# PHASE TRANSITIONS IN FOODS

# FOOD SCIENCE AND TECHNOLOGY

*International Series*

A complete list of the books in this series appears at the end of the volume.

# PHASE TRANSITIONS
# IN FOODS

## Yrjö H. Roos

*University of Helsinki*
*Faculty of Agriculture and Forestry*
*Department of Food Chemistry and Technology*
*Helsinki, Finland*

**ACADEMIC PRESS**

San Diego   New York   Boston
London   Sydney   Tokyo   Toronto

Copyright © 1995 by ACADEMIC PRESS, INC.

Academic Press, Inc.
A Division of Harcourt Brace & Company
525 B Street, Suite 1900, San Diego, California 92101-4495

*United Kingdom Edition published by*
Academic Press Limited
24-28 Oval Road, London NW1 7DX

Library of Congress Cataloging-in-Publication Data

Roos, Yrjo.
    Phase transitions in foods / by Yrjo Roos.
        p.        cm. -- (Food science and technology international series)
    Includes bibliographical references and index.
    ISBN 0-12-595340-2 (acid-free)
    1. Food--Composition.   2. Phase transformations (Statistical
physics)  I. Title.  II. Series.
    TP372.5.R66   1995
    664--dc20                                               95-5987
                                                            CIP

PRINTED IN THE UNITED STATES OF AMERICA
95  96  97  98  99  00  MM  9  8  7  6  5  4  3  2  1

*To my beloved wife, Sari
and our son, Johan Henrik*

# Contents

*Preface*   xi

## 1. Introduction to Phase Transitions

I.  Introduction   1
II.  Thermodynamics   2
    A. Basic Terminology   2
    B. First Law of Thermodynamics   4
    C. Second Law of Thermodynamics   6
III.  Characterization of Phase Transitions   9
    A. Phase Diagrams   9
    B. Gibbs Energy of Phases   11
    C. Classification of Phase Transitions   13
    D. Effects of Composition on Transition
       Temperatures   17
   References   18

## 2. Physical State and Molecular Mobility

I.   Introduction   19
II.  Crystallization and Melting   20
     A. Nucleation and Crystal Growth   20
III. The Physical State of Amorphous Materials   26
     A. Mechanical Properties   26
     B. Characterization of the Physical State   29
     C. Glass Transition Theories   31
IV.  Molecular Mobility and Plasticization   34
     A. Mechanical Properties   35
     B. Plasticization and Molecular Weight   39
     C. Crystallization of Amorphous Compounds   42
     References   47


## 3. Methodology

I.   Introduction   49
II.  Determination of the Physical State   50
     A. Crystallinity   50
     B. Molecular Mobility   53
III. Physical Properties and Transition Temperatures   60
     A. Volumetric Changes   60
     B. Changes in Enthalpy   63
     C. Mechanical and Dielectric Properties   66
     References   69


## 4. Water and Phase Transitions

I.   Introduction   73
II.  Properties of Water   74
     A. Phase Behavior of Water   74
     B. Water in Solutions   77
III. Water in Foods   86
     A. Sorption Behavior   87
     B. Water Plasticization   91
     C. Ice Formation and Freeze-Concentration   97
     References   103

## 5. Food Components and Polymers

 I. Introduction 109
 II. Carbohydrates 110
   A. Sugars 110
   B. Starch 119
 III. Proteins 133
   A. Denaturation 134
   B. Glass Transition 137
 IV. Lipids 142
   A. Polymorphic Forms 142
   B. Melting of Fats and Oils 145
   References 149

## 6. Prediction of the Physical State

 I. Introduction 157
 II. Prediction of Water Plasticization 158
   A. Plasticization Models 159
   B. Effects of Composition 165
 III. Mechanical Properties and Flow 170
   A. Viscosity of Amorphous Foods 171
   B. Viscoelastic Properties 179
   References 188

## 7. Time-Dependent Phenomena

 I. Introduction 193
 II. Time-Dependent Properties of the Physical State 194
   A. Glass Formation 194
   B. Relaxation Phenomena in Amorphous Foods 198
 III. Collapse Phenomena 202
   A. Stickiness and Caking 202
   B. Collapse 206
 IV. Crystallization and Recrystallization 210
   A. Crystallization of Amorphous Sugars 210
   B. Ice Formation and Recrystallization 224
   C. Retrogradation of Starch 235
   References 240

## 8. Mechanical Properties

    I.   Introduction   247
   II.   Stiffness   248
          A. Modulus Curves of Food Materials   248
          B. Stiffness and Food Properties   253
  III.   Mechanical Properties and Crystallinity   262
          A. Food Lipids   262
          B. Crystallinity in Amorphous Foods   264
          References   267

## 9. Reaction Kinetics

    I.   Introduction   271
   II.   Principles of Reaction Kinetics   272
          A. Reaction Order   272
          B. Temperature Dependence   276
  III.   Kinetics in Amorphous Foods   283
          A. Low-Moisture Foods   284
          B. Frozen Foods   302
          References   308

## 10. Food Processing and Storage

    I.   Introduction   313
   II.   Food Processing   314
          A. Dehydration and Agglomeration   314
          B. Melt Processing and Extrusion   330
  III.   Food Formulation and Storage   333
          A. Prediction of Stability   333
          B. Food Formulation   336
          References   345

*Index*   349

# *Preface*

Temperature, time, and water content have enormous effects on the physical state and quality of food and biological materials. These materials are often metastable and they undergo phase and state transitions in various processes and during storage. Kinetics of various changes is related to these transitions and therefore to molecular mobility. Understanding phase and state transitions and the relationships between molecular mobility and stability is often the basis of proper control of food processing and storage conditions. Temperature and water content are variables that govern rates of both desired and detrimental changes in quality. This book describes phase and state transitions of food and biological materials and discusses their dependence on temperature, time, and water content.

It was recognized quite early that changes in the physical state of lactose in dried milk and freeze-concentrated ice cream solids were responsible for loss of quality. However, the tremendous effects of phase and state transitions on food properties and quality were not realized and widely accepted until 1980's. Harry Levine and Louise Slade have been the pioneers in describing foods as metastable systems and the role of water as the ubiquitous plasticizer of food solids. Especially, the glass transition temperature of amorphous biological materials has been admitted to be one of the main attributes in controlling shelf life of low-moisture and frozen foods. In addition to the description of phase transitions, such as crystallization and melting, this book considers the effects of glass transition on the time-dependent changes in state and quality of food and biological materials.

This book is intended to those interested in physical chemistry of food, pharmaceutical, and other biological materials. There have been no other books of similar type that summerize literature on phase and state transitions and their effects on mechanical and physicochemical properties. This book gives an introduction to thermodynamics and definitions of phase and state transitions. The relationships between the physical state and molecular mobility are discussed as well as the common methodology is introduced. The importance of properties of water and especially water plasticization are given particular attention. Transitions, such as protein denaturation, starch gelatinization, and those of lipids, are discussed with numerical values for transition temperatures. The glass transition is given the main emphasis and the contents of the book stress predictions of changes in the physical state, time-dependent phenomena, and mechanical properties. Definition of reaction kinetics is included and the effect of diffusional limitations on quality changes is given particular attention. The book also discusses the role of phase and state transitions in food processing and storage. I hope that my view on phase transitions in foods will be found useful by teachers and students of food engineering and technology courses. The book has succeeded if it is also found by professionals in academia and industry in the areas of biological, food, and pharmaceutical sciences to be a valuable source of information on phase and state transitions.

Various people have contributed to the contents of this book through discussions and helping me to understand the complexity of biological systems. In particular, I appreciate the time taken by Jorge Chirife, Marcus Karel, Theodore P. Labuza, Harry Levine, and Micha Peleg for reading the manuscript. I also wish to express my gratitude for their valuable comments and suggestions that were extremely helpful in the final revision of the contents. The help of the graduate students of the course "Water in Foods, Drugs and Biological Systems," FScN 5555 at the University of Minnesota taught by Professor Theodore P. Labuza, is also appreciated.

The time needed for writing this book was taken from my family. I wish to thank Sari for her warm understanding and support.

# CHAPTER 1

# Introduction to Phase Transitions

## I. Introduction

Phase transitions are changes in the physical state of materials, which have significant effects on their physical properties. Chemically pure compounds such as water or many organic and inorganic compounds in foods have exact phase transition temperatures. There are three basic physical states, which are the solid, the liquid, and the gaseous states. The term *transition* refers to the change in the physical state that is caused by a change in temperature or pressure. Heating of solid foods is often used to observe temperatures at which changes in thermal or physical properties, e.g., in heat capacity, viscosity, or textural characteristics, occur.

Water is one of the most important compounds in nature and also in foods. It may exist in all of the three basic states during food processing, storage, and consumption. The effect of water on the phase behavior of food solids is of utmost importance in determining processability, stability, and quality. Well-known examples include transformation of liquid water into ice (freezing) or water into vapor (evaporation). These transitions in phase are the main physical phenomena that govern food preservation by freezing and dehydration. Engineering and sensory characteristics of food materials are often defined by the complicated combination of the physical state of component compounds. The main constituents of food solids are carbohydrates, proteins, water, and fat. These materials may exist in the liquid state and the solid crystalline or amorphous noncrystalline state. Many of the component compounds, e.g., sugars, fats, and water, when they are chemically pure, crystallize below their equilibrium melting temperature.

Stability is an important criterion in food preservation. Materials in thermodynamic equilibrium are stable, i.e., they exist in the physical state that is determined by the pressure and temperature of the surroundings. However, most biological materials are composed of a number of compounds and they often exist in a thermodynamically nonequilibrium, amorphous state. Such materials exhibit many time-dependent changes that are not typical of pure compounds and they may significantly affect the shelf life of foods. The physical state of food solids is often extremely sensitive to water content, temperature, and time. This chapter introduces the basic terminology of thermodynamics and phase transitions, and describes the common thermodynamic principles that govern the physical state of foods.

## II.  Thermodynamics

Thermodynamics describes the physical state of materials in terms of basic state variables. Thermodynamics gives and defines the basis for description and understanding of the physical state at equilibrium and various transitions, which may occur due to changes in the quantities of the state variables that define the equilibrium. General principles of thermodynamics, which are reported here according to Alberty (1983), can be found in most textbooks of physical chemistry.

### A.  Basic Terminology

Thermodynamics describes relationships between various *systems* that are in equilibrium, i.e., no changes in the physical state of the systems are observed as a function of time. Such systems are often pure compounds at a given temperature and pressure. Thermodynamics may also be used to characterize differences in physical properties between various equilibrium states and the driving forces towards equilibrium. Thus, a change in temperature or pressure may introduce a driving force for a system to approach another equilibrium state. Biological materials and foods are often metastable systems and they exhibit time-dependent changes as they approach equilibrium. Temperature changes that occur over a phase transition temperature result in a change of phase. The basic quantitative concepts of thermodynamics are *temperature*, *internal energy*, and *heat*. These concepts are used to describe the physical state of thermodynamic systems. Obviously, the most important parameter that has to be taken into account in the definition of the physical state of any food is temperature.

Thermodynamic systems can be separated from their surroundings with boundaries. Boundaries may allow the occurrence of heat transfer between the system and its surroundings, e.g., a food product may gain heat from its surroundings. However, boundaries of an *isolated system* prevent all interactions of the system with its surroundings and there can be no transfer of energy or matter between the system and its surroundings. Foods are seldom isolated systems, although the principle can be applied when insulators are used to avoid rapid cooling of hot foods or warming of cold drinks. A system may also be *open* or *closed*. An open system may have transfer of both energy and matter with the surroundings. A closed system may only have transfer of energy with surroundings. In food processing food systems may be open systems, e.g., they can be heated by steam infusion, which includes transfer of matter and energy from the surroundings to the food. Foods may also be closed systems, which is true when they are hermetically sealed in containers. Systems that are uniform in all properties are called homogeneous and those containing more than one phase are heterogeneous. When systems are at equilibrium they exist in *definite states* and their properties have definite values. The equilibrium state is defined by *state variables*, which are pressure, temperature, and volume. Two state variables, i.e., pressure and temperature or temperature and volume, can be used to define the physical state of pure materials.

Thermodynamic quantities that are proportional to the amount of material are *extensive functions of state*. Such quantities include total energy, volume, number of moles, and mass, which depend on the amount of material in the system. Properties which are independent of the amount of material, e.g., pressure and temperature, are called *intensive functions of state*. The ratio of two extensive functions of state is always an intensive function of state. Intensive properties are also obtained if extensive properties are divided by the amount of material. At a thermodynamic equilibrium extensive functions of state and the intensive functions of state of each phase are constant. The state variables define the physical state of food materials. Especially temperature can be used to control their stability. However, phase transitions in foods are complicated and their exact thermodynamic characterization may be difficult due to the number of component compounds, coexisting phases, and metastability.

## B. First Law of Thermodynamics

The physical state of thermodynamic systems can be related to the state variables. According to the *zeroth law of thermodynamics* the temperature of two or more systems at *thermal equilibrium* is the same. Equation (1.1), which defines the well known relationships between pressure, $p$, temperature, $T$, and volume, $V$, is called the *equation of state*. According to the gas law, which is based on the equation of state, the product, $pV$, of ideal gases follows the same function. Therefore, a change in $p$, $T$, or $V$ results in a change of one or both of the other variables. The gas law, which is given in equation (1.2), where $n$ is the number of moles, defines the *gas constant, R*. At the temperature of absolute zero (0 K, -273.16°C) molecular mobility ceases and also the pressure approaches 0 Pa.

$$T = f(p,V) \tag{1.1}$$

$$pV = nRT \tag{1.2}$$

A process that occurs in a system that is thermally isolated from the surroundings, i.e., there is no heat exchange between the system and the environment, is called an *adiabatic* process. Such systems have a given amount of internal energy, $U$. A change in the state of the system requires work, $W$, on the environment that is equal to the change in the amount of internal energy, $\Delta U$. The internal energy of a system may also be altered by heat exchange between the system and the environment. If no work on the environment is done $\Delta U$ is equal to the quantity of heat, $q$, exchanged. A change in the internal energy can be calculated with equation (1.3), which defines that heat absorbed by a system either increases the amount of internal energy or is used to do work on the environment.

$$\Delta U = q + W \tag{1.3}$$

The definition of internal energy can be used to derive the *first law of thermodynamics*. The first law of thermodynamics states that a thermodynamic system has a property of internal energy. The internal energy is a function of state variables that can be changed due to energy exchange between the system and its surroundings.

## 1. Enthalpy

*Enthalpy*, or *heat content*, is a thermodynamic quantity that is defined by the sum of internal energy and pressure-volume work which has been done on the system. It is obvious that most common processes occur at the atmospheric pressure and therefore also at a constant pressure. The pressure-volume work is defined by equation (1.4). Equation (1.4) states that the heat absorbed by a system becomes a function of state. The amount of heat absorbed at a constant pressure can then be obtained from equation (1.5), where the subscripts 1 and 2 refer to the initial and the final state, respectively. Equation (1.5) suggests that at a constant pressure a change in temperature involves the amount of heat that is defined by the difference in the sums of internal energy and pressure-volume work.

$$W = -p\Delta V \tag{1.4}$$

$$q = (U_2 + pV_2) - (U_1 + pV_1) \tag{1.5}$$

Enthalpy, $H$, which is defined by equation (1.6), is also a function of state. The difference in enthalpy, $\Delta H$, between two states that are referred to with the subscripts 1 and 2 is given by equation (1.7). It can be shown that the change in enthalpy at a constant pressure is equal to the change in the amount of heat, $q$, between the two states.

$$H = U + pV \tag{1.6}$$

$$q = \Delta H = H_2 - H_1 \tag{1.7}$$

The heat absorbed or released in chemical reactions or due to phase transitions is studied in *thermochemistry*. A negative value for $q$ shows that a process is *exothermic* and the system releases heat to the surroundings. In *endothermic* changes heat must flow from the surroundings to the system and $q$ has a positive value.

## 2. Heat capacity

The internal energy of a system with a constant mass is a function of pressure, temperature, and volume. If the volume of the system is constant, there is no pressure-volume work and a change in temperature results in a change in the amount of internal heat. The heat capacity at a constant volume, $C_v$, is defined by equation (1.8). Equation (1.8) shows that heat capacity at a constant volume is a measure of the change in the internal heat that is caused by the change in temperature.

$$C_v = \frac{dU}{dT} \tag{1.8}$$

Heat capacity is often determined at a constant pressure. At a constant pressure a change in temperature results also in a change in volume. The change in enthalpy involves the change in both internal energy and pressure-volume work. If the pressure of the system is constant, a change in temperature results in a change in enthalpy that is equal to the heat exchange between the system and its surroundings. Therefore, the heat capacity at a constant pressure is defined by equation (1.9). It should be noticed that the heat capacities at constant pressure and constant volume are not the same, but $C_p > C_v$ since the change in enthalpy includes both the change in internal energy and the energy change due to pressure-volume work.

$$C_p = \frac{dH}{dT} \tag{1.9}$$

It has been found that the molar $C_p$ and its enlargement with raising temperature often increases with increasing molecular complexity. However, the heat capacity in most foods is due to water. Most changes in state occur at a constant pressure and calorimetric experiments may be conducted to obtain changes in enthalpy as a function of temperature. The results can be used for the determination of $C_p$, which is an important property in food processing and engineering.

## C. Second Law of Thermodynamics

Most natural processes are spontaneous and they occur to the direction of an equilibrium. The *second law of thermodynamics* provides criteria for predicting the probability of thermodynamic processes. It can be used to evaluate whether changes in the physical state occur spontaneously. The second law of thermodynamics includes basis for understanding spontaneous changes such as the well known fact that no heat is transferred from a colder system to a warmer system without the occurrence of other simultaneous changes in the two systems or in the environment. It may also be shown that spontaneous changes involve changes in energy and the directions of the changes are defined by the second law of thermodynamics.

## 1. Entropy

It is obvious that the total amount of energy within an isolated system is constant, but it may become unavailable in irreversible processes. The amount of unavailable energy within a system is known as *entropy, S*. Entropy is a function of state that is defined by equation (1.10).

$$dS = \frac{dq}{T} \qquad (1.10)$$

Irreversible processes in isolated systems are spontaneous and they produce entropy. Energy in reversible processes of isolated systems cannot become unavailable and therefore the entropy within the system remains constant. All natural processes are irreversible and the entropy of natural systems increases as they are changed towards equilibrium. In various reversible processes such as first-order phase transitions, e.g., in melting and evaporation, the pressure of the system is constant. Since a first-order transition at a constant pressure includes no pressure-volume work and it occurs at a constant temperature, the latent heat of the transition is equal to the change in enthalpy. Therefore, the change in entropy, $\Delta S$, can be related to enthalpy according to equation (1.11).

$$\Delta S = \frac{\Delta H}{T} \qquad (1.11)$$

$$dU = TdS - pdV \qquad (1.12)$$

$$dS = \frac{C_v}{T} dT \qquad (1.13)$$

$$dS = \frac{C_p}{T} dT \qquad (1.14)$$

Combination of the first and second laws of thermodynamics can be used to establish relationships between entropy, internal energy, pressure, temperature, and volume. The relationships as shown by equation (1.12) suggest that entropy is a function of internal energy and volume. Equation

(1.12) is an important thermodynamic equation that may be used to calculate entropy for various processes. It may be shown that an increase in temperature increases entropy. Equations (1.13) and (1.14) can be used to calculate changes in entropy that occur as a function of temperature at a constant volume or at a constant pressure, respectively.

In addition to the zeroth, first, and second laws of thermodynamics the physical state of materials at absolute zero can be related to the state variables by definition of the *third law of thermodynamics*. The third law of thermodynamics is given by the statement that the entropy of each pure element or substance in a perfect crystalline form is zero at absolute zero (0 K). It is important to notice that the entropy of an amorphous solid or a supercooled liquid, or almost any food, may be considered to be higher than zero at absolute zero.

### 2. Helmholtz free energy

According to the second law of thermodynamics changes in an isolated system are spontaneous if $dS > 0$, and the system is in equilibrium if $dS = 0$. The Helmholtz free energy, $A$, is a state function that defines the direction of changes in a closed system at a constant temperature and a constant volume. The Helmholtz free energy is defined by equation (1.15), which shows that the Helmholtz free energy is an extensive function of state that is defined by entropy, internal energy, and temperature.

$$A = U - TS \qquad (1.15)$$

The Helmholtz free energy of a closed system at a constant temperature and a constant volume can be shown to decrease for changes that occur spontaneously. Therefore, changes in closed systems may occur spontaneously if $dA < 0$, the systems are in equilibrium or the changes are reversible if $dA = 0$, and the changes are forced if $dA > 0$.

### 3. Gibbs energy

Gibbs energy, $G$, is analogous to the Helmholtz free energy for changes that occur in closed systems at a constant temperature and a constant pressure. Most changes in foods occur at the atmospheric pressure and therefore at a constant pressure. The Gibbs energy can be used to show whether changes occur spontaneously or if they are forced. The Gibbs energy is an extensive function of state that is defined by equation (1.16). Equation (1.16) shows that Gibbs energy is defined by entropy, internal energy, pressure-volume work, and temperature. Since enthalpy at a constant pressure is equal to the sum of internal energy and pressure-volume work, the definition of Gibbs energy is also given by equation (1.17), which defines Gibbs energy to be a function of enthalpy, entropy, and temperature.

$$G = U + pV - TS \qquad (1.16)$$

$$G = H - TS \qquad (1.17)$$

The Gibbs energy of a closed system at a constant temperature and a constant pressure can be shown to decrease for changes that occur spontaneously. Therefore, changes in closed systems are spontaneous if $dG < 0$, the systems are in equilibrium or the changes are reversible if $dG = 0$, and the changes are forced if $dG > 0$.

## III.  Characterization of Phase Transitions

Phase transitions in foods are often a result of changes in composition or temperature during processing or storage. Knowledge of transition temperatures and of the thermodynamic quantities is particularly important in understanding such processes as evaporation, dehydration, and freezing. These processes are governed by the transition of water into the gaseous or crystalline state. Foods are complex materials that contain at least one component compound and water. The dependence of the physical state of various materials as a function of temperature can be characterized with phase diagrams. Changes that are observed at the transition temperatures can be used for the description of the effect of the transition on physical properties.

### A. Phase Diagrams

A phase can be defined to be a physically and chemically homogenous state of a material that is clearly separated from other matter. A phase transition can be observed from a change in internal energy, $U$, volume, $V$, number of moles, $n$, or mass. The change in phase is a result of a change in temperature or pressure. An equilibrium, e.g., between ice and water, requires that the two phases have the same temperature and pressure. Also the chemical

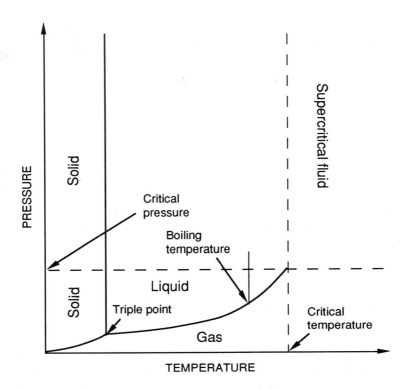

**Figure 1.1** A schematic phase diagram showing the equilibrium curves between various physical states and their dependence on pressure and temperature.

potential, $\mu$, and Gibbs energy, $G$, of the material are the same in both phases. The basic three physical states of chemically pure substances are solid, liquid, and gaseous states. These states are equilibrium states and the change of one state to another state occurs at exact temperature and pressure conditions that are specific to each material.

The relationships between the physical state, pressure, temperature, and volume can be shown in three-dimensional phase diagrams. Such diagrams show surfaces that indicate values for the state variables at equilibrium conditions. It can be shown that at most equilibrium situations two phases may coexist. The possible combinations of coexisting phases are solid and liquid, solid and gas, and liquid and gas. All three phases may coexist only at the *triple point*, which has exact temperature and pressure conditions that are specific for each material. Two-dimensional projections of the three-dimensional phase diagrams are often more useful in practical applications. As shown in Figure 1.1 the two-dimensional phase diagrams show equilibrium lines for pressure and temperature for each phase. One of the most important of such two-dimensional phase diagrams for foods is that of water.

Molecular organization of the liquid and gaseous phases are similar, but different from that of the highly ordered crystalline solid phase. The equilibrium curve between the liquid and gaseous states ends at the *critical state*. At the *critical point* the temperature and pressure of both phases become equal and the two phases can no longer be separated. Thus, the *critical temperature* is the highest temperature at which the liquid state may exist. The pressure at the critical point is called *critical pressure* and the corresponding volume is the *critical volume*. The critical conditions of carbon dioxide are important in supercritical extraction, which is often applied as a food processing method.

## B. Gibbs Energy of Phases

An equilibrium between various phases exists only when no driving force is present for the molecules to change phase. The driving force for a phase transition is the chemical potential, $\mu$, and the conditions that allow equilibrium can be defined by chemical potential, pressure, and temperature. It may be shown that at equilibrium the chemical potential is equal for all phases. However, if the chemical potentials between various phases differ a spontaneous change in the direction from high to low potential occurs. The stable phase is that with the lowest chemical potential.

In a one-component system the molar Gibbs energy is equal to chemical potential. At equilibrium two or three phases of a single component may have the same molar Gibbs energy. Therefore, phases coexist at melting temperature, $T_m$, boiling temperature, $T_b$, and triple point. A schematic representation of the Gibbs energy of various phases as a function of temperature at a constant pressure is shown in Figure 1.2. Below $T_m$ the solid phase has the lowest Gibbs energy, between $T_m$ and $T_b$ the lowest Gibbs energy is in the liquid phase, and above $T_b$ the gaseous phase has the lowest Gibbs energy. It may be shown that the slopes of the lines are defined by equation (1.18).

$$\left( \frac{\partial G}{\partial T} \right)_p = -S \tag{1.18}$$

The slopes of the lines in Figure 1.2 are negative, since the entropies of each phase are positive, and follow the order $S_g > S_l > S_s$, where the subscripts $g$, $l$, and $s$ refer to the gaseous, liquid, and solid states, respectively.

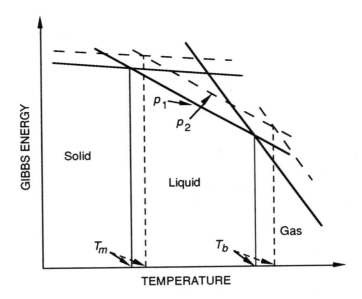

**Figure 1.2** A schematic representation of Gibbs energy of phases for a pure compound as a function of temperature at a constant pressure. The Gibbs energy decreases with increasing temperature and is the same for two phases at the intercept of the lines, i.e., the Gibbs energy is the same for the solid and liquid states at the melting temperature, $T_m$, and for the liquid and gaseous states at the boiling temperature, $T_b$. At various temperatures the stable phase has the lowest Gibbs energy. An increase in pressure from $p_1$ to $p_2$ increases Gibbs energy and also $T_m$ and $T_b$.

Therefore, the slopes of the lines differ and the decrease of Gibbs energy is steepest for the gaseous phase. At the intercepts of the lines the Gibbs energies are equal and the two phases may coexist. Between the intercepts the most stable phase has the lowest Gibbs energy.

An increase in pressure increases Gibbs energy. The effect of pressure on the Gibbs energy at a constant temperature is defined by equation (1.19).

$$\left(\frac{\partial G}{\partial p}\right)_T = V \tag{1.19}$$

The volume in the gaseous state is significantly larger than the volume in the liquid or solid state. Therefore, as shown in Figure 1.2 an increase in pressure increases the boiling temperature and often also the melting temperature. However, pressure has a much larger effect on the boiling temperature than on the melting temperature due to the higher molar volume of the gaseous state. The effect of pressure on the boiling temperature is applied in several food processes, e.g., in dehydration, evaporation, and sterilization.

## C. Classification of Phase Transitions

Classification of phase transitions is important in establishing criteria for defining general principles that govern various effects of phase transitions on material properties that are changed at the transition temperatures. Gibbs energy or chemical potential can be used in the classification of phase transitions. Such classification of phase transitions into first-order, second-order, and higher-order transitions was made by Ehrenfest (1933). The classification reported by Ehrenfest (1933) is based on observed discontinuities that occur in the state functions at the transition temperatures.

### 1. First-order transitions

The classification of phase transitions is often based on changes in chemical potential or Gibbs energy (Wunderlich, 1981). At equilibrium the chemical potentials of two phases are equal. However, as was shown in Figure 1.2 a change in chemical potential or Gibbs energy at a transition temperature shifts the equilibrium state to that with the lower chemical potential and Gibbs energy. According to the classification of phase transitions by Ehrenfest (1933) the first derivative of chemical potential and Gibbs energy that were defined by equations (1.18) and (1.19) show discontinuity at the first-order transition temperature.

Most phase transitions occur at a constant pressure. Equation (1.17) defined relationships between enthalpy, entropy, Gibbs energy, and temperature at a constant pressure. It is obvious that two phases with the same Gibbs energy at the same temperature, e.g., a solid and a liquid or a liquid and a gaseous phase, have different enthalpies and therefore different entropies. Moreover, it can be shown that if the two phases have different enthalpies, they also have different volumes. Therefore, changes in enthalpy, entropy, and volume are typical of first-order phase transitions. The common techniques used in the determination of phase transition temperatures such as calorimetry and dilatometry are based on the determination of the change in enthalpy or volume, respectively. Since Gibbs energy for the two phases at a first-order transition temperature is the same for both phases, the first derivative of Gibbs energy shows discontinuity at the transition temperature. Therefore, the quantities of $H$, $S$, and $V$ show a step change at the transition

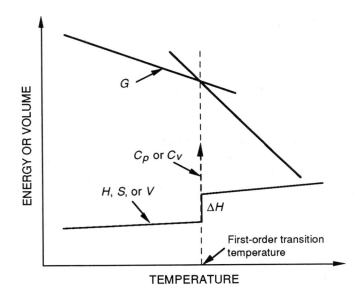

**Figure 1.3** Changes in thermodynamic quantities at a first-order phase transition. Gibbs energy, $G$, is the same for both phases at the transition temperature. Enthalpy, $H$, entropy, $S$, and volume, $V$, show a step change at the transition temperature. The heat capacity, $C_p$ or $C_v$, has an infinite value at a first-order phase transition temperature.

temperature as shown in Figure 1.3. The heat capacity is obtained from the second derivative of the Gibbs energy and it has an infinite value at a first-order transition temperature.

    Phase transitions that occur between the three basic physical states, i.e., between the solid, liquid, and gaseous states, are first-order transitions. These transitions include melting and crystallization, which occur between the solid and liquid states. Vaporization and condensation, which are transitions between the liquid and gaseous states, and sublimation and ablimation, which are phase changes between the solid and gaseous states without the presence of the liquid state, are also first-order transitions.

2. Second-order and higher-order transitions

Second-order phase transitions according to the classification of Ehrenfest (1933) are those for which the second derivative of the chemical potential or Gibbs energy shows a discontinuous change at the transition temperature. At the second-order transition the thermodynamic quantities of enthalpy,

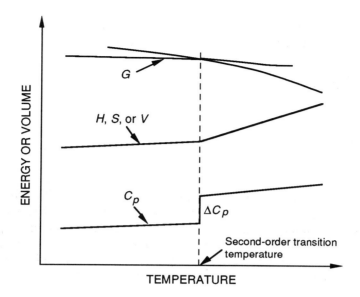

**Figure 1.4** Changes in thermodynamic quantities at a second-order phase transition. Enthalpy, *H*, entropy, *S*, Gibbs energy, *G*, and volume, *V*, of the two phases are the same at the transition temperature. The second derivatives of the Gibbs energy are discontinuous at the transition temperature, and therefore a step change is observed in heat capacity, $C_p$.

entropy, and volume of the two phases are the same at the transition temperature. Therefore, a second-order transition has no latent heat of the phase change, but there is a discontinuity in the heat capacity. The heat capacity is different in the two phases, but does not become infinite at the transition temperature, which occurs at the first-order transition temperature.

The effects of a second-order transition on enthalpy, entropy, Gibbs energy, heat capacity, and volume are shown in Figure 1.4. The second derivatives of Gibbs energy are defined by equations (1.20), (1.21), and (1.22), where $\alpha$ is the thermal expansion coefficient and $\beta$ is compressibility.

$$\left(\frac{\partial^2 G}{\partial T^2}\right) = \frac{-C_p}{T} \tag{1.20}$$

Equation (1.20) suggests that at a constant pressure a second-order transition results in a discontinuity in heat capacity. Equations (1.21) and (1.22) show that there is a discontinuity in the thermal expansion coefficient and isothermal compressibility at a second-order phase transition. Therefore, experimentally determined changes in heat capacity and thermal expansion can be used in locating second-order transition temperatures.

$$\left(\frac{\partial^2 G}{\partial p \partial T}\right) = V\alpha \tag{1.21}$$

$$\left(\frac{\partial^2 G}{\partial p^2}\right) = -V\beta \tag{1.22}$$

Third-order and higher-order transitions are those for which the third or higher derivatives of the chemical potential or Gibbs energy become discontinuous at the transition temperature. At a third-order transition temperature both phases have the same heat capacity, but the change in heat capacity as a function of temperature in the two phases is different. The third- or higher-order transitions have not been reported for food materials.

3. Effects of pressure on transition temperatures

An equilibrium between two phases of a one-component system requires that the chemical potential, pressure, and temperature are the same in both phases. A change in temperature at a constant pressure or a change of pressure at a constant temperature results in a nonequilibrium state and disappearance of one of the phases. However, if both pressure and temperature are changed, but the chemical potential of both phases are kept equal, the two phases may coexist. Such changes in pressure and temperature that maintain equilibrium between two phases can be obtained with the Clausius-Clapeyron equation, which is given in equation (1.23), where $\Delta H_l$ is the latent heat of the transition and $\Delta V$ is the difference in volume between the two phases.

$$\frac{dp}{dT} = \frac{\Delta H_l}{T\Delta V} \tag{1.23}$$

The Clausius-Clapeyron equation can be used to calculate the effect of pressure on the boiling temperature, e.g., on the boiling temperature of water, which is important in evaporation and dehydration. The equation may also be used to calculate the effect of pressure on the melting temperature. The Clausius-Clapeyron equation may be written into the form of equation (1.24), which can be used to obtain the latent heat of vaporization or sublimation. According to equation (1.24) a plot of ln $p$ against $1/T$ is linear. However, linearity is often valid only over a narrow temperature range, since the properties of the vapor phase differ from those of a perfect gas and $\Delta H_l$ changes with temperature.

$$\ln \frac{p_2}{p_1} = \frac{-\Delta H_l}{R} \left( \frac{1}{T_1} - \frac{1}{T_2} \right) \tag{1.24}$$

It should be noticed that an increase in pressure causes a decrease in the total volume, which may increase both first- and second-order transition temperatures. According to Sperling (1986) an increase in pressure may result in vitrification of amorphous materials during extrusion.

### D. Effects of Composition on Transition Temperatures

The effects of composition on phase transition temperatures are important in defining various effects of food composition on observed phase transition temperatures and the physical state. The effect of composition on vapor pressure of water in foods is one of the most important examples that affect food behavior in both processing and storage.

### 1. Raoult's law

Raoult's law defines the relationship between composition and vapor pressure of dilute solutions. A liquid and its vapor phase have the same temperature and pressure at equilibrium. The equilibrium between the two phases also requires that the chemical potentials of the two phases are equal. Raoult's law, which is given in equation (1.25), states that the vapor pressure of a liquid, $p$, in a mixture is a linear function of its mole fraction, $x$, in the liquid phase and the vapor pressure of the pure liquid, $p_0$, at the same temperature.

$$p = x p_0 \tag{1.25}$$

Raoult's law is important in defining properties of water in food materials. Raoult's law can be applied to evaluate effects of food composition on the melting and boiling temperatures of water in foods. Several salts, however, may lower the vapor pressure of water in solutions more than is predicted by Raoult's law, which affects water activity, $a_w$, of the salt solutions accordingly (Benmergui *et al.*, 1979).

## 2. Henry's law

Raoult's law applies to the depression of the vapor pressure of solvents in dilute solutions and therefore it is approached as the mole fraction of the solvent becomes unity. Henry's law is analogous to Raoult's law, but it defines the vapor pressure of the solute in a dilute solution. Therefore, as the mole fraction of a solute approaches zero its partial pressure is defined by Henry's law. Henry's law is given by equation (1.26), which states that the vapor pressure, $p$, is a linear function of the mole fraction of the solute, $x$, and a constant, $K$.

$$p = Kx \qquad\qquad (1.26)$$

It should be noticed that Henry's law can also be applied to predict solubility of gases in liquids.

## References

Alberty, R.A. 1983. *Physical Chemistry*. John Wiley & Sons, New York.

Benmergui, E.A., Ferro Fontan, C. and Chirife, J. 1979. The prediction of water activity in aqueous solutions in connection with intermediate moisture foods. I. $a_w$ prediction in single aqueous electrolyte solutions. *J. Food Technol. 14*: 625-637.

Ehrenfest, P. 1933. *Proc. Acad. Sci., Amsterdam 36*: 153. Cited in Wunderlich (1981).

Sperling, L.H. 1986. *Introduction to Physical Polymer Science*. John Wiley & Sons, New York.

Wunderlich, B. 1981. The basis of thermal analysis. In *Thermal Characterization of Polymeric Materials*, ed. E.A. Turi. Academic Press, New York, pp. 91-234.

# CHAPTER 2

# *Physical State and Molecular Mobility*

## I.  Introduction

The physical state of food materials is defined by the physical state of component compounds and their phase behavior. In foods the physical state and stability are related to both first- and second-order phase transitions and molecular mobility in various phases. The physical state and molecular mobility are affected by temperature and also by composition of food solids. Various solid foods have a low water content and their molecular mobility is also low. It is also known that amorphous polymers have a higher molecular mobility than crystalline polymers (Barker and Thomas, 1964), which probably applies also to molecular mobility in foods. Molecular mobility is often related to such phenomena as diffusion, viscosity, electrical conductivity, and glass transition.

Food solids may be divided into lipids and water solubles that are composed mostly of carbohydrates and proteins. The physical state of lipids is often related to the transitions between various polymorphic forms and the liquid state. Carbohydrates and proteins may exist in the noncrystalline amorphous state, in the crystalline state, or in solution. The physical state and phase transitions of the water solubles are significantly affected by water content. As the water content increases the molecular mobility of water solubles may also increase (Lillford, 1988), which results in a decreasing viscosity and probably also in lower stability. Liquid materials have fairly high molecular mobility, which decreases with decreasing temperature.

Changes in the physical state of food materials include first-order phase transitions such as crystallization and melting, and the second-order phase change that occurs at the glass transition of an amorphous material. The amorphous state is typical of water-soluble food components at low water contents and in frozen foods (White and Cakebread, 1966; Levine and Slade, 1986; Slade and Levine, 1991). The emphasis of this chapter is on the basic relationships between the physical state and properties of food solids, and on the effects of temperature on dynamic phenomena.

## II.   Crystallization and Melting

Crystallization and melting are first-order phase transitions that occur between the solid and liquid states. The latent heat released by crystallization is equal to the amount of heat required by melting at the same temperature. Crystallization may occur from a melt or from a solution. Crystallization from a melt may occur at temperatures below the equilibrium melting temperature, $T_m$, and from a solution due to supersaturation. In amorphous foods with low water contents crystallization of solids may occur from the melt at temperatures above the glass transition temperature, $T_g$, but below $T_m$ or in frozen foods from the supersaturated solution formed due to freeze-concentration. Crystallization is often a three-step process that includes nucleation (formation of nuclei), propagation (crystal growth), and maturation (crystal perfection and/or continued growth) (Levine and Slade, 1990).

### A. Nucleation and Crystal Growth

Numerous applications of crystallization technology exist in the food industry. Crystallization is an important phase transition of frozen foods and in the manufacturing of sugars, salts, and various fats and spreads. Several materials may also crystallize during food processing and storage.

Crystallization may be either a desired or a hazardous phenomenon depending on the product and its quality attributes. Examples include a wide range of processes and products from ice formation during freezing to sugar crystallization in confectionery or food powders to various recrystallization phenomena.

## 1. Nucleation

Nucleation is the process that precedes crystallization and it involves the formation of the incipient crystalline phase. Nucleation results from the metastable state that occurs after supersaturation due to solvent removal or a decrease in temperature of a solution or a melt (Van Hook, 1961). Nucleation may be either homogeneous or heterogeneous. Heterogeneous nucleation is more likely to occur in foods due to the presence of impurities. Homogeneous and heterogeneous nucleation processes are considered to be primary nucleation mechanisms, i.e., crystalline centers of the crystallizing compounds are not present in the nucleating systems (Hartel and Shastry, 1991).

*a. Homogeneous Nucleation.* Homogeneous nucleation occurs spontaneously, i.e., the molecules of the material arrange together and form the nuclei. Formation of a new phase requires energy due to its higher solubility or vapor pressure, which may cause its disappearance (Van Hook, 1961). The vapor pressure variation of small droplets is given by the Kelvin equation, which is given in equation (2.1), where $p$ is vapor pressure of a droplet, $p_\infty$ is vapor pressure at a plane surface, $\sigma$ is surface tension, $v$ is molar volume, $R$ is the gas constant, and $r$ is the radius of the droplet.

$$\ln \frac{p}{p_\infty} = \frac{2\sigma v}{rRT} \tag{2.1}$$

$$\ln \frac{L}{L_\infty} = \frac{2\sigma M}{r\rho RT} \tag{2.2}$$

The Kelvin equation is one of the most important equations in the quantitative theories of crystallization (Van Hook, 1961) and it can be written into the form of the Ostwald-Freundlich equation, which is given in equation (2.2), where $L$ is solubility, $L_\infty$ is the solubility over extended plane surfaces, i.e., the typical solubility of the material, $M$ is molecular weight, $\rho$ is density, and $r$ is the radius of the nucleus (Van Hook, 1961; Brennan *et al.*, 1990).

The quantities of $p/p_\infty$ or $L/L_\infty$ in equations (2.1) and (2.2), respectively, are measures of the extent of supersaturation or supercooling, $\alpha$. The critical radius, $r_c$, for crystal formation can be obtained from equation (2.3). Equation (2.3) defines that spontaneous nucleation may occur when nuclei with a size of $r > r_c$ are formed.

$$r_c = \frac{2\sigma M}{\rho RT \ln \alpha} \tag{2.3}$$

It has been observed that the rate of nucleation increases with an increase in supersaturation. The supersaturated state may result from a decrease of temperature, further removal of solvent at a constant temperature, or both. The rate of nucleation may be assumed to follow the Arrhenius-type temperature dependence according to equation (2.4), where $v$ is the rate of nuclei formation, $k$ is the frequency factor, and $E_n$ is the minimum energy required for nucleus formation (Van Hook, 1961; Brennan *et al.*, 1990).

$$v = ke^{-\frac{E_n}{RT}} \tag{2.4}$$

Van Hook (1961) reported that the work required for the formation of a spherical nucleus with the critical size is one-third of the work that is required for the forming of its own surface. According to Van Hook (1961) the rate of nucleation is defined by equation (2.5). The rate of nucleation can be shown to have a dramatic increase above a critical value of supersaturation or supercooling as shown in Figure 2.1.

$$v = ke^{\left(-\frac{16\pi M^2 \sigma^3}{3d^2 R^3 T^3 \ln^2 \alpha}\right)} \tag{2.5}$$

As defined by the critical size of a nucleus the requirement for the occurrence of homogeneous nucleation is that molecules form clusters due to molecular collisions. The size of such clusters must become sufficiently large and exceed the energy barrier for nucleation (Hartel and Shastry, 1991). Homogeneous nucleation requires a fairly large extent of supercooling or supersaturation and its occurrence even in chemically pure materials is rare. Most crystallization processes are not homogeneous due to the presence of foreign surfaces and particles that are in contact with the material.

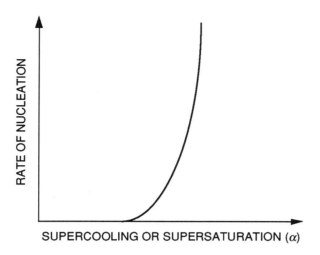

**Figure 2.1** Effect of supercooling or supersaturation, $\alpha$, on the rate of nucleation. As the driving force, $\alpha$, for nucleation increases above a critical value nucleation occurs with a rate that increases dramatically with increasing $\alpha$. After Van Hook (1961).

*b. Heterogeneous Nucleation.* Heterogeneous nucleation is the main path in the crystallization processes that occur in food materials. Heterogeneous nucleation involves the presence of foreign bodies or impurities that may act as the nucleating sites. The foreign bodies reduce the energy needed for the formation of the critical nuclei and therefore facilitate crystallization (Van Hook, 1961). The extent of supercooling or supersaturation is also lower in the heterogeneous than in the homogeneous nucleation process. According to Hartel and Shastry (1991) sugar syrups that have been filtered may require higher supersaturations for nucleation than syrups containing more impurities. The mechanisms of nucleation and various factors affecting the formation of nuclei in foods are not well understood. However, information on nucleation in supercooled solutions exists to some extent (Van Hook, 1961; Hartel and Shastry, 1991), but nucleation phenomena in supercooled amorphous food materials have not been studied. It should be noticed that nucleation may be reduced in the presence of other compounds, e.g., in foods that are composed of mixtures of several solutes.

*c. Secondary Nucleation.* Secondary nucleation differs from homogeneous and heterogeneous nucleation processes in that it occurs in the presence of existing crystals of the nucleating compound. According to Hartel and Shastry (1991) secondary nucleation requires external forces, e.g., agitation of saturated solutions. Secondary nucleation may result from mechanical reduction of the size of the existing crystals, which may occur due to shear forces. The small crystal pieces may grow greater and exceed the critical size of the stable nuclei. Secondary nucleation phenomena are common in sugar crystallization processes (Hartel and Shastry, 1991).

## 2. Crystal growth

The nucleation step in crystallization is followed by crystal growth. Crystal growth requires that molecules are able to diffuse to the surface of the growing nuclei. The rate of the process is very sensitive to the extent of supersaturation or supercooling, temperature, and the presence of impurities. The effects of impurities are particularly important to overall crystallization rates in food materials.

The diffusion of molecules on the growing nuclei surface is the main requirement of crystal growth. According to Van Hook (1961) the growth rate of a nucleus, $J$, is defined by equation (2.6), where $E_d$ is the activation energy of diffusion and $(A)T$ is the work required to form the surface of the nucleus.

$$J = ce^{-\frac{E_d}{RT}} e^{-\frac{(A)T}{RT}} \qquad (2.6)$$

Crystal growth depends on several factors, but when nuclei grow as spheres the velocity of crystallization follows equation (2.7), where $Q_{cr}$ is the velocity of crystallization, $c$ is a constant, $v$ is the molar volume, and $\Delta H_{cr}$ is the latent heat of crystallization.

$$Q_{cr} = ce^{\left\{ -\frac{1}{RT}\left[ E_d + \frac{4}{3}\pi\sigma\left( \frac{2v\sigma T}{\Delta H_{cr}(T_0-T)} \right)^2 \right] \right\}} \qquad (2.7)$$

Equations (2.6) and (2.7) may be used for establishing a growth curve similar to that shown in Figure 2.2. Van Hook (1961) reported increasing rates of crystallization with increasing supersaturation for sucrose when supersaturation was low. Crystallization became delayed at high levels of supersaturation as well as in the presence of impurities.

In addition to diffusion the growth of existing nuclei requires that several other requirements have been accomplished. Hartel and Shastry (1991) reported that such requirements in sugar crystallization may include (1) diffusion of the compound from bulk to the solid surface; (2) mutarotation of

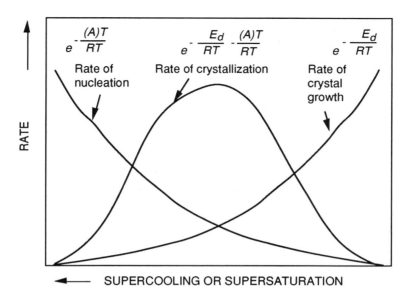

$$e^{-\frac{(A)T}{RT}} \qquad\qquad e^{-\frac{E_d}{RT} - \frac{(A)T}{RT}} \qquad\qquad e^{-\frac{E_d}{RT}}$$

Rate of nucleation    Rate of crystallization    Rate of crystal growth

RATE

← SUPERCOOLING OR SUPERSATURATION

**Figure 2.2** A schematic representation of the effect of supercooling or supersaturation on rate of nucleation, crystal growth, and crystallization. At high levels of supercooling or supersaturation the rate of nucleation is high but the rate of diffusion is low, which decreases the rate of crystallization. At low levels of supercooling the rate of diffusion is high but the rate of nucleation is low. Therefore, the maximum rate of crystallization occurs at intermediate levels of supercooling or supersaturation. After Van Hook (1961).

molecules to the anomeric form for growth; (3) removal of hydration water; (4) diffusion of water away from the crystal surface; (5) proper orientation of the molecules; (6) surface diffusion of the molecules to appropriate sites; (7) incorporation of the growth unit into the crystal lattice; and (8) removal of latent heat. These steps are important in defining the rate-limiting steps for crystallization and they are used in various crystal growth theories. According to Hartel and Shastry (1991) the diffusion theory assumes that molecular diffusion to the surface is the rate-limiting step of crystal growth. Adsorption layer theories assume formation of an adsorption layer of molecules on the crystal surface in which formation of a growth unit, diffusion of the growth unit to an appropriate site, or incorporation of the growth unit to the lattice may be the rate-limiting factors.

## III.   The Physical State of Amorphous Materials

A number of synthetic polymers but also most food materials exist in an amorphous state, which is a nonequilibrium state at temperatures below the equilibrium melting temperature of the material. Physical properties of the amorphous state are related to the mechanical behavior and stability of various solid and highly viscous food materials. It is well known in polymer science that melting of crystalline polymers results in the formation of an amorphous melt, which can be supercooled to a viscoelastic, "rubbery" state or to a solid, glassy state. The state transition that occurs between the rubbery and glassy states is known as the second-order glass transition.

### A.  Mechanical Properties

The physical state of amorphous materials defines their mechanical properties. Mechanical properties of glassy materials resemble those of solid materials, but the molecules have no definite order such as solids that exist in the equilibrium, crystalline state. A dramatic change in mechanical properties is observed as amorphous materials are plasticized and transformed to the supercooled liquid-like state. The characterization of the physical state of amorphous materials requires definition of glass formation and modulus, which can be used in determination of changes in the physical state.

1. Glass formation and glass transition

Quenching of a liquid to the form of a glass is the most common method of producing amorphous supercooled materials and glasses. Elliot *et al.* (1986) listed a number of other methods that may be used to produce amorphous materials. The methods included rapid evaporation of solvent molecules, which occurs in food processing, among a number of other methods, which allow a sufficiently rapid cooling or solvent removal to avoid the formation of the equilibrium crystalline state. A common feature of amorphous materials is that they contain excess free energy and entropy in comparison to their crystalline counterparts at the same temperature and pressure conditions.

At low temperatures amorphous materials are vitreous, i.e., they are glassy, hard, and brittle (Sperling, 1986). An increase in temperature may cause the transformation of such materials into the less viscous, "rubbery" state at the glass transition temperature, $T_g$. The glass transition temperature of polymers is defined to be the temperature at which the material softens due to the onset of long-range coordinated molecular motion (Sperling, 1986). It is important to notice that the physical behavior of foods may differ from that of polymers. A general rule for polymers is that cross-linked amorphous polymers exhibit rubber elasticity and linear amorphous polymers exhibit flow at temperatures above $T_g$ (Sperling, 1986). Amorphous polymers show no X-ray patterns that are typical of crystalline materials and they show no melting at the first-order melting transition temperature.

The physical state of amorphous materials below melting temperature, $T_m$, is thermodynamically a nonequilibrium state. Therefore, various properties of amorphous materials depend on temperature and time. The physical properties and applicability of synthetic, noncrystalline polymers are strongly related to their glass transition temperature. The structure of amorphous materials often "stabilizes" as a function of time when they are annealed at a temperature slightly below $T_g$ (Elliot *et al.*, 1986). Because of the metastability of the amorphous state such materials tend to crystallize when sufficient time is allowed at temperatures above $T_g$. The importance of the $T_g$ of amorphous food materials to the processability and storage stability has been recognized and emphasized particularly by Levine and Slade (e.g., 1986; Slade and Levine, 1991, 1993).

## 2. Young's modulus

Young's modulus can be used to evaluate the elasticity of various materials. According to Hook's law the ratio of the stress to the strain produced within the elastic limits of any body is constant. Young's modulus, $E$, is a measure of material stiffness and it is defined by equation (2.8), where $\sigma$ and $\varepsilon$ refer to tensile stress and strain, respectively.

$$\sigma = E\varepsilon \qquad (2.8)$$

The tensile stress is the force per unit area that causes the strain. The strain is the difference between the final length of the body, $L$, and the initial length of the body, $L_0$, divided by the initial length of the body according to equation (2.9).

$$\varepsilon = \frac{(L - L_0)}{L_0} \qquad (2.9)$$

Young's modulus of glassy polymers just below the glass transition temperature is almost constant, being $3 \times 10^9$ Pa (Sperling, 1986). Compliance may be used to measure softness instead of stiffness and it can be defined to be equal to $1/E$.

### 3. Shear modulus

Young's modulus applies to tensile stress and strain. In addition to elongation or compression an external stress may cause shear. The viscoelastic properties of amorphous materials are often related to their behavior in shear and the changes in the physical state are observed from changes in the shear modulus. The shear modulus, $G$, may be defined to be the ratio of the shear stress, $f$, to the shear strain, $s$, according to equation (2.10).

$$G = \frac{f}{s} \tag{2.10}$$

Shear strain and shear stress are related to viscosity, $\eta$. The relationship is given by equation (2.11), where $t$ is time.

$$f = \eta \left( \frac{ds}{dt} \right) \tag{2.11}$$

Both Young's modulus and the shear modulus are high for glassy materials and they decrease dramatically at temperatures above $T_g$.

### 4. Storage and loss moduli

Cyclical or repetitive motions of stress and strain may be used to obtain mechanical moduli. The modulus caused by such motions is defined by complex Young's modulus, $E^*$, or by complex shear modulus, $G^*$. The complex Young's modulus is defined by equation (2.12), where $E'$ refers to *storage modulus* and $E''$ is the *loss modulus*.

$$E^* = E' + E'' \sqrt{-1} \tag{2.12}$$

A similar relationship can be established for the shear modulus, which is given by equation (2.13).

$$G^* = G' + G'' \sqrt{-1} \tag{2.13}$$

The storage modulus as defined by either $E'$ or $G'$ is a measure of the amount of energy that is stored elastically during a deformation, which is caused by the tensile or shear stress, respectively. The loss modulus as defined by either $E''$ or $G''$ is related to the amount of energy that is converted to heat during the deformation. Since viscosity is related to the amount of energy that is converted to heat, the loss modulus is also a measure of viscosity. The *loss tangent*, tan $\delta$, can be defined to be the ratio of $E''$ and $E'$. Storage and loss moduli are often determined as a function of temperature and they show a characteristic change at the $T_g$. The quantities of $E''$ or $G''$ and tan $\delta$ have a maximum at the glass transition of an amorphous material.

### B. Characterization of the Physical State

Physical properties of amorphous materials are dependent on their composition, temperature, and time. There are no exact definitions for the various physical states similar to those of the crystalline, solid, and gaseous states. A fully amorphous material can exist in the glassy state or in the viscous, supercooled, liquid state. The transition between these states occurs over the glass-rubber transition region. The term *glass transition* refers to the temperature at which ordinary inorganic glasses begin to soften and flow. Although the transition occurs over a temperature range rather than at an exact temperature, it has the thermodynamic properties of a second-order phase transition at very slow rates of temperature change (Sperling, 1986). The modulus curve of amorphous polymers is often divided into various flow regions according to Ferry (1980).

### 1. The glassy state

Amorphous materials in the glassy state are practically solid liquids that have an extremely high viscosity of $\eta > 10^{12}$ Pa s. Such viscosities are difficult to measure. Typical properties of glassy materials are brittleness and transparency. Their molecular motions are restricted to vibrations and short-range rotational motions (Sperling, 1986). Changes that occur in the glassy state are extremely slow and they are often referred to as *physical aging*. Glassy foods are often considered to be stable and the glassy state is of great importance to textural characteristics of crispy foods such as potato chips, cookies, and extruded snacks (Levine and Slade, 1986; Slade and Levine, 1991).

### 2. Glass transition temperature range

Transformation of glassy materials into the viscous, supercooled, liquid state occurs over the glass transition temperature range. Glass transition results in a dramatic drop of the modulus from about $10^9$ to $10^6$ Pa. The drop in modulus occurs over a temperature range of 10 to 30°C. However, it is important to notice that over the glass transition temperature range, a few degrees change in temperature may have a significant effect on stiffness. The change in stiffness is particularly important to the structure and stability of amorphous food materials. There is also a dramatic change in molecular mobility within the glass transition temperature range. According to Sperling (1986) the $T_g$ may also be defined to be the onset temperature of long-range, coordinated molecular motion, which does not occur below $T_g$ due to the freezing of the molecules to the solid state at $T_g$.

## 3. Rubbery plateau region

The glass transition temperature range of several polymers is followed by a temperature range over which the modulus remains constant. This temperature range is known as the rubbery plateau region at which polymers have an almost constant modulus of $2 \times 10^6$ Pa and they show long-range rubber elasticity (Sperling, 1986). At temperatures within the rubbery plateau region elastomers can be stretched and they regain the original length when the force is released. The temperature range of the rubbery plateau is a function of molecular weight and it is affected by the linearity of the molecules. The width of the plateau of linear molecules increases with increasing molecular weight and they exhibit a slow drop in modulus. Cross-linking improves rubber elasticity. These factors may significantly affect also the physical properties of amorphous food components. The modulus also increases with increasing crystallinity (Sperling, 1986). It should be noticed that the existence of the rubbery plateau is not typical of inorganic amorphous materials or low molecular weight amorphous compounds such as sugars or water.

## 4. Rubbery flow region

The rubbery plateau temperature region of polymers is followed by a rubbery flow region. Over the rubbery flow temperature range amorphous polymers may exhibit both rubber elasticity and flow depending on the experimental time scale (Sperling, 1986). Rubbery flow is a property of linear polymers. Cross-linked polymers remain in the rubbery plateau region until they decompose at high temperatures.

5. Liquid flow region

Liquid flow region follows the rubbery flow region as temperature is increased. At the liquid flow region synthetic polymers flow readily and they may behave like molasses (Sperling, 1986). The modulus of semicrystalline polymers depends on crystallinity. The amorphous portions of such materials go through the glass transition at $T_g$. However, the crystalline regions remain hard at temperatures $T_g < T < T_m$ and a rapid drop in modulus is observed as the crystalline regions melt at $T_m$. Such behavior may occur in polymeric food components such as starch, which is composed of partially crystalline carbohydrate polymers (Slade and Levine, 1991). It should be noticed that the $T_g$ of a number of synthetic polymers and also of food components occurs at about 100°C below $T_m$. Therefore, most amorphous materials approach the equilibrium liquid state at about $T_g + 100°C$.

### C. Glass Transition Theories

The basic theories of the occurrence of glass transition are the free volume theory, the kinetic theory, and the thermodynamic theory. The free volume theory assumes that molecular motion depends on the presence of holes, vacancies, or voids that allow molecular movement. The holes between molecules provide the free volume that is needed for molecular rearrangement. The kinetic theory defines glass transition as the temperature at which the relaxation time for the segmental motion of polymer chains approaches that of the experimental time scale. The thermodynamic theory uses the concept of equilibrium and the thermodynamic requirements for a real second-order transition.

1. Free volume theory

The free volume theory is based on the change in volume expansion coefficient that occurs at the glass transition. It assumes that the fractional free volume, $f$, in the glassy state is constant and it increases above $T_g$ according to equation (2.14), where $f_g$ is the fractional free volume at $T_g$ and $\alpha_f$ is the thermal expansion coefficient of free volume.

$$f = f_g + \alpha_f (T - T_g) \tag{2.14}$$

The fractional free volume is defined by equation (2.15) and the specific free volume is given by equation (2.16), where $v$ is macroscopic volume, $v_f$

is specific free volume, and $v_0$ is the volume occupied by molecules (Tant and Wilkes, 1981).

$$f = \frac{v_f}{v} \qquad\qquad (2.15)$$

$$v_f = v - v_0 \qquad\qquad (2.16)$$

$$v_f = K + (\alpha_R - \alpha_G)T \qquad\qquad (2.17)$$

Fox and Flory (1950) suggested that the specific free volume of polystyrene with an infinite molecular weight above $T_g$ followed equation (2.17), where $K$ is a constant that is related to the free volume at 0 K, and $\alpha_R$ and $\alpha_G$ are volume expansion coefficients in the rubbery and glassy states,

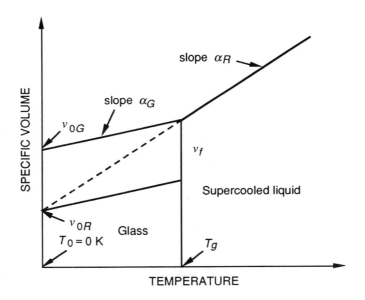

**Figure 2.3** A schematic representation of the determination of free volume. The specific volume, $v$, in the glassy state, which has the value $v_{0G}$ at 0 K, is temperature-dependent as defined by the thermal expansion coefficient, $\alpha_G$. Thermal expansion above the glass transition temperature, $T_g$, is defined by the thermal expansion coefficient, $\alpha_R$. The extrapolated specific volume of the supercooled liquid at 0 K is $v_{0R}$. The specific free volume, $v_f$, is given by the difference in the extrapolated specific volumes of the supercooled liquid and glassy states at 0 K. After Sperling (1986).

respectively. Below $T_g$ the temperature-volume relationship was found to be independent of molecular weight. The determination of specific free volume is shown in Figure 2.3. Reported values for specific free volume at $T_g$ vary depending on polymer from about 1 to 13% (Sperling, 1986). Tant and Wilkes (1981) pointed out that the free volume theory has been useful in the description of transport properties such as viscosity and diffusivity in liquids. The free volume theory can also be used to relate viscoelastic motion to temperature and time and to explain thermal expansion at temperatures below and above the glass transition. The inadequacies of the free volume theory were discussed by Perez (1994).

## 2. Kinetic theory

It has been recognized that experimentally determined glass transition temperatures depend on the experimental time scale. The kinetic theory of glass transition considers the time-dependent characteristics of the glass transition and the time-dependent molecular relaxations that take place over the glass transition temperature range.

According to Sperling (1986) the kinetic theory assumes that matter may have holes with a molar volume, $v_h$, or it may exist in a no-hole situation with molar excess energy, $\varepsilon_h$. The activation energy for the disappearance of a hole is $\varepsilon_j$. The partition function of the holes is $Q^h$ and that in the activated state is $Q^{\Xi}$. Removal of holes decreases volume. In the glassy state the number of holes and their spatial positions become frozen in and the molecules may not move to holes (Sperling, 1986). The equilibrium number of holes, $N_h^*$, is given by equation (2.18), where $N_0$ and $v_0$ are the number of moles of repeating units and the molar volume per repeating unit, respectively. The relaxation time for the disappearance of holes, $\tau_h$, is given by equation (2.19), where $h$ and $k$ are the Planck and Boltzmann constants, respectively.

$$N_h^* = N_0 (v_0 / v_h) e^{-\varepsilon_h / RT} \tag{2.18}$$

$$\tau_h = \frac{h}{kT} \frac{Q^h}{Q^{\Xi}} e^{\varepsilon_j / RT} \tag{2.19}$$

The change in the number of holes that go through the glass transition yields the change in heat capacity, $\Delta C_p$. Thus, experimental determinations of $\Delta C_p$ can be used to calculate $\varepsilon_h$. The value for $v_0/v_h$ may also be derived from experimental values for density and cohesive energy.

3. Thermodynamic theory

All amorphous materials have a glass transition with properties typical of a second-order transition. The thermodynamic theory assumes that the glass transition attains a true equilibrium at an infinitely long experiment. The true $T_g$ may be observed at a temperature that is about 50°C lower than that observed with common times of experiments (Sperling, 1986). The thermodynamic theory of the glass transition requires that a glassy phase with an entropy that is negligibly higher than that of the crystalline phase can be formed.

4. Other theories

The free volume, kinetic, and thermodynamic theories are the most applied in the evaluation of the changes that are observed at the glass transition temperature. These theories have both advantages and disadvantages in relating viscoelastic properties, expansion coefficients, heat capacity, molecular mobility, composition, and a number of other properties of amorphous materials to $T_g$. Several other theories of glass transition have also been reported. The mode coupling theories use a generalized viscosity term that is expressed in terms of products of correlation functions, which introduce coupling effects between different modes of vibration (Perez, 1994). Other theories are based on the hierarchical correlated molecular motion and computer simulation of quasi-point defects in disordered condensed matter (Perez, 1994).

## IV.  Molecular Mobility and Plasticization

Molecular mobility of amorphous compounds becomes apparent at the glass transition temperature. As the temperature increases above $T_g$ the molecular mobility increases, which is observed from decreasing viscosity and increasing flow. In such cases temperature has a plasticizing effect on the material. However, plasticizers, which often have a lower molecular weight than the bulk material, may decrease the glass transition temperature and therefore cause plasticization. It is also well known that a diluent may cause a depression of both the melting temperature and glass transition temperature of a polymer. The extent of the depression of $T_m$ and $T_g$ depends on the amount of the diluent and its compatibility with the polymer. Food solids are often water soluble or miscible compounds. It is well known that food solids often become soft if the water content is increased. Thus, water can be considered as a plasticizer of food materials. Water has also been found to act as a $T_g$ decreasing plasticizer especially in natural polymers such as cellulose and collagen, and to a lesser degree in synthetic polyamides (Batzer and Kreibich, 1981). The decrease of $T_g$ becomes more pronounced with increasing $T_g$ of the polymer at low water contents (Batzer and Kreibich, 1981).

## A. Mechanical Properties

According to Sperling (1986) the molecular mobility in the rubbery and liquid states of polymers involves 10 to 50 carbon atoms, while the molecular mobility in the glassy state is restricted to vibrations, rotations, and motions by relatively short segments of the polymer chains. The molecular mobility increases dramatically at temperatures above $T_g$. Relaxation times of mechanical properties are related to molecular mobility, which is related to temperature. Relationships between temperature and relaxation times are used in the time-temperature superposition principle (Ferry, 1980). The time-temperature superposition principle applies to the mechanical properties of various amorphous materials including food components and food solids.

## 1. Relaxation times

According to Williams *et al.* (1955) a single empirical function may be used to describe the temperature dependence of mechanical properties of amorphous polymers above their glass transition temperature. The ratio, $a_T$, of relaxation times $\tau$ and $\tau_s$ of configurational rearrangements at a temperature, $T$, to that at a reference temperature, $T_s$, respectively, as defined by equation (2.20) reflects the temperature dependence of mobility.

$$a_T = \frac{\tau_T}{\tau_{T_s}} \qquad (2.20)$$

The temperature dependence of the relaxation times of mechanical properties are often predicted by the Williams-Landel-Ferry (WLF) equation.

*a. The WLF Equation.* Williams *et al.* (1955) used literature data for $a_T$ that were derived from measurements of dynamic mechanical, stress relaxation, and dielectric properties and viscosity to show that $a_T$ followed the empirical equation (2.21), where $C_1$ and $C_2$ are constants and $T_s$ is a reference temperature. The only food component for which data were included was glucose. Williams *et al.* (1955) used viscosity data for glucose over a wide temperature range above $T_g$. They found that plots of $a_T$ against $T$ could be matched for a number of inorganic, organic, and synthetic glass-forming compounds using horizontal and vertical translations. The average $T_s - T_g$ value was found to be $50 \pm 5°C$ and the constants $C_1$ and $C_2$ were 8.86 and 101.6, respectively.

$$\log a_T = \frac{-C_1(T - T_s)}{C_2 + (T - T_s)} \tag{2.21}$$

Equation (2.21) is the original form of the well known WLF equation. Williams *et al.* (1955) suggested that the WLF equation was applicable over the temperature range from $T_g$ to $T_g + 100°C$. They pointed out that the use of $T_g$ as the reference temperature instead of $T_s$ was possible if the "universal" constant values were $C_1 = 17.44$ and $C_2 = 51.6$. However, the use of $T_s$ at $T_g + 50°C$ was considered to be more accurate.

The free volume theory assumes that the temperature dependence of viscosity is related to free volume. Doolittle (1951) showed that viscosity was related to free volume according to equation (2.22), where $A$ and $B$ are constants, $v_0$ is occupied volume, and $v$ is specific free volume.

$$\ln \eta = \ln A + B(v_0 / v) \tag{2.22}$$

Williams *et al.* (1955) were able to show that $\ln a_T$ is obtained also from equation (2.23), where $f_g$ is the fractional free volume at the glass transition temperature and $\alpha_2$ is the difference of the thermal expansion coefficient below and above $T_g$.

$$\ln a_T = \frac{-f_g^{-1}(T - T_g)}{\left(\dfrac{f_g}{\alpha_2}\right) + (T - T_g)} \tag{2.23}$$

Equation (2.23) has the form of the WLF equation. Therefore, it can be shown that the universal constants suggest that the values for $\alpha_2$ and $f_g$ are $4.8 \times 10^{-4} \, °C^{-1}$ and 0.025, respectively.

*b. WLF Constants.* Although Williams *et al.* (1955) found that the data for several glass forming inorganic, organic, and organic polymeric compounds followed the WLF model with the universal constants, their use has been criticized. The universal constants may be used only if sufficient data for calculating specific constants are unavailable. The WLF equation can be written also into the form of a straight line that is given by equation (2.24).

$$\frac{1}{\ln a_T} = \frac{1}{-C_1} - \frac{C_2}{C_1(T - T_s)} \qquad (2.24)$$

The relaxation times of mechanical properties at $T_g$ are extremely long and the experimental determination of $\tau_g$ is often impossible. However, the reference temperature, $T_s$, can be located at another temperature above $T_g$ at which experimental data for $\tau_s$ are available. Equation (2.24) suggests that a plot of $1/\ln a_T$ against $1/(T - T_s)$ gives a straight line. It is obvious that the WLF constants can be derived from the slope and the intercept of the straight line. However, the constants $C_1$ and $C_2$ are dependent on the location of $T_s$ and it is often practical to use the glass transition temperature as the reference temperature. A shift in the reference temperature requires that correct values for the constants at the other reference temperature are available. Corrected values for the constants can be derived with equations (2.25) and (2.26), where $C'_1$ and $C'_2$ are the corrected values for the constants $C_1$ and $C_2$, respectively, at a temperature $T'_s = T_s - \delta$ (Ferry, 1980; Peleg, 1992).

$$C'_1 = \frac{C_1 C_2}{C_2 - \delta} \qquad (2.25)$$

$$C'_2 = C_2 - \delta \qquad (2.26)$$

Ferry (1980) pointed out that equation (2.26) suggests that the relationship given in equation (2.27) exists between $C_2$ and temperature, where $T_\infty$ is a fixed temperature at which $\log a_T$ becomes infinite.

$$T'_s - C'_2 = T_s - C_2 = T_\infty \qquad (2.27)$$

The relationship given by equation (2.27) can be considered in the WLF equation, which may be rewritten into the form of equation (2.28).

$$\log a_T = \frac{C'_1 (T - T'_s)}{T - T_\infty} \qquad (2.28)$$

According to equation (2.28) a plot of $\log a_T$ against $(T - T'_s)/(T - T_\infty)$ is linear through the origin and it has the slope $C'_1$. Ferry (1980) pointed out that $T_\infty$ is often at about $T_g - 50°C$. However, computer programs may be used to fit experimental data to equation (2.28) and to derive a more appropriate $T_\infty$, which may be used to obtain $C'_2$ from equation (2.27). A shift of the reference temperature to $T'_s = T_g$ allows prediction of the relaxation times at temperatures above $T_g$ and a comparison of the universal constants with those derived from the experimental data (Peleg, 1992).

## 2. Viscosity

The glassy state is an isoviscous state with a viscosity of about $10^{12}$ Pa s (e.g., White and Cakebread, 1966; Sperling, 1986). Relaxation times of mechanical changes in the glassy state are extremely long, but as the temperature is increased to above the glass transition temperature dramatic decreases in the relaxation times of mechanical changes and in viscosity are observed.

Viscosity is a time-dependent quantity that can be related to the relaxation times above glass transition (Sperling, 1986). The temperature dependence of the relaxation times and therefore viscosity above $T_g$ is defined by equation (2.29), where $\tau$ and $\eta$, and $\tau_s$ and $\eta_s$, refer to relaxation time and viscosity at $T$ and $T_s$, respectively. Thus, the WLF equation can be used to relate viscosity or any other temperature-dependent mechanical property at temperatures above $T_g$ to a reference temperature, $T_s$, or to $T_g$ according to equation (2.30).

$$\ln a_T = \ln\left(\frac{\tau}{\tau_s}\right) = \ln\left(\frac{\eta}{\eta_s}\right) \tag{2.29}$$

$$\ln\frac{\eta}{\eta_s} = \frac{-C_1(T-T_s)}{C_2+(T-T_s)} \tag{2.30}$$

Other equations are also commonly used to predict temperature dependence of viscosity of supercooled liquids. These equations include the Vogel-Tammann-Fulcher (VTF) equation (2.31) and the power-law equation (2.32), where $A$, $D$, and $r$ are constants.

$$\eta = \eta_s e^{DT_s/(T-T_s)} \tag{2.31}$$

$$\eta = A(T-T_g)^r \tag{2.32}$$

Viscosity data above $T_g$ may follow the VTF and power-law equations in addition to the WLF equation, but the free volume-based WLF theory has probably been the most applicable in predicting temperature-dependent mechanical properties of amorphous materials. However, the VTF and WLF equations are interconvertible and various $D$ values correspond to various WLF constants. Angell *et al.* (1994) used the $D$ value to classify liquids into fragile and strong liquids. The liquid relaxation behavior is observed from the $D$ value. Strong liquids that resist thermal degradation have a large $D$ value while those with small $D$ are fragile liquids. Angell *et al.* (1994) also pointed out that strong liquids exhibit only small or undetectable changes in heat capacity at $T_g$, whereas fragile liquids show large changes in heat capacity at $T_g$.

## 3. Dynamic mechanical properties

Mechanical loss measurements can be used to probe the time scales of molecular motions in the region of the glass transition (Allen, 1993). The relaxation measurements apply an oscillating stress or strain and the response of the sample to the external force. The energy transformed to heat is that lost during the experiment. The measurements can be done isothermally to determine the loss factor as a function of frequency or by keeping the frequency constant to determine the response (*damping*) to the frequency as a function of temperature (Allen, 1993).

The time and frequency dependence of viscoelastic functions reflects the response of the system to the duration of the experiments. The response is dependent on the time required for the slowest rearrangement of molecules, i.e., relaxation time. In a mechanical measurement a maximum energy loss is observed at the temperature where the average of molecular motions is equivalent to the applied frequency. In the glassy state the storage modulus, $G'$, is $10^9$-$10^{11}$ Pa. At the glass transition temperature range $G'$ drops to $10^3$ to $10^5$ Pa. At the rubbery plateau region $G'$ shows only slight dependence on frequency, but the loss modulus, $G''$, shows a minimum. In the glassy state $G''$ is considerably smaller than $G'$, since the response to external stress is very small. At the glass transition $G'$ increases with frequency (Allen, 1993).

## B. Plasticization and Molecular Weight

Plasticization and molecular weight affect both first-order and second-order transition temperatures. In amorphous materials plasticizers may be considered as compounds that increase the free volume and therefore depress the transition temperatures. Molecular weight can also be related to the free volume of homopolymers and to the plasticizing effect of diluents.

1. Melting temperature

Flory (1953) reported relationships between equilibrium melting temperature of a polymer and the amount of a diluent. The relationship between $T_m$ and volume fraction of the diluent is defined by the Flory-Huggins equation.

The Flory-Huggins equation is given in equation (2.33), where $T_m^0$ is the equilibrium melting temperature, $T_m$ is melting temperature in the presence of the diluent, $R$ is the gas constant, $\Delta H_u$ is the latent heat of melting per repeating unit, $V_u/V_1$ is the ratio of the molar volume of the repeating unit of the polymer and that of the diluent, $\chi_1$ is the polymer-diluent interaction parameter, and $v_1$ is the volume fraction of the diluent.

$$\frac{1}{T_m} - \frac{1}{T_m^0} = \frac{R}{\Delta H_u} \frac{V_u}{V_1} (v_1 - \chi_1 v_1^2) \tag{2.33}$$

A common assumption is that $\chi_1$ in equation (2.33) is zero. Thus, a plot of $1/T_m$ against $v_1$ gives a straight line with intercept at $1/T_m^0$.

2. Glass transition and molecular weight

The molecular weight of polymers can be related to their free volume. Common synthetic polymers and a number of natural carbohydrate polymers and polymer mixtures, but not native proteins, have a distribution of molecular weights. Therefore, the molecular weight can be given as the number average molecular weight, $M_n$, or as the weight average molecular weight, $M_w$, and the ratio $M_w/M_n$ can be used as a measure of the molecular weight distribution. The number average molecular weight is given by equation (2.34), where $n_i$ is the number of molecules with the molecular weight $M_i$. The weight average molecular weight is defined by equation (2.35).

$$M_n = \frac{\sum_i n_i M_i}{\sum_i n_i} \tag{2.34}$$

$$M_w = \frac{\sum_i n_i M_i^2}{\sum_i n_i M_i} \tag{2.35}$$

Fox and Flory (1950) prepared ten polystyrene fractions with molecular weights ranging from 2970 to 85,000. The glass transition temperature for each fraction was determined from volume-temperature curves that were obtained from dilatometric measurements. The $T_g$ was found to increase rapidly with increasing molecular weight up to 25,000 and then to level off for materials with higher molecular weights. Fox and Flory (1950) found that a plot of $T_g$ against the reciprocal molecular weight gave a linear relationship. They concluded that the transition temperature increased with increasing molecular weight giving a linear relationship between the transition temperature and the reciprocal of the number average molecular weight. The Fox and Flory equation that is given in equation (2.36), where $T_g(\infty)$ is the limiting value of $T_g$ at high molecular weight and $K$ is a constant, has been shown to apply for a number of synthetic polymers and biopolymers, including starch hydrolysis products (maltodextrins) (Roos and Karel, 1991a).

$$T_g = T_g(\infty) - \frac{K}{M_n} \qquad (2.36)$$

It is obvious that the addition of low molecular weight components to polymers or the decrease of the size of polymer molecules decreases their glass transition temperature. An increase in the size of polymer molecules that occurs in thermosetting results in an increase of $T_g$.

3. Glass transitions of mixtures

The free volume of a liquid (plasticizer) with a low molecular weight such as water in foods is very large in comparison to that of a polymer at the same temperature and pressure (Levine and Slade, 1986; Slade and Levine, 1991). Knowledge of the effects of composition on the glass transition temperature is of significant importance in the evaluation of the technological properties of polymers.

Couchman and Karasz (1978) suggested a thermodynamic theory, which related the glass transition temperature of binary blends of miscible polymers to composition. The $T_g$ dependence of composition was suggested to follow equation (2.37), where $x_1$, $T_{g1}$, and $\Delta C_{p1}$ and $x_2$, $T_{g2}$, and $\Delta C_{p2}$ refer to the mole fraction, glass transition temperature, and change in heat capacity at glass transition of the components 1 and 2, respectively. It should be noticed that equation (2.37) was identical with the empirical Gordon and Taylor (1952) equation with a constant, $k = \Delta C_{p2}/\Delta C_{p1}$.

$$T_g = \frac{x_1 T_{g1} + (\Delta C_{p1} / \Delta C_{p2}) x_2 T_{g2}}{x_1 + (\Delta C_{p1} / \Delta C_{p2}) x_2} \qquad (2.37)$$

The change in heat capacity that occurs over the glass transition temperature range is also dependent on composition. The $\Delta C_p$ of sugars plasticized by water have been found to increase linearly with increasing weight fraction of water (Roos and Karel, 1991b; Hatley and Mant, 1993).

## C. Crystallization of Amorphous Compounds

Crystallization mechanisms of amorphous food components have not been well established. Obviously, the driving force for crystallization from the amorphous supercooled state is the degree of supercooling and the rates of nucleation and crystal growth may be assumed to follow those in supersaturated solutions. Crystallization behavior of amorphous food components may vary significantly depending on various factors, and in particular on the specific properties of the components and on the degree of polymerization. Polymers and sugars with a high value for the $T_m/T_g$ ratio tend to crystallize rapidly (Roos, 1993).

### 1. Nucleation and crystal growth

Low molecular weight amorphous food components may crystallize at temperatures higher than the glass transition temperature but below the melting temperature. Such compounds as amorphous sugars crystallize to the thermodynamically favorable form although the presence of other food components may delay crystallization. Crystallization behavior of amorphous food component polymers is more complicated and it may be assumed to correspond to that of synthetic polymers.

According to Flory (1953) crystallization occurs in polymers that have a sufficiently ordered chain structure. Flory (1953) stated that polymer crystals may have regularity in three dimensions. The unit cells of the lattice may be considered to be composed of the repeating polymer units instead of single molecules. At the unit cell level the crystal structure is analogous to the crystal structure of monomeric compounds. However, the occurrence of complete crystallization of polymers is unlikely and the crystallites are usually embedded in a residual amorphous matrix (Flory, 1953). The structure of the crystallites may vary depending on the crystallization conditions.

Nucleation and crystallization of polymers occur slowly in comparison with low molecular weight compounds. The slow crystallization is due to the restrictions of molecular motions that are required for molecular arrangements to form the crystalline structure. Flory (1953) pointed out that the extent of crystallization of polymers may depend on supercooling and more perfect crystallites are obtained when crystallization occurs at a temperature close to the melting temperature. However, the overall rate of crystallization increases with increasing supercooling (Sperling, 1986).

Polymer crystallization has been studied with several methods, including microscopy. Microscopy may be used to follow the growth of the radius of spherulites at isothermal conditions (Sperling, 1986). The increase in radius often exhibits linearity when plotted against time. The linear growth rate is dependent on the crystallization temperature. Plots showing the linear rate as a function of temperature give information on the crystallization rate in relation to the glass transition and melting temperature. It may be shown that the rate increases with decreasing temperature below $T_m$, i.e., with increasing supercooling, until a peak occurs and the rate starts to decrease as shown in Figure 2.4. Such decrease of the rate of crystallization may be accounted for by decreasing diffusion and increasing viscosity. At temperatures close to the $T_g$ the rate approaches zero as the material is transformed into the solid, glassy state (Sperling, 1986). Such behavior may be assumed to be also typical of biopolymers. Based on experimental crystallization studies

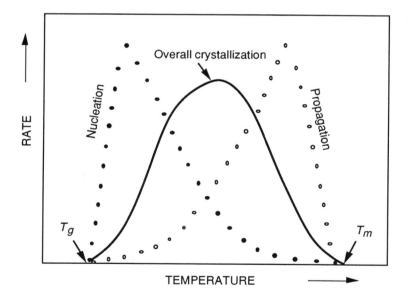

**Figure 2.4** Crystallization kinetics in amorphous or partially crystalline polymers. The rate of nucleation is high at temperatures close to the glass transition, $T_g$, due to the extensive supercooling, but kinetic restrictions decrease the rate of propagation. Propagation increases with increasing temperature due to increasing diffusion until the liquid state becomes the equilibrium state above the melting temperature, $T_m$. The overall rate of crystallization exhibits a maximum between $T_g$ and $T_m$. After Levine and Slade (1990).

of starch gels Marsh and Blanshard (1988) assumed that crystallization in starch shows a maximum between $T_g$ and $T_m$.

Crystallization kinetics of amorphous food component polymers was discussed in detail by Levine and Slade (1990). Crystallization of amorphous noncrystalline and partially crystalline materials occurs due to supercooling below the equilibrium melting temperature. Such crystallization may occur at temperatures above $T_g$ but below $T_m$, which obviously applies also to the crystallization of low molecular weight amorphous food components such as sugars (Roos and Karel, 1990, 1992). A schematic representation of crystallization kinetics in partially crystalline polymers is shown in Figure 2.4. Nucleation and crystal growth cease at $T_m$ due to the fact that the thermodynamic equilibrium shifts to the liquid state. At temperatures below $T_g$ crystallization is unlikely due to the extremely high viscosity and slow diffusion. The rate of nucleation in the vicinity of $T_m$ is slow and it increases rapidly with decreasing temperature below $T_m$ until the temperature approaches $T_g$. At temperatures close to $T_m$ crystal growth is rapid, but the increasing viscosity with decreasing temperature delays propagation, which becomes extremely slow in the vicinity of $T_g$ (Levine and Slade, 1990). In agreement with the nucleation and crystal growth theories the number of nuclei formed at temperatures close to $T_g$ is high, but propagation is slow due to diffusional limitations.

## 2. Crystallization kinetics

Crystallization of amorphous compounds may significantly affect food quality and kinetics of other deteriorative changes. Such phenomena as sandiness in ice cream, caking and browning of milk powders, or starch retrogradation are often related to crystallization of amorphous carbohydrates. Knowledge of factors affecting crystallization can be applied both in controlled crystallization processes during food manufacturing and in shelf life predictions.

$$\alpha = 1 - e^{-V_t} \qquad\qquad (2.38)$$

Crystallinity in amorphous materials is often produced as a function of time. A plot of crystallinity against time of crystallization has a sigmoid shape, showing that the rate at which crystallization occurs is initially slow, then it increases, and at the end of the process the rate is again reduced (Avrami, 1939). Crystallization kinetics are particularly important for evaluating time dependence of the change in phase that may occur in amorphous food components, e.g., sugars and starch. Crystallization kinetics are often modeled with the Avrami equation (Avrami, 1940). The Avrami equation is applied in modeling crystallization kinetics of amorphous polymers (Sperling, 1986). It may be shown that the extent of crystallization is related to the volume of the crystalline material, $V_t$, at time, $t$, according to equation (2.38), where $\alpha$ is crystallinity (Sperling, 1986).

Crystallization may occur due to growth of nuclei that have been formed during cooling of the amorphous material to the crystallization temperature or due to nucleation and crystal growth at the crystallization temperature. The growth of spherical nuclei may proceed at a constant rate, $g$, and it can be shown that the volume increase of the crystalline material during a time period from $t$ to $t + dt$ is defined by equation (2.39), where $N$ is the number of nuclei with radius, $r$, at time, $t$.

$$dV_t = 4\pi r^2 N dr \tag{2.39}$$

$$V_t = \int_0^t 4\pi g^2 t^2 N g \, dt \tag{2.40}$$

$$V_t = \frac{4}{3}\pi g^3 N t^3 \tag{2.41}$$

Considering that $r = gt$ and integration of equation (2.39) results in equation (2.40) and the volume of the crystalline fraction at time $t$ is given by equation (2.41). If the assumption is that the rate of crystallization is defined by the formation of nuclei, which is followed by propagation of spheres that are nucleated at time $t_i$, the volume of the crystalline fraction is defined by equation (2.42). Moreover, equation (2.42) assumes that nucleation occurs linearly with time and therefore the relationship $N = It$ applies provided that the rate of nucleation is 1 (Sperling, 1986). The volume fraction of the crystalline material is then obtained from equation (2.43). Equation (2.43) is obtained by integration of equation (2.42) and it may be used to evaluate the morphology of the crystals formed.

$$dV_t = 4\pi g^2 (t - t_i)^2 \, It g \, dt \tag{2.42}$$

$$V_t = \frac{2}{3}\pi g^3 I t^4 \tag{2.43}$$

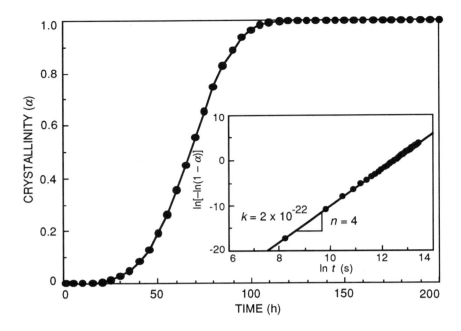

**Figure 2.5** Avrami plot for a hypothetical crystallization process occurring in an amorphous material. The Avrami constants, $n$ and $k$, can be derived from experimental data with linear regression as shown in the inset figure.

Equations (2.41) and (2.43) give the crystalline fraction as a function of time. Therefore, equation (2.38) can be rewritten into the form of equation (2.44), where $k$ and $n$ are constants. Equation (2.44) is equal to the Avrami equation (Avrami, 1940), which suggests that the increase of crystallinity as a function of time is described by a sigmoid curve in a plot of $\alpha$ against $t$ similar to that shown in Figure 2.5. The Avrami equation can also be rewritten into the form of equation (2.45). Equation (2.45) suggests that a plot of $\ln[-\ln(1 - \alpha)]$ against $\ln t$ gives a straight line. The constants, $n$ and $k$, correspond to the slope and intercept of the line, respectively. Thus, the constants can be derived from experimental crystallization data as shown in the inset of Figure 2.5.

$$\alpha = 1 - e^{-kt^n} \tag{2.44}$$

$$\ln[-\ln(1 - \alpha)] = \ln k + n \ln t \tag{2.45}$$

The above derivation of the Avrami equation according to Sperling (1986) suggests that the value for *n* is 3 or 4. However, if crystallization is diffusion controlled other values may apply (Sperling, 1986). Typical values for *n* of polymers are within the range of 2 to 4.1 and the values for both *k* and *n* can be used in analyzing the crystallization mechanism of an amorphous polymer (Sperling, 1986). Sperling (1986) emphasized that the Avrami equation is valid only for low degrees of crystallinity and that the value for *n* may decrease with increasing crystallinity, which should be considered in applying the equation in modeling crystallization kinetics of amorphous food components and other biological materials.

# References

Allen, G. 1993. A history of the glassy state. Chpt. 1 in *The Glassy State in Foods*, ed. J.M.V. Blanshard and P.J. Lillford. Nottingham University Press, Loughborough, pp. 1-12.

Angell, C.A., Bressel, R.D., Green, J.L., Kanno, H., Oguni, M. and Sare, E.J. 1994. Liquid fragility and the glass transition in water and aqueous solutions. *J. Food Eng. 22*: 115-142.

Avrami, M. 1939. Kinetics of phase change. I. General theory. *J. Chem. Phys. 7*: 1103-1112.

Avrami, M. 1940. Kinetics of phase change. II. Transformation-time relations for random distribution of nuclei. *J. Chem. Phys. 8*: 212-224.

Barker, R.E. and Thomas, C. R. 1964. Glass transition and ionic conductivity in cellulose acetate. *J. Appl. Phys. 35*: 87-94.

Batzer, H. and Kreibich, U.T. 1981. Influence of water on thermal transitions in natural polymers and synthetic polyamides. *Polym. Bull. 5*: 585-590.

Brennan, J.G., Butters, J.R., Cowell, N.D. and Lilly, A.E.V. 1990. *Food Engineering Operations*, 3rd ed. Elsevier, London.

Couchman, P.R. and Karasz, F.E. 1978. A classical thermodynamic discussion of the effect of composition on glass transition temperatures. *Macromolecules 11*: 117-119.

Doolittle, A.K. 1951. Studies in Newtonian flow. II. The dependence of the viscosity of liquids on free space. *J. Appl. Phys. 22*: 1471-1475.

Elliott, S.R., Rao, C.N.R. and Thomas, J.M. 1986. The chemistry of the noncrystalline state. *Angew. Chem. Int. Ed. Engl. 25*: 31-46.

Ferry, J.D. 1980. *Viscoelastic Properties of Polymers*, 3rd ed. John Wiley & Sons, New York.

Fox, T.G. and Flory, P.J. 1950. Second-order transition temperatures and related properties of polystyrene. I. Influence of molecular weight. *J. Appl. Phys. 21*: 581-591.

Flory, P.J. 1953. *Principles of Polymer Chemistry*. Cornell University Press, Ithaca, New York.

Gordon, M. and Taylor, J.S. 1952. Ideal copolymers and the second-order transitions of synthetic rubbers. I. Non-crystalline copolymers. *J. Appl. Chem. 2*: 493-500.

Hartel, R.W. and Shastry, A.V. 1991. Sugar crystallization in food products. *Crit. Rev. Food Sci. Nutr. 30*: 49-112.

Hatley, R.H.M. and Mant, A. 1993. Determination of the unfrozen water content of maximally freeze-concentrated carbohydrate solutions. *Int. J. Biol. Macromol. 15*: 227-232.

Levine, H. and Slade, L. 1986. A polymer physico-chemical approach to the study of commercial starch hydrolysis products (SHPs). *Carbohydr. Polym. 6*: 213-244.

Levine, H. and Slade, L. 1990. Influences of the glassy and rubbery states on the thermal, mechanical, and structural properties of doughs and baked products. Chpt. 5 in *Dough Rheology and Baked Product Texture*, ed. H. Faridi and J.M. Faubion. Van Nostrand Reinhold, New York, pp. 157-330.

Lillford, P.J. 1988. The polymer/water relationship - its importance for food structure. Chpt. 6 in *Food Structure - Its Creation and Evaluation*, ed. J.M.V. Blanshard and J.R. Mitchell. Butterworths, London, pp. 75-92.

Marsh, R.D.L. and Blanshard, J.M.V. 1988. The application of polymer crystal growth theory to the kinetics of formation of the $\beta$-amylose polymorph in a 50% wheat-starch gel. *Carbohydr. Polym. 9*: 301-317.

Peleg, M. 1992. On the use of the WLF model in polymers and foods. *Crit. Rev. Food Sci. Nutr. 32*: 59-66.

Perez, J. 1994. Theories of liquid-glass transition. *J. Food Eng. 22*: 89-114.

Roos, Y. 1993. Melting and glass transitions of low molecular weight carbohydrates. *Carbohydr. Res. 238*: 39-48.

Roos, Y. and Karel, M. 1990. Differential scanning calorimetry study of phase transitions affecting the quality of dehydrated materials. *Biotechnol. Prog. 6*: 159-163.

Roos, Y. and Karel, M. 1991a. Water and molecular weight effects on glass transitions in amorphous carbohydrates and carbohydrate solutions. *J. Food Sci. 56*: 1676-1681.

Roos, Y. and Karel, M. 1991b. Nonequilibrium ice formation in carbohydrate solutions. *Cryo-Lett. 12*: 367-376.

Roos, Y. and Karel, M. 1992. Crystallization of amorphous lactose. *J. Food Sci. 57*: 775-777.

Slade, L. and Levine, H. 1991. Beyond water activity: Recent advances based on an alternative approach to the assessment of food quality and safety. *Crit. Rev. Food Sci. Nutr. 30*: 115-360.

Slade, L. and Levine, H. 1993. The glassy state phenomenon in food molecules. Chpt. 3 in *The Glassy State in Foods*, ed. J.M.V. Blanshard and P.J. Lillford. Nottingham University Press, Loughborough, pp. 35-101.

Sperling, L.H. 1986. *Introduction to Physical Polymer Science*. John Wiley & Sons, New York.

Tant, M.R. and Wilkes, G.L. 1981. An overview of the nonequilibrium behavior of polymer glasses. *Polym. Eng. Sci. 21*: 874-895.

Van Hook, A. 1961. *Crystallization. Theory and Practice*. Reinhold, New York.

White, G.W. and Cakebread, S.H. 1966. The glassy state in certain sugar-containing food products. *J. Food Technol. 1*: 73-82.

Williams, M.L., Landel, R.F. and Ferry, J.D. 1955. The temperature dependence of relaxation mechanisms in amorphous polymers and other glass-forming liquids. *J. Am. Chem. Soc. 77*: 3701-3707.

# CHAPTER 3

# *Methodology*

## I.  Introduction

The physical states of food materials include crystalline and liquid states, and almost any combination of these two phases and volatile compounds. A food material may be a completely liquid or solid material, or it may contain a solid, crystalline phase and a liquid phase, which is typical of food fats. Most food solids are amorphous or partially crystalline materials. Native starch is an example of a partially crystalline material, which is composed of amorphous amylose and partially crystalline amylopectin.

It is obvious that only food materials that are completely liquid or solid or composed of separate phases can exist in a thermodynamic equilibrium. A change in temperature of such materials may shift the equilibrium between the liquid and solid states. In amorphous food materials a change in temperature may result in glass transition. It is important to notice that changes between the amorphous and crystalline states may also occur as a function of time. Since food materials exist usually at a constant pressure, the most important parameters that affect their physical state, physicochemical properties, and transition temperatures are temperature, time, and water content.

The physical state of food materials and transitions between various states can be determined with a number of methods that are commonly used to study the physical state and phase transitions of matter. The most important transitions that are related to food properties, quality, and stability occur between the solid and liquid states. These transitions include both the first-order crystallization and melting transitions and the second-order glass transition. Therefore, the methods used to study the physical state and physical properties and to detect both the first- and second-order transitions are discussed in this chapter.

## II. Determination of the Physical State

Most methods used in the characterization of the solid state are based on the detection of its structural properties such as the crystalline structure, which may be derived from X-ray diffraction studies. The amorphous states have no ordered structure and the physical state of amorphous materials is related to molecular mobility, which can be observed with such techniques as electron spin resonance (ESR) and nuclear magnetic resonance (NMR) spectroscopy.

### A. Crystallinity

The crystalline state is the most ordered state of molecular arrangements. There are three fundamental laws of crystallinity, which reflect the well-organized physical state of crystals. The fundamental laws are the law of constancy of interfacial angles, the law of rationality of indices, and the law of symmetry. The methods used to study crystallinity of various materials and polymers include microscopy, X-ray diffraction, electron diffraction, infrared absorption, and Raman spectra (Sperling, 1986). These methods and in particular microscopy and X-ray diffraction techniques have been used in observing crystallinity in food materials.

1. Microscopy

Microscopic methods, i.e., optical microscopy and electron microscopy, are extremely useful in observing the physical state of food components such as starch and the microscopic structure of various food materials (Morris and Miles, 1994). Microscopic methods can also be used to study crystal morphology of food components and of ice in frozen foods. Microscopic methods are particularly useful in the observation of changes that may occur in crystal size and crystallinity in foods during storage.

    *a. Optical Microscopy.* Several optical microscopic methods are available. They are based either on the observation of light transmission or reflection. Polarizing light microscopy uses light transmission and it can be used to observe crystallinity. A typical example is given by native starch granules that exhibit crystallinity under polarizing light microscopy, which disappears during gelatinization. Optical microscopes can also be used with a hot or cold stage that allows temperature control of the samples and examination of the temperature dependence of observed changes. Reflected light microscopy allows examination of surface properties of particles. A typical resolution limit of optical microscopy is about 1 $\mu$m.

    *b. Electron Microscopy.* Electron microscopy includes such techniques as conventional transmission electron microscopy (CTEM), high resolution electron microscopy (HREM), and scanning electron microscopy (SEM), which may be of either transmission or reflection design (Hermansson and Langton, 1994).

    The resolution limits of electron microscopy techniques vary from approximately 2 Å for HREM to about 100 Å for SEM. Reflection SEM is the main technique that has been used to observe the physical state and surface morphology of food materials and the microstructure of polymers. The SEM instruments use an electron beam that is scanned systematically over the material studied. The material emits X-rays and secondary electrons, which can be used to obtain information on chemical properties and to produce an image of the sample surface, respectively.

    SEM provides a valuable technique for the examination of surface properties and structure of a number of food materials. It has proved to be a particularly useful technique in the determination of the morphology of spray-dried powders and changes in structure of other amorphous, low-moisture foods (Gejl-Hansen and Flink, 1977; Alexander and King, 1985). However, it should be noticed that amorphous food materials are extremely sensitive to humidity and temperature, and therefore special care must be taken in sample preparation and handling.

## 2. X-ray diffraction

X-ray diffraction is probably the most important technique in observing properties of crystalline solid materials, polymers, and undoubtedly food materials. The technique has been traditionally used in observation and characterization of crystallinity and crystal structure of polymers (Sperling, 1986) but also of food materials such as starch (e.g., Zobel, 1988) and lipids (e.g., Chapman, 1965) and biopolymers (Clark, 1994).

X-rays are electromagnetic radiation with a wavelength of about 1 Å. They are produced when high energy particles such as electrons collide with matter. In X-ray diffraction studies X-rays are emitted from a radiation source to a sample of the material studied. The radiation is diffracted from the surfaces of the material. Such diffraction is analogous to the diffraction of light by an optical grating. According to Bragg's law the distance, $d$, between successive identical planes of atoms in a crystal is defined by equation (3.1), where $\lambda$ is the X-ray wavelength, $\theta$ is the Bragg angle between the X-ray beam and the lattice planes, and $n$ is any whole number.

$$d = \frac{n\lambda}{2\sin\theta} \tag{3.1}$$

Bragg's law states that when reflected beams are in phase they interfere effectively. The basic principle of powder X-ray diffraction methods is shown in Figure 3.1. Diffraction of the X-rays occurs when the lattice planes are at the Bragg angle. Any set of lattice planes may then produce a surface of a cone that is formed by the diffracted radiation. The diffracted radiation can

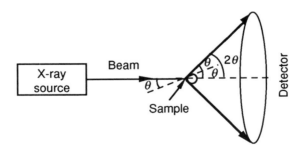

**Figure 3.1** A schematic representation of the formation of a X-ray diffraction cone. X-rays are produced in a radiation source and they are transmitted to a powdered sample that has a random arrangement of crystals. Crystals that have planes oriented at the Bragg angle, $\theta$, to the beam diffract the radiation effectively. The diffracted radiation can be detected using photographic film or a diffractometer.

be detected using photographic film or a diffractometer. The X-ray diffraction pattern is produced on the film or the detected intensity of the diffracted radiation is plotted against $2\theta$.

X-ray diffraction patterns are often considered as "fingerprints" of crystalline materials (West, 1987). X-ray diffraction patterns may also be used to obtain qualitative and quantitative information of the crystalline structure. In food applications X-ray diffraction patterns are extremely useful in the characterization of crystalline states and crystal structure in such materials as starch, amylose-lipid complexes, and sugars, and in the detection of crystallinity in partially amorphous food components. Obviously, amorphous food materials have no characteristic X-ray diffraction patterns.

Electron and neutron diffraction methods are closely related to the X-ray methods. Electron diffraction can be used to study very small crystallites in polymers (Sperling, 1986). Instead of X-rays, the electron and neutron diffraction methods use electrons and neutrons as the radiation, respectively. Thus, the methods produce an electron or neutron diffraction pattern. A general description of the X-ray, electron, and neutron diffraction techniques can be found, e.g., in Willard *et al.* (1981), West (1987), and Clark (1994).

## B. Molecular Mobility

Molecular mobility is significantly affected by temperature and phase transitions. Various spectroscopic methods are available for the examination of different types of molecular mobility. The methods apply the well known principle that materials may absorb or emit energy, which is often electromagnetic radiation. Some of the spectroscopic methods may also be used to evaluate crystallinity in polymers (Sperling, 1986).

### 1. Spectroscopic techniques

Common spectroscopic techniques determine absorption or emission of electromagnetic radiation by the materials studied. The energy involved in such radiation can be related to its frequency, or wavelength. A plot showing the intensity of the absorbed or emitted radiation against the energy of the radiation gives the spectra that can be related to several types of mobility.

Spectroscopic techniques use frequencies with the lower limit at the radiowave region up to the frequency range of X-ray radiation. The energy of electromagnetic radiation is given by equation (3.2).

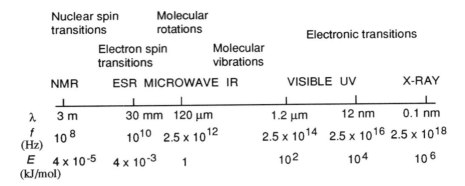

**Figure 3.2** Regions of electromagnetic spectrum that are used in spectroscopic techniques and the effects of radiation at various wavelengths, $\lambda$, frequencies, $f$, and energy, $E$, on mobility.

$$E = hf \qquad\qquad (3.2)$$

Equation (3.2) states that the energy of electromagnetic radiation, $E$, is the product of the Planck's constant, $h$ (6.6242 x $10^{-34}$ J s), and frequency, $f$. The frequency of the radiation is related to the wavelength, $\lambda$, and to the velocity of light, $c$ (2.99776 x $10^8$ m/s), by $c = f\lambda$. The effects of various frequencies on observed changes in matter are shown in Figure 3.2. It is obvious that at short wavelengths and high frequencies the radiation has a high energy, which may cause changes at the electronic level of atoms. The energy of the radiation decreases with decreasing frequency and increasing wavelength, resulting in various changes in vibrations, rotations, or transitions at the electronic, nuclear, or molecular level.

Spectroscopic methods that have been used in the determination of molecular mobility in food materials include nuclear magnetic resonance (NMR) and electron spin resonance (ESR) spectroscopy. Infrared (IR) and Raman spectroscopy are frequently used in the determination of crystallinity and cross-linking in synthetic polymers (Sperling, 1986). According to West (1987) ultraviolet (UV) and visible (VIS) spectroscopy may be used to obtain information on the local structure of amorphous materials. Other spectroscopic techniques used to study glass transitions and their effects in food incredients and products have included Brillouin scattering, Fourier transform infrared (FTIR), phosphorescence, and Mossbauer spectroscopy (Slade and Levine, 1995).

## 2. Infrared and Raman spectroscopy

IR and Raman spectra may be used to obtain information on crystallinity in food lipids (Chapman, 1965) and synthetic polymers (Sperling, 1986). However, their use in the analysis of crystallinity in other food materials has been limited.

West (1987) pointed out that atoms in solids vibrate at frequencies of $10^{12}$ to $10^{13}$ Hz, which interfere with the electromagnetic radiation over the infrared range. The IR and Raman spectra are plots of intensity of IR absorption and scattering as a function of frequency or wavenumber, respectively. In infrared spectroscopy the frequency of the radiation is varied and the intensity of absorbed or transmitted radiation is determined. Raman spectroscopy uses monochromatic, preferably laser, light and its scatter by the sample material is determined. The scattered light may be Rayleigh scatter that has the same energy and wavelength as the incident light or Raman scatter that has either longer or shorter wavelength than the incident light.

The infrared spectra of semicrystalline polymers have "crystallization-sensitive bands" and the intensity of the bands may vary with the degree of crystallinity (Sperling, 1986). Polarized infrared spectra of oriented semicrystalline polymers give information about the molecular and crystal structure, while Raman spectroscopy may provide complementary information on S—S linkages in vulcanized rubber or conjugated linkages between carbon atoms (Sperling, 1986). Moreover, Raman spectroscopy may be used to study solid materials as well as liquids such as biopolymers that are solubilized in water, e.g., the technique can be used to obtain differences in disulfide linkages between solid biopolymers and those solubilized in water (Marshall, 1978). Raman spectroscopy has also been applied in the determination of crystallinity and rate of retrogradation in starch-water systems (Bulkin *et al.*, 1987).

## 3. Nuclear magnetic resonance spectroscopy

NMR involves the magnetic spin energy of atomic nuclei. Paramagnetic nuclei that have a nonzero nuclear spin such as $^1H$ and $^{13}C$ are influenced by an externally applied magnetic field at which they absorb radiation at the radiofrequency range of the electromagnetic spectra. Several variations of the technique such as pulsed NMR, pulsed field gradient NMR, cross-polarization magic angle spinning (CP MAS) NMR, and NMR imaging have been developed. A detailed description of the NMR spectroscopic techniques can be found in, e.g., Mehring (1983), Derome (1987), and Ernst *et al.* (1992).

NMR spectroscopy is used to measure absorption of radiofrequency energy by nuclear spins in the presence of an externally applied magnetic field. Nonzero spins as such are characterized by a magnetic field that is surrounding each nucleus and has a strength and direction, which are known as magnetic moment, $\mu$. In the presence of an external magnetic field, $H_0$, the nonzero spin nuclei become aligned or opposed with the field. The energy of the orientations are different and the nuclei aligned parallel to the field have a lower energy than those in the opposite position. The energy difference between the two states is defined by $\Delta E = 2\mu H_0$. The number of the nuclei at each energy state is dependent on the spin value of the nuclei. The spectra can be produced to show energy absorption against frequency of the electromagnetic radiation at fixed $H_0$ or against $H_0$ at fixed resonance frequency. In pulsed NMR a homogenous magnetic field is applied to the sample, which causes orientation of the nonzero spin nuclei. A pulse of electromagnetic radiation at the resonant frequency is applied at a correct angle to $H_0$, which results in the alignment of all probed nuclei in a single direction. The pulse is defined in terms of the angle of magnetic rotation from $H_0$. Discontinuation of the pulse allows the nuclei to relax in their original positions, which occurs through longitudinal or spin-lattice and transverse or spin-spin relaxation. Longitudinal relaxation is referred to with time constant, $T_1$, which is a measure of energy exchange between nuclei and surrounding molecular environment and of the rate at which magnetization of the $z$ vector component returns to the equilibrium state. Transverse relaxation time constant, $T_2$, is related to the rate of decay of the magnetization in the $xy$ plane and entropic processes of signal decay. The time constants may have the same value, but $T_2$ is often less than $T_1$ (Schmidt and Lai, 1991). Oscillation of magnetization of the nuclei in the $xy$ plane results in an induced voltage in a receiver coil. A plot of voltage intensity of magnetization amplitude against time is called free induction decay (FID). The FID is related to the reciprocal of frequency and Fourier transformation can be used to obtain common NMR spectra showing the amplitude against frequency. NMR imaging uses the NMR spectroscopy to produce three-dimensional images of samples often by probing the [1]H nuclei (Schmidt and Lai, 1991; McCarthy and McCarthy, 1994).

NMR techniques are important in the determination of the physical state and molecular characteristics of almost any material, including polymers and food materials. Wide line NMR is one of the main techniques in the determination of solid fat contents in food lipids and it may be used to observe unfrozen water in frozen foods. The time constants are related to molecular mobility and physical state as shown in Figure 3.3. The main excellence of the NMR techniques is that they can differentiate motions of nuclei that are attached to or located near various chemical groups. In polymers the [1]H NMR

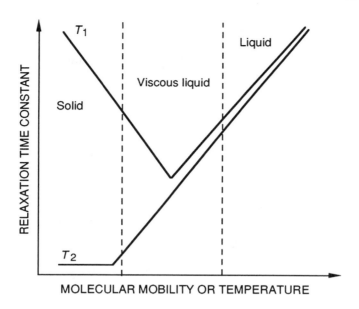

**Figure 3.3** Schematic representation of relationship between physical state, molecular mobility, and NMR relaxation time constants, $T_1$ and $T_2$. $T_1$ shows a minimum for the most efficient relaxations that occur in the liquid state. The $T_2$ has a limiting value for very slow relaxations in the solid state and it increases with increasing mobility above the limiting value. After Ablett et al. (1993).

line width has been shown to decrease within the glass transition temperature range (Sperling, 1986). The change in the NMR line width occurs due to the changes in mobility of the polymer molecules. At temperatures below $T_g$ interactions between magnetic dipoles of different hydrogen nuclei are strong, which results in a broad signal. At temperatures above $T_g$ the distribution of proton orientations becomes more random, which sharpens the NMR signal (Sperling, 1986). NMR was used by Abblett et al. (1993) to determine changes in molecular mobility, which occurred over the glass transition temperature range of food materials.

The use of NMR techniques in the determination of molecular mobility and self-diffusion of molecules in various foods and food components in relation to water in foods was reviewed by Schmidt and Lai (1991). NMR and $^{13}C$ solid state NMR as well as NMR imaging have proved to be promising and useful techniques in studies of crystallinity, molecular mobility, and physical state and phase transitions in foods (Ablett et al., 1993; Gidley et al., 1993; Willenbücher et al., 1993; Kulik et al., 1994; McCarthy and McCarthy, 1994).

4. Electron spin resonance spectroscopy

ESR spectroscopy is based on the absorption of electromagnetic radiation, which occurs in a magnetic field due to magnetic resonance of unpaired electrons such as those of spin-labeled radicals. As a spectroscopic technique ESR is closely related to NMR. However, the intensity of the electron magnetic resonance absorption is substantially higher than the absorption in NMR (Marshall, 1978). Moreover, the ESR spectra are usually shown as the first derivative of the absorption.

The ESR technique may be used to observe mobility of probes incorporated in the material studied. Such probes are often nitroxide radicals and the ESR absorption of the unpaired electron of the probe can be detected. The nitroxide electron is magnetically anisotropic. It absorbs a different amount of energy when the N—O bond is aligned parallel than when the bond is perpendicular to the applied magnetic field (Marshall, 1978). Obviously, the ESR line shape changes due to the rotation of the radical. Therefore, the ESR line shape is sensitive to the rate of rotational reorientation of the spin-labeled nitroxide probes. The ESR line shape can be considered to be a result of the rates of reorientation of the spin-labeled probes. The difference in the energy between the two energy levels of the unpaired electrons is defined by equation (3.3), where $g$ and $\beta$ are constants and $H$ is the strength of the magnetic field. The $g$ factor is a dimensionless constant and equal to 2.002319 for an unbound electron. The $g$ factor is dependent on the chemical environment, but the value of the odd electron of organic compounds is often close to that of the free electron (Willard *et al.*, 1981).

$$\Delta E = g\beta H \qquad\qquad\qquad (3.3)$$

Absorption of energy by electrons occurs in an electromagnetic field when $\Delta E = hf$, which allows reorientation of the electrons antiparallel to the magnetic field. Therefore, the absorption of energy in ESR as well as in NMR is a function of $H$. The first derivative of the absorbed energy is shown in ESR spectra at a fixed frequency as a function of magnetic field. Incorporation of stable spin-labeled probes into food matrices may be used to study molecular mobility and the effect of the physical state on observed mobility. The information obtained reflects molecular mobility of the probe and the polarity of its environment (Simatos *et al.*, 1981). The absorption lines are sharp and of equal height when molecular motion is fast, e.g., in liquid state with a correlation time, $\tau_c$, of less than $10^{-11}$ s. In conventional ESR correlation times of $10^{-11} < \tau_c < 10^{-7}$ s can be determined. Saturation transfer ESR allows determination of slower motions of $10^{-7} < \tau_c < 10^{-3}$ s, which may occur in powders and solid materials (Le Meste *et al.*, 1991). The sharpness of the absorption lines decreases with decreasing mobility. Low mobility is characterized by broad ESR lines, which, however, remain at their original positions (Hemminga *et al.*, 1993). Therefore, ESR can be used to observe changes in molecular mobility that occur at phase transition temperatures. In the solid state molecular mobility is low and the probes are highly immobilized. A change in phase due to melting of crystalline structure or due to glass transition increases the rotational freedom of the probes.

Molecular mobility is related to viscosity and diffusion. It may be shown that molecular motions occur in three dimensions and the rotations observed with ESR can be related to rotational diffusion. The rotational diffusion coefficient can be derived from the ESR correlation time. The relationship between the correlation time and the rotational diffusion coefficient, $D_{rot}$, is given by equation (3.4).

$$D_{rot} = \frac{1}{6\tau_c} \qquad (3.4)$$

Molecular motions may be either translational or rotational. Both translational diffusion coefficients, $D_{trans}$ and $D_{rot}$, are dependent on the friction coefficients, $f_{trans}$ and $f_{rot}$, respectively. The relationships are given by equations (3.5) and (3.6), where $k$ is the Boltzmann constant ($k = R/N$, 1.38048 x 10-23 J/K, N = Avogadro's number), and $T$ is absolute temperature.

$$D_{trans} = \frac{kT}{f_{trans}} \qquad (3.5)$$

$$D_{rot} = \frac{kT}{f_{rot}} \qquad (3.6)$$

The friction coefficients are functions of viscosity, $\eta$. The friction coefficients for spherical molecules are given by equations (3.7) and (3.8), where $r$ is the radius of the diffusing molecules.

$$f_{trans} = 6\pi\eta r \qquad (3.7)$$

$$f_{rot} = 8\pi\eta r^3 \qquad (3.8)$$

$$\tau_c = \frac{V\eta C}{kT} \qquad (3.9)$$

It is obvious that the ratio $D_{trans}/D_{rot}$ is dependent on the size of the diffusing molecule. According to Le Meste *et al.* (1991) the above equations are applicable to macromolecules. When correlation times are defined in terms of viscosity and volume, $V$, for small diffusing molecules the relationships are defined by equation (3.9), where $C$ is an additional coupling parameter.

Molecular mobility and diffusion of reactants in food matrices are important in defining rates of various deteriorative changes. Le Meste *et al.* (1991) found that *rotational* mobilization of molecules as observed by ESR correlated with water activities that were assumed to allow sufficient solute mobility for chemical reactivity. It may also be expected that molecular mobilization temperatures that are observed by ESR correlate with phase transition temperatures. Le Meste *et al.* (1991) found that at low water contents the mobilization temperatures differed considerably from glass transition temperatures that were determined with differential scanning calorimetry (DSC). However, Bruni and Leopold (1991) used ESR to study glass transitions in soybean seeds. The temperature at which a change in $D_{rot}$ occurred was taken as $T_g$. Hemminga *et al.* (1993) observed a high correlation between the temperatures at which $\tau_c$ of sucrose-water and maltodextrin-water mixtures decreased and the glass transition temperature. Some of the mixtures were studied in the freeze-concentrated state and it should be noticed that an increase in molecular mobility of freeze-concentrated solutions is probably affected by an increasing amount of unfrozen water above the transition temperature.

## III.  Physical Properties and Transition Temperatures

Changes in physical state are observed from changes in the thermodynamic quantities, which can be measured with a number of techniques such as calorimetric techniques, dilatometry, and thermal analysis. The various spectroscopic methods are also applicable for the determination of transition temperatures. In addition to the changes in thermodynamic quantities and molecular mobility, phase transitions affect physicochemical properties including viscosity, viscoelastic properties, and dielectric properties.

### A. Volumetric Changes

Phase transitions result in dimensional and volumetric changes that can be observed with such techniques as dilatometry and thermal mechanical analysis (TMA). Dilatometry is used to measure the volume or length of a material as a function of temperature. TMA differs from dilatometry in that the measurement of length or volume is made under an applied tension or load (Wunderlich, 1990).

## 1. Changes at transition temperatures

A change in phase at a constant pressure involves a change in volume. The change in volume in first-order transitions occurs at a constant temperature. In second-order transitions there is a change in the thermal expansion coefficient, $\alpha$, at the transition temperature.

The change in volume can be observed from dimensional changes or from a change in the thermal expansion coefficient that accompanies the phase transition. The thermal expansion coefficient defines a change in volume that is caused by a change in temperature according to equation (3.10).

$$\alpha = \frac{1}{V}\left(\frac{\partial V}{\partial T}\right)_p \tag{3.10}$$

Volume in various phases and the effects of phase transitions on volume and thermal expansion are shown in Figure 3.4. The volume in the solid state is lowest for the crystalline material. A step change occurs at the first-order melting transition at $T_m$ as the crystalline material is transformed into the liquid state. Therefore, the thermal expansion coefficient has an infinite value at $T_m$. The volume in the glassy state is higher than that of the crystalline state. A step change in $\alpha$ can be observed at $T_g$ as the solid glass is transformed into the supercooled liquid state. It should be noticed that no change in $\alpha$ is observed at $T_m$ as a supercooled liquid is heated to above $T_m$.

## 2. Dilatometry

Dilatometry can be used to measure volume changes as a function of temperature. Dilatometry has been a common method in the determination of fat melting and solid fat content. The method can also be used to observe volume changes in polymers that occur due to glass transition (Sperling, 1986).

Dilatometers are often made of glass and they consist of a capillary that is used for calibration and recording volume changes. The sample is placed in a bulb that is connected to a separate bulb, which contains a liquid such as mercury. The liquid is used to fill the instrument with sample. The apparatus is

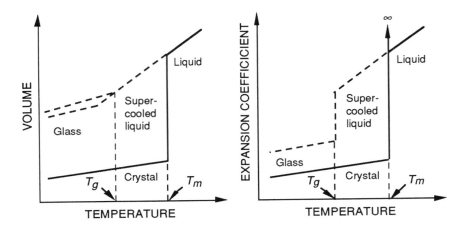

**Figure 3.4** A schematic representation of changes in volume and thermal expansion coefficient, $\alpha$, at first-order and second-order phase transition temperatures. The volume of the solid, crystalline state is dependent on temperature and a function of temperature as defined by $\alpha$. At the first-order melting transition temperature, $T_m$, an isothermal increase in volume occurs due to melting of the crystals and $\alpha$ has an infinite value. The amorphous, glassy solid, depending on the cooling rate, may have various volumes that are higher than that of the crystalline state. A change in $\alpha$ occurs at the second-order glass transition temperature, $T_g$, as the solid is transformed into the supercooled liquid state.

evacuated and the expansion of the liquid and the sample is recorded. Thus, the volume changes of the sample can be obtained at various temperatures provided that the thermal properties of the liquid are known.

Dilatometric techniques in the determination of phase transitions of lipids have been replaced by NMR and thermal analysis. According to Sperling (1986) dilatometric data of polymers agree well with modulus-temperature studies especially when the same experimental time scales are used.

3. Thermal mechanical analysis

TMA can be used to measure volume expansion, compression, and tension. The measurements are made by heating samples at a constant rate and the change in volume, penetration depth, or force and length are obtained as a function of temperature.

TMA is common in the determination of phase transitions and mechanical properties of polymers. It may also be used to determine the elastic modulus in bending (Wunderlich, 1990), which is useful in the characterization of changes in mechanical properties that occur in amorphous materials above $T_g$.

Food materials are extremely sensitive to water and a change in water content may cause a dramatic change in mechanical properties. Such changes may occur in TMA due to exchange of water between the sample and its surroundings. However, the method may be used to observe mechanical changes at low temperatures and in frozen materials (Le Meste and Huang, 1992; Le Meste *et al.*, 1992).

## B. Changes in Enthalpy

Enthalpy at a constant pressure includes internal energy and pressure-volume work. The total enthalpy of a material is unknown, but changes in enthalpy can be measured with calorimetric techniques. Calorimetric methods that are used to determine phase transitions include differential thermal analysis (DTA) and DSC. The use of DTA and DSC in food applications was reviewed by Biliaderis (1983) and Lund (1983). A general description of the techniques and their use in the determination of phase transitions in polymers was given by Wunderlich (1990).

### 1. Enthalpy and phase transitions

Enthalpy changes occur at first-order transition temperatures and there is a change in heat capacity at second-order transitions. These changes can be determined with calorimetry, which gives information on transition temperatures and quantitative data on changes in energy.

The effect of phase transitions on enthalpy, $H$, is shown in Figure 3.5. Changes in enthalpy and heat capacity, $C_p$, at transition temperatures correspond to those of volume and thermal expansion coefficient, respectively. Enthalpy is lowest in the crystalline solid state. Enthalpy of a single phase increases with increasing temperature by the amount defined by $C_p$. At a first-order transition temperature, e.g., melting temperature, the enthalpy change occurs isothermally and it is equal to the latent heat, $\Delta H_l$, of the transition. Heat capacity at a first-order transition temperature has an infinite value. Enthalpy of a solid glass is higher than that of the crystalline material at the same temperature and it may vary depending on the rate of "freezing" of molecules to the solid state during glass formation. A change in heat capacity occurs at a second-order transition temperature, e.g., at glass transition temperature. Therefore, the increase in enthalpy with increasing temperature is higher in the supercooled liquid state than in the glassy state. It should be

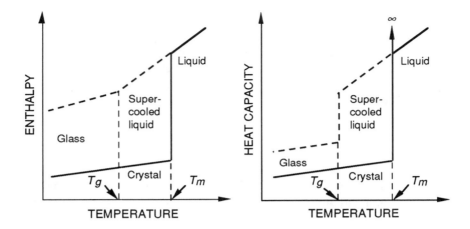

**Figure 3.5** A schematic representation of changes in enthalpy and heat capacity, $C_p$, at first-order and second-order phase transition temperatures. Enthalpy of the solid, crystalline state is dependent on temperature and a function of temperature as defined by $C_p$. At a first-order melting transition temperature, $T_m$, an isothermal increase in enthalpy occurs due to the latent heat of melting of the crystals and $C_p$ has an infinite value. The enthalpy of the amorphous, glassy solid is higher than that of the crystalline material. A change in $C_p$ occurs at the second-order glass transition temperature, $T_g$, as the solid is transformed into the supercooled liquid state.

pointed out that the change in heat capacity or thermal expansion coefficient occurs over a temperature range of 10 to 30°C, but not isothermally at a single transition temperature.

2. Differential scanning calorimetry

DTA and DSC are closely related methods, which are probably the most common techniques in the determination of phase transitions in inorganic, organic, polymeric and also food materials. DTA measures temperature of a sample and a reference as a function of temperature. A phase transition causes a temperature difference between the sample and the reference, which is recorded. DSC may use the same principle, but the temperature difference between the sample and the reference is used to derive the difference in the energy supplied. DSC may also measure the amount of energy supplied to the sample and the reference. A modulated DSC uses oscillating temperature during heating and cooling of samples (Boller *et al.*, 1994).

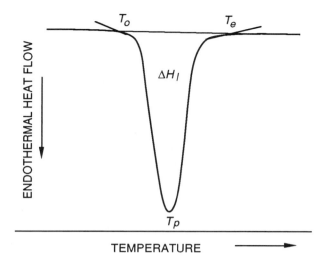

**Figure 3.6** A schematic DSC thermogram showing an endothermal, first-order phase transition, e.g, melting. The onset of the transition occurs at $T_o$, which is the transition temperature. In broad melting transitions the peak temperature of the endotherm, $T_p$, and the endset temperature, $T_e$, may also be determined. $T_o$ and $T_e$ are obtained from the intercept of tangents drawn at the point at which deviation from the baseline occurs. Peak integration is used to obtain the latent heat of the transition, $\Delta H_l$.

DTA and DSC are used to detect endothermal and exothermal changes that occur during a dynamic measurement as a function of temperature or isothermally as a function of time. The thermograms obtained show the heat flow to the sample and DSC data can be used to calculate enthalpy changes and heat capacities. First-order phase transitions produce peaks and a step change in heat flow occurs at second-order transitions. As shown in Figure 3.6 thermograms showing first-order transitions can be analyzed to obtain transition temperatures. The latent heat of the transition is obtained by peak integration. Thermograms showing second-order transitions can be used to derive transition temperatures and changes in heat capacity as shown for glass transition in Figure 3.7. The transition occurs over a temperature range of 10 to 30°C. Both the onset and midpoint temperatures of the glass transition temperature range are commonly referred to as $T_g$.

Applications of differential scanning calorimetry in the determination of phase transitions in foods include such changes as crystallization and melting of water, lipids, and other food components, protein denaturation, and gelatinization and retrogradation of starch. The samples are usually placed in pans

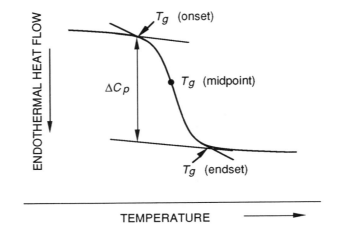

**Figure 3.7** Determination of second-order and glass transition temperatures, $T_g$, and change in heat capacity, $\Delta C_p$, that occurs over the glass transition temperature range from DSC thermograms. The endothermal step change in heat flow during heating of glassy materials occurs due to $\Delta C_p$ at the second-order transition temperature.

that can be hermetically sealed. Therefore, the method can be used to observe phase transitions and to determine transition temperatures without changes in water content. A constant water content is extremely important in the determination of phase transitions of food materials. Water has an enormous effect on transition temperatures and its impact on food behavior cannot be overemphasized.

## C. Mechanical and Dielectric Properties

Mechanical and dielectric properties of food materials are related to physical state and they are significantly affected by phase transitions. It is obvious that physical properties of solid, liquid, and gaseous materials differ. Changes in the physical state of water in freezing and melting has a significant effect on the mechanical and dielectric properties of foods. Mechanical and dielectric properties are also affected by glass transitions, which often govern the physical state of low-moisture and frozen foods.

1. Changes at glass transition temperature

The main importance of mechanical and dielectric properties in relation to phase transitions in foods is their sensitivity to indicate changes in modulus and dielectric properties that occur in amorphous food materials above $T_g$.

Molecular mobility in amorphous materials is in large part governed by glass transition, which is often considered as an $\alpha$ transition and the onset temperature of long-range motions in amorphous macromolecules. At temperatures below $T_g$ other transitions that are referred to as $\beta$, $\gamma$, etc., with decreasing transition temperature may occur. These transitions are due to local mode relaxations of polymer chains and rotations of terminal groups or side chains (Ferry, 1980; Sperling, 1986; Kalichevsky et al., 1993). The transitions are observed from changes in mechanical and dielectric properties, i.e., storage modulus, $E'$ or $G'$, loss modulus, $E''$ or $G''$, dielectric constant, $\varepsilon'$, dielectric loss constant, $\varepsilon''$, and mechanical and dielectric loss, tan $\delta$. Therefore, the methods observe relaxation times of molecular rotations and energy loss, which are related to viscosity.

The effect of glass transition on viscosity and dielectric constant is shown in Figure 3.8. Determination of mechanical properties has been used in several studies to observe changes that occur in amorphous biological and food materials at various temperatures and the effect of glass transition on mechanical properties (e.g., Kalichevsky et al., 1993; Kokini et al., 1994; Leopold et al., 1994; Williams, 1994).

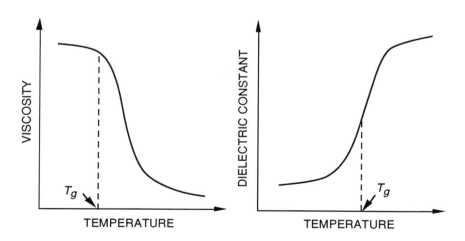

**Figure 3.8** Effect of glass transition, $T_g$, on viscosity and dielectric constant. Increasing molecular mobility above $T_g$ affects mechanical and dielectric properties.

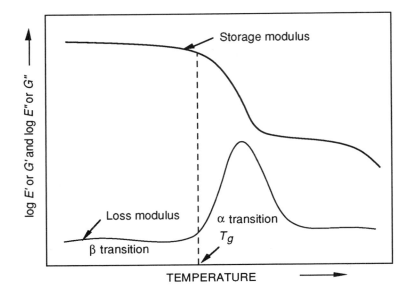

**Figure 3.9** Changes in storage modulus ($E'$ or $G'$) and loss modulus ($E''$ or $G''$) in amorphous materials as a function of temperature. The storage modulus is high in the glassy state and it decreases dramatically at glass transition, $T_g$. The loss modulus may show peaks below $T_g$ due to transitions in molecular rotations. A significant peak in loss modulus occurs over the $T_g$ temperature range.

### 2. Dynamic mechanical thermal analysis and mechanical spectroscopy

Dynamic mechanical thermal analysis (DMTA) and mechanical spectroscopy can be used to obtain mechanical modulus as a function of temperature. Mechanical properties can be obtained from samples during periodic changes in stress or strain that are applied in samples under bending, shear, tension, or torsion (Ross-Murphy, 1994).

The effect of physical state on mechanical moduli of polymers was extensively discussed by Ferry (1980). The behavior of amorphous food materials has been shown to be of similar type (e.g., Kalichevsky *et al.*, 1993; Kokini *et al.*, 1994). The effect of glass transition on storage modulus and loss modulus is shown in Figure 3.9. Storage modulus has a high value of $10^{6.5}$ to $10^{9.5}$ Pa (Kalichevsky *et al.*, 1993) in the glassy state. It suffers a dramatic decrease at the glass transition. The temperature location of the drop in storage modulus is dependent on the frequency, $f$, of the applied stress or strain. Arrhenius plots showing $\log f$ against $1/T$ have suggested that the apparent activation energy is 200 to 400 kJ/mol (Kalichevsky *et al.*, 1993). The loss modulus and tan $\delta$ show a pronounced peak at the transition temperature.

Cocero and Kokini (1991) reported that the loss modulus peak of glutenin in shear at 1 Hz occurred at about the same temperature as the endset temperature of glass transition that was determined with DSC at 5°C/min. Roos and Karel (1991) found that the onset of the loss modulus peak of glucose and sucrose solutions occurred close to the endset of $T_g$ measured with DSC.

DMTA and mechanical spectroscopy have been applied in the analysis of mechanical properties of such food materials as polysaccharides (Kalichevsky *et al.*, 1993; Chinachoti, 1994), proteins (Kokini *et al.*, 1994), and sugars (Roos and Karel, 1991; Blond *et al.*, 1994). However, samples may suffer changes in water contents at high temperatures, which must be avoided due to the dramatic effect of water on mechanical properties of food materials.

3. Dielectric properties

Changes in dielectric properties of amorphous materials can be studied with dielectric thermal analysis (DETA). The dielectric constant, $\varepsilon'$, dielectric loss constant, $\varepsilon''$, and tan $\delta$ can be measured by placing a sample between parallel plate capacitors and alternating the electrical field. Polar groups in the sample respond to the alternating electrical field and an absorption maximum is obtained at the frequency that equals the molecular motion. Generally the shapes of the $\varepsilon'$, $\varepsilon''$, and tan $\delta$ curves are similar to those of corresponding mechanical properties (Sperling, 1986). The method has been applied in the analysis of some food-related materials such as amorphous sugars (Noel *et al.*, 1992) and amylopectin (Kalichevsky *et al.*, 1992). Dielectric spectroscopy and thermal stimulated current/relaxation map analysis (TSC/RMA) spectroscopy have also been applied in the characterization of food and related materials (Leopold *et al.*, 1994; Slade and Levine, 1995).

# References

Ablett, S., Darke, A.H., Izzard, M.J. and Lillford, P.J. 1993. Studies of the glass transition in malto-oligomers. Chpt. 9 in *The Glassy State in Foods*, ed. J.M.V. Blanshard and P.J. Lillford. Nottingham University Press, Loughborough, pp. 189-206.

Alexander, K. and King, C.J. 1985. Factors governing surface morphology of spray-dried amorphous substances. *Drying Technol. 3*: 321-348.

Biliaderis, C.G. 1983. Differential scanning calorimetry in food research - A review. *Food Chem. 10*: 239-265.

Blond, G., Ivanova, K. and Simatos, D. 1994. Reliability of dynamic mechanical thermal analyses (DMTA) for the study of frozen aqueous systems. *J. Rheol. 38*: 1693-1703.

Boller, A., Jin, Y. and Wunderlich, B. 1994. Heat capacity measurement by modulated DSC at constant temperature. *J. Thermal Anal. 42*: 307-330.

Bruni, F. and Leopold, A.C. 1991. Glass transitions in soybean seed. *Plant Physiol. 96*: 660-663.

Bulkin, B.J., Kwak, Y. and Dea, I.C.M. 1987. Retrogradation kinetics of waxy-corn and potato starches; a rapid, Raman-spectroscopic study. *Carbohydr. Res. 160*: 95-112.

Chapman, D. 1965. *The Structure of Lipids by Spectroscopic and X-Ray Techniques.* Methuen and Co., London.

Chinachoti, P. 1994. Probing molecular and structural thermal events in cereal-based products. *Thermochimica Acta 246*: 357-369.

Clark, A.H. 1994. X-ray scattering and diffration. Chpt. 3 in *Physical Techniques for the Study of Food Biopolymers*, ed. S.B. Ross-Murphy. Chapman & Hall, New York, pp. 65-149.

Cocero, A.M. and Kokini, J.L. 1991. The study of the glass transition of glutenin using small amplitude oscillatory rheological measurements and differential scanning calorimetry. *J. Rheol. 35*: 257-270.

Derome, A.E. 1987. *Modern NMR Techniques for Chemistry Research.* Pergamon Press, New York.

Ernst, R.R., Bodenhausen, G. and Wokaun, A. 1992. *Principles of Nuclear Magnetic Resonance in One and Two Dimensions.* Oxford University Press, Oxford.

Ferry. J.D. 1980. *Viscoelastic Properties of Polymers*, 3rd ed. John Wiley & Sons, New York.

Gejl-Hansen, F. and Flink, J.M. 1977. Freeze-dried carbohydrate containing oil-in-water emulsions: Microstructure and fat distribution. *J. Food Sci. 42*: 1049-1055.

Gidley, M.J., Cooke, D. and Ward-Smith, S. 1993. Low moisture polysaccharide systems: Thermal and spectroscopic aspects. Chpt. 15 in *The Glassy State in Foods*, ed. J.M.V. Blanshard and P.J. Lillford. Nottingham University Press, Loughborough, pp. 303-331.

Hemmings, M.A., Roozen, M.J.G.W. and Walstra, P. 1993. Molecular motions and the glassy state. Chpt. 7 in *The Glassy State in Foods*, ed. J.M.V. Blanshard and P.J. Lillford. Nottingham University Press, Loughborough, pp. 157-187.

Hermansson, A.-M. and Langton, M. 1994. Electron microscopy. Chpt. 6 in *Physical Techniques for the Study of Food Biopolymers*, ed. S.B. Ross-Murphy. Chapman & Hall, New York, pp. 277-341.

Kalichevsky, M.T., Jaroszkiewicz, E.M., Ablett, S., Blanshard, J.M.V. and Lillford, P.J. 1992. The glass transition of amylopectin measured by DSC, DMTA and NMR. *Carbohydr. Polym. 18*: 77-88.

Kalichevsky, M.T., Blanshard, J.M.V. and Marsh, R.D.L. 1993. Applications of mechanical spectroscopy to the study of glassy biopolymers and related systems. Chpt. 6 in *The Glassy State in Foods*, ed. J.M.V. Blanshard and P.J. Lillford. Nottingham University Press, Loughborough, pp. 133-156.

Kokini, J.L., Cocero, A.M., Madeka, H. and de Graaf, E. 1994. The development of state diagrams for cereal proteins. *Trends Food Sci. Technol. 5*: 281-288.

Kulik, A.S., de Costa, C. J.R. and Haverkamp, J. 1994. Water organization and molecular mobility in maize starch investigated by two-dimensional solid-state NMR. *J. Agric. Food Chem. 42*: 2803-2807.

Le Meste, M. and Huang, V. 1992. Thermomechanical properties of frozen sucrose solutions. *J. Food Sci. 57*: 1230-1233.

Le Meste, M., Voilley, A. and Colas, B. 1991. Influence of water on the mobility of small molecules dispersed in a polymeric system. In *Water Relationships in Foods*, ed. H. Levine and L. Slade. Plenum Press, New York, pp. 123-138.

Le Meste, M., Huang, V.T., Panama, J., Anderson, G. and Lentz, R. 1992. Glass transition of bread. *Cereal Foods World 37*: 264-267.

Leopold, A.C., Sun, W.Q. and ernal-Lugo, I. 1994. The glassy state in seeds: analysis and function. *Seed Sci. Res. 4*: 267-274.

Lund, D.B. 1983. Applications of differential scanning calorimetry in foods. Chpt. 4 in *Physical Properties of Foods*, ed. M. Peleg and E.B. Bagley. AVI Publishing Co., Westport, CT, pp. 125-143.

Marshall, A.G. 1978. *Biophysical Chemistry. Principles, Techniques, and Applications*. John Wiley & Sons, New York.

McCarthy, M.J. and McCarthy, K.L. 1994. Quantifying transport phenomena in food processing with nuclear magnetic resonance imaging. *J. Sci. Food Agric. 65*: 257-270.

Mehring, M. 1983. *Principles of High Resolution NMR in Solids*. 2nd ed. Springer-Verlag, Berlin.

Morris, V.J. and Miles, M.J. 1994. Birefringent techniques. Chpt. 5 in *Physical Techniques for the Study of Food Biopolymers*, ed. S.B. Ross-Murphy. Chapman & Hall, New York, pp. 215-275.

Noel, T.R., Ring, S.G. and Whittam, M.A. 1992. Dielectric relaxations of small carbohydrate molecules in the liquid and glassy states. *J. Phys. Chem. 96*: 5662-5667.

Roos, Y. and Karel, M. 1991. Nonequilibrium ice formation in carbohydrate solutions. *Cryo-Lett. 12*: 367-376.

Ross-Murphy, S.B. 1994. Rheological methods. Chpt. 7 in *Physical Techniques for the Study of Food Biopolymers*, ed. S.B. Ross-Murphy. Chapman & Hall, New York, pp. 343-392.

Schmidt, S.J. and Lai, H.-M. 1991. Use of NMR and MRI to study water relations in foods. In *Water Relationships in Foods*, ed. H. Levine and L. Slade. Plenum Press, New York, pp. 405-452.

Simatos, D., Le Meste, M., Petroff, D. and Halphen, B. 1981. Use of electron spin resonance for the study of solute mobility in relation to moisture content in model food systems. In *Water Activity: Influences on Food Quality*, ed. L.B. Rockland and G.F. Stewart. Academic Press, New York, pp. 319-346.

Slade, L. and Levine, H. 1995. Glass transitions and water-food structure interactions. *Adv. Food Nutr. Res. 38*. In press.

Sperling, L.H. 1986. *Introduction to Physical Polymer Science*. John Wiley & Sons, New York.

West, A.R. 1987. *Solid State Chemistry and Its Applications*. John Wiley & Sons, New York.

Willard, H.H., Merritt, L.L., Dean, J.A. and Settle, F.A. 1981. *Instrumental Methods of Analysis*, 6th ed. D. Van Nostrand, New York.

Williams, R.J. 1994. Methods for determination of glass transitions in seeds. *Ann. Botany 74*: 525-530.

Willenbücher, R.W., Tomka, I. and Müller, R. 1993. Thermally induced structural transitions in the starch-water system. Chpt. 26 in *The Glassy State in Foods*, ed. J.M.V. Blanshard and P.J. Lillford. Nottingham University Press, Loughborough, pp. 491-497.

Wunderlich, B. 1990. *Thermal Analysis*. Academic Press, San Diego, CA.

Zobel, H.F. 1988. Starch crystal transformations and their industrial importance. *Starch 40*: 1-7.

# CHAPTER 4

# Water and Phase Transitions

## I. Introduction

Water is the most important diluent of food solids. Water in foods has several effects on physical properties, food behavior in processing, microbial growth, stability, palatability, and phase transitions. The importance of water in foods is due to its presence in almost all foods, the dominance of water as a food component, and the physical properties and peculiarities of water in comparison with other food components.

The phase behavior of water as such is extraordinary. Water has fairly high melting and boiling temperatures as well as high latent heats of first-order transitions. In foods with high water contents the phase behavior of water is fairly similar to, although different from, that of pure water. Ice formation and separation of water from food solids occur below 0°C and boiling of water occurs at temperatures close to 100°C. The phase behavior of water and its thermodynamic characteristics govern and define processing conditions that are often necessary for food preservation by dehydration and freezing. Removal of water affects the phase behavior of remaining concentrated solids that begin to govern the physical state and transitions as water content is reduced. However, it should be noticed that the phase behavior of lipids defines the physical state of fats and oils and their mixtures. The importance of the phase behavior of other food components and especially the amorphous state and the interactions of water with amorphous food solids has been recognized in particular due to the efforts of Levine and Slade (e.g., 1986; Slade and Levine, 1991).

This chapter describes physical properties of pure water and its phase behavior within food matrices. The emphasis is on the physical state of water in food materials and its sorption properties in food solids, which are important to water plasticization and phase behavior of amorphous food components.

## II. Properties of Water

Pure water is an essential substance for all life. Water is a well-characterized substance and the peculiarities of its physical properties in comparison with other compounds of similar molecular structure and its importance and significance in foods have been described elsewhere (e.g., Fennema, 1973, 1985). The most important characteristics of water include its ability to act as a solvent and plasticizer for a large number of compounds and food components as well as the fact that it may exist in the solid, liquid, and gaseous states at common food processing and storage temperatures.

### A. Phase Behavior of Water

Ice formation may occur in biological materials, including foods, when temperature is depressed to below the freezing temperature or pressure is reduced to below the triple point pressure of water. Such conditions are typical of freezing, frozen storage, and freeze-drying of foods. Food materials have no exact freezing temperatures, although onset temperatures of freezing and ice melting can be defined. Usually supercooling precedes ice formation, which may occur at fairly low temperatures. Some food components, e.g., salts and sugars, may also crystallize due to freeze-concentration and decreased solubility at low temperatures. At high temperatures the boiling temperature and properties of steam set limits to various food processing conditions. Evaporation and evaporative cooling are other important phenomena in food processing.

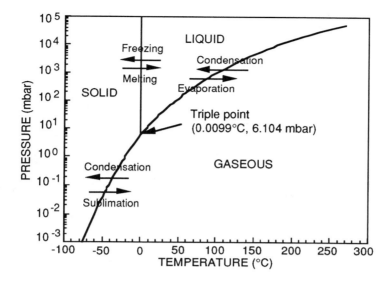

**Figure 4.1** Phase diagram for water, showing relationships between equilibrium states and their dependence on temperature and pressure. All three phases may coexist only at the triple point.

## 1. Phase behavior of pure water

The pressure-temperature phase diagram for water is shown in Figure 4.1. The two-dimensional phase diagram is a simple representation of temperature-pressure relationships that govern phase transitions of pure water and gives guidelines to phase behavior of water in foods with high water contents. Necessary thermodynamic data for the solid, liquid, and gaseous states of water may be obtained from the works of Perry and Green (1984) and Weast (1986) among other handbooks that report data on transition temperatures and thermodynamic quantities. The most important physicochemical data for water are given in Table 4.1 with some comparisons to other compounds that have a corresponding chemical structure and molecular size.

Water is not present in foods in the chemically pure state, although its characteristics are those that determine phase behavior of water in foods. The purest forms of water in foods probably occur in the crystalline state and in the gaseous state. Freezing of water in foods separates ice from the solutes and food solids as almost pure ice crystals, which, when removed, allow the use of freeze-concentration as a food concentration technique. Freeze-drying is another food processing method that is based on the removal of ice

**Table 4.1**  Properties of water in comparison with compounds having a corresponding chemical structure and molecular size.

| Property | Compound | | | | |
|---|---|---|---|---|---|
| | Ammonia (NH$_3$) | Hydrofluoric acid (HF) | Hydrogen sulfide (H$_2$S) | Methane (CH$_4$) | Water (H$_2$O) |
| Molecular weight | 17.03 | 20.01 | 34.08 | 16.04 | 18.01528 |
| Melting point (°C) | -77.7 | -83.1 | -85.5 | -182.6 | 0.000 |
| Boiling point (°C) | -33.35 | 19.54 | -60.7 | -161.4 | 100.000 |
| Critical | | | | | |
| $T$ (°C) | 132.5 | 188 | 100.4 | -82.1 | 374.15 |
| $p$ (bar) | 114.0 | 64.8 | 90.1 | 46.4 | 221.5 |

*Source*: Weast (1986)

crystals from frozen foods. In comparison to freeze-concentration freeze-drying uses the pressure-temperature relationships between the physical states of water. In freeze-drying an artificial depression of pressure can be used to maintain the pressure of the sublimating ice below the triple point and to achieve direct transformation of solid ice into water vapor. Other important processes that apply phase behavior of water to improve food quality and stability include vacuum dehydration and evaporation processes.

## 2. Supercooled amorphous water

The equilibrium solid form of water is ice, which may exist in several pressure-dependent crystalline structures (e.g., Fennema, 1973). Extremely pure water may be supercooled to fairly low temperatures before homogeneous nucleation occurs. Water may also solidify as an amorphous glass, which may have various forms (Jenniskens and Blake, 1994). However, the formation of amorphous water requires special conditions and the existence of amorphous water at normal atmospheric conditions is unlikely.

One of the first studies reporting thermal properties and transition temperatures for amorphous water was that of Sugisaki *et al.* (1968). Sugisaki *et al.* (1968) used a vapor-condensation calorimeter for the measurement of heat capacity of amorphous water, and of cubic and hexagonal ice. Vapor-deposited glassy water was found to have a glass transition with onset at

**Table 4.2** Glass transition temperature, $T_g$, change of heat capacity at the glass transition, $\Delta C_p$, and crystallization temperature, $T_{cr}$, of amorphous water determined with various techniques.

| $T_g$ (°C) | $\Delta C_p$ | | $T_{cr}$ (°C)[a] | Method[b] | Reference |
|---|---|---|---|---|---|
| | (J/g°C) | (J/mol°C) | | | |
| -138 | 1.94 | 35 | - | vd | Sugisaki et al. (1968) |
| -137±1 | - | - | - | extrapolation | Rasmussen and MacKenzie (1971) |
| -134±2 | 1.05-1.39 | 19-25 | - | extrapolation | Angell and Tucker (1980) |
| -137±1 | - | - | - | hq | Johari et al. (1987) |
| -137±1 | 0.11±0.01 | 1.9±0.2 | -123 | vd | Hallbrucker et al. (1989a) |
| -144±1 | 0.11±0.01 | 2.0±0.2 | -121 | pa | Hallbrucker et al. (1989b) |
| -143±1 | 0.1 | 1.8 | -123 | pa | Johari et al. (1990) |

[a] Onset temperature of crystallization
[b] Hyperquenched water, hq; pressure-amorphized ice, pa; vapor-deposited amorphous water, vd

-138°C. The glass transition was followed by an immediate crystallization observed from the release of the latent heat. Various techniques can be used to produce amorphous water. Amorphous water produced with various methods have exhibited identical or slightly different properties when properly annealed (Angell, 1983). Common techniques given in Table 4.2 include hyperquenching of liquid water (hq), vapor deposition (vd), and pressure-induced amorphization (pa) of ice (Angell, 1983; Hallbrucker et al., 1989a,b; Angell et al., 1994). Thermal properties of the various forms of amorphous water have usually been determined with DSC. The glass transition temperatures reported agree fairly well, showing that the $T_g$ is located at around -135°C, but most values reported for the change of the heat capacity over the $T_g$ temperature range disagree (Table 4.2). Several studies have also used extrapolation methods to derive the glass transition temperature of amorphous water. The extrapolation methods have used binary, glass-forming mixtures of water and organic compounds (Rasmussen and MacKenzie, 1971) or solutions of water and inorganic salts (Angell and Tucker, 1980). The $T_g$ for amorphous water has been derived from data extrapolated with the composition dependence of the $T_g$.

## B. Water in Solutions

The physical state of a food material is often quantitatively defined by the physicochemical properties of pure water. Properties of water in solutions and in foods exhibit differences that depend on the effect of dissolved compounds on the phase behavior of water. These changes can be derived for dilute solutions from thermodynamics. However, the complexity of foods often makes it difficult or impossible to make predictions of compositional effects on phase behavior.

## 1. Freezing temperature depression

Equilibrium freezing temperatures of solutions are always lower than freezing temperatures of pure solvents. Water in foods is the most important solvent and the depressing effect of solutes on the freezing temperature of water in foods has been well established.

The effect of solutes on the vapor pressure of water in dilute solutions is defined by Raoult's law, which was given in equation (1.24). Raoult's law states that the vapor pressure of a solvent in solution is directly proportional to the mole fraction of the solvent at a constant temperature. Vapor pressure of water and ice and the effect of a solute on vapor pressure as a function of temperature is shown in Figure 4.2. As shown in Figure 4.2 the vapor pressures of ice and water are equal at the equilibrium freezing

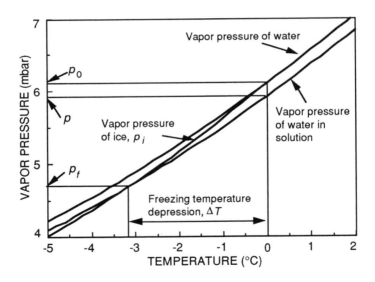

**Figure 4.2** Freezing temperature depression of water in a solution according to Raoult's law when the solute mole fraction is 0.97 (e.g., 4.8% NaCl solution). Vapor pressure of pure water, $p_0$, is higher than that of water in the solution, $p$, but equals the vapor pressure of ice, $p_i$, at the equilibrium freezing temperature. Ice formation in the solution occurs at a lower temperature as the vapor pressure of water in solution at the freezing temperature, $p_f$, becomes equal to $p_i$ at the same temperature.

temperature. Below the freezing temperature the vapor pressure of ice is lower than that of water and the equilibrium state is the solid state. Ice formation in a solution with decreased vapor pressure requires that the vapor pressure of ice becomes lower than that of water. Therefore, ice formation starts at a lower temperature as the vapor pressure of ice becomes lower than vapor pressure in solution.

The freezing temperature depression that is derived from vapor pressure depression as defined by Raoult's law applies only to solutions with low solute concentrations and ideal behavior (e.g., Fennema, 1973). Phase transitions of foods at low temperatures are complicated and often poorly understood, which applies even to ice formation and equilibrium freezing temperature. Several methods have been used to calculate freezing temperature depression of water in foods due to its importance in freezing and freeze-concentration. It should be noticed that ice formation changes solute concentration in a solution. Separation of water from a solution as ice results in freeze-concentration of solutes and the vapor pressure of unfrozen water decreases. The increasing solute concentration and decreasing vapor pressure cause a further depression of the equilibrium freezing temperature. Ice formation in foods occurs at relatively low temperatures and foods stored at -18°C are not likely to be totally frozen. It may be shown that the freezing temperature depression of water in foods follows equation (4.1), where $a_w$ is water activity, $T_0$ is the freezing temperature of pure water, $R$ is the gas constant, and $\Delta \overline{H}_m$ is the average heat of melting over the temperature range from $T$ to $T_0$ (Fennema, 1973; Schwartzberg, 1976).

$$\ln a_w = \frac{M_w \Delta \overline{H}_m (T - T_0)}{RT_0 T} \tag{4.1}$$

Schwartzberg (1976) reduced equation (4.1) to equation (4.2), where $\Delta H_0$ is the latent heat of melting at $T_0$, $w_w$ is the weight fraction of water, $w_s$ is the weight fraction of solids, $E$ and $b$ are constants, and $M_w$ is the molecular weight of water.

$$\frac{E w_s}{(w_w - b w_s)} = \frac{M_w \Delta H_0 (T_0 - T)}{RT_0^2} \tag{4.2}$$

Equation (4.2) allows the use of weight fractions of solids and water instead of the respective mole fractions. However, equation (4.2) was obtained with several assumptions. The most important assumption considered $\Delta \overline{H}_m / (T - T_0)$ to be equal to $\Delta H_0 / T_0^2$ and that $b$ was a temperature-independent constant ($\Delta H_0 = 6008.2$ J/mol $= 333.55$ J/g). Since both constants $E$ and $b$ are specific to each solute or food composition, equation (4.2) can be rewritten into the form of equation (4.3), where $k$ is defined by equation (4.4). Equation (4.3) allows the use of experimental data with linear regression to obtain numerical values for the constants $E$ and $b$.

$$\frac{1}{w_s} = (1+b) + \frac{E}{k}\left(\frac{1}{T_0 - T}\right) \qquad (4.3)$$

$$k = \frac{M_w \Delta H_0}{RT_0^2} \qquad (4.4)$$

Equation (4.3) can be used to fit experimental data on freezing temperature depression as shown for sucrose in Figure 4.3. Experimental data are needed for solute weight fraction and equilibrium freezing temperature.

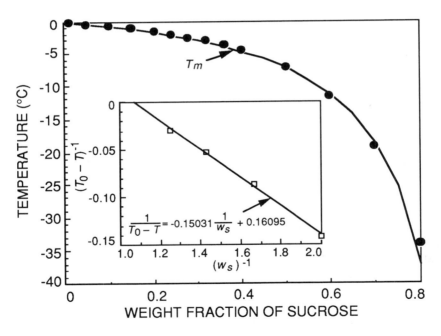

**Figure 4.3** Equilibrium freezing temperature, $T_m$, of sucrose solutions. The inset figure shows fitting of equation (4.3) to experimental data and values for $1 + b$ and $E/k$ that were used to predict the $T_m$ curve. Experimental data are from Young and Jones (1949), Weast (1986), and Roos and Karel (1991a).

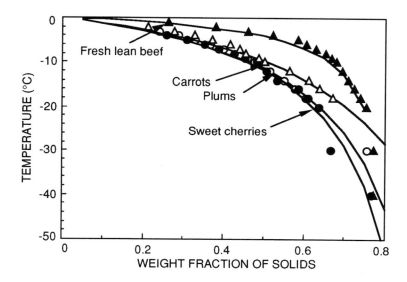

**Figure 4.4** Estimated values of freezing temperature depression in foods based on data for unfrozen water contents (Heldman, 1992) as a function of weight fraction of solids. The solid lines show the predicted freezing temperature curves obtained with equation (4.3). The weight fraction of solids gives the amount of solids in the unfrozen phase as a function of temperature.

Equation (4.3) is likely to overestimate freezing temperature depression if data close to the freezing temperature of pure water are used in modeling. Data over the whole concentration range are needed, since the true values for $b$ and $E/k$ are temperature-dependent. It is obvious that $k$ decreases with decreasing temperature, since the average value for $\Delta\overline{H}_m$ decreases with decreasing experimental temperature. It may also be observed that both $1 + b$ and $E/k$ decrease with the decreasing temperature range that is used in modeling. The best fit of equation (4.3) to experimental data over the whole concentration range is obtained when data collected below -10°C are used, as shown in Figure 4.3.

Equation (4.3) is useful in fitting experimental data on unfrozen water contents of foods and to predict freezing temperature depression as a function of food solids or to predict amounts of unfrozen water in foods at various storage conditions. As shown in Figure 4.4 freezing temperature depression of food solids is related to composition. Foods with high amounts of low molecular weight sugars and therefore lower effective molecular weights show considerable freezing temperature depression and contain higher amounts of unfrozen water at lower temperatures than foods composed of polymeric compounds. Such foods as sweet fruits are likely to contain fairly high amounts of unfrozen water at temperatures used in frozen storage while the unfrozen water content of meat products decreases significantly at temperatures close to the initial freezing temperature (Figure 4.4). Sufficient data to calculate freezing temperature depression and to predict unfrozen water contents of various foods were reported by Heldman (1992).

Another approach to obtain relationships between food composition and freezing temperature depression, although derived from the same theoretical relationships, was reported by Heldman and Singh (1981) and in a modified form by Chen (1985). According to the approach reported by Heldman and Singh (1981) the mole fraction of water is obtained with equation (4.5) when the freezing temperature of water in the food, i.e., the initial freezing temperature of the fresh product, is known.

$$\ln X_w = \frac{\Delta H_0}{R}\left(\frac{1}{T_0} - \frac{1}{T}\right) \tag{4.5}$$

Equation (4.6), where $w_s$ is the weight fraction of solids $(1 - w_w)$, can then be used to solve an "effective molecular weight" of solids, $M_s$, in the food for the estimation of unfrozen water fractions at other temperatures with the same equations.

$$X_w = \frac{w_w / M_w}{w_w / M_w + w_s / M_s} \tag{4.6}$$

The freezing temperature depression of dilute solutions may be obtained with a simplified equation (4.7), where $x_s$ is the molality (mol/kg water) of solids and $L_f$ is the latent heat of fusion (333.55 kJ/kg) (Heldman and Singh, 1981). In some studies unfrozen water contents have been derived from enthalpy data or measured using DSC (Pham, 1987; Pongsawatmanit and Miyawaki, 1993).

$$\Delta T = \frac{RT_0^2 M_w x_s}{1000 L_f} \tag{4.7}$$

Simatos and Blond (1993) calculated activity coefficients for glucose and sucrose with modified UNIQUAC (Universal Quasi Chemical) and UNIFAC (Universal Functional Activity Coefficient) models (Larsen *et al.*, 1987) and suggested the use of equation (4.8) to obtain the equilibrium freezing temperature, $T_m$. The heat capacity difference between ice and water at $T_0$ is also taken into account in $T_m$ prediction. Although the model predicted freezing temperatures for glucose and sucrose solutions well, the use of the model for the prediction of freezing temperature depressions of foods due to difficulties in calculating activity coefficients may not be feasible.

$$-\ln a_w = \frac{\Delta H_0}{R}\left(\frac{1}{T_m} - \frac{1}{T_0}\right) + \frac{\Delta C_p}{R}\ln\frac{T_0}{T_m} - \frac{\Delta C_p}{R}\left(\frac{T_0}{T_m} - 1\right) \quad (4.8)$$

Freezing temperature depressions and unfrozen water predictions are used in the evaluation of optimal temperatures for freeze-concentration of various liquid foods. Freezing temperature depression and the unfrozen water content are also related to proper freeze-drying conditions and stability of frozen foods. The freezing temperature depression and unfrozen water are mainly due to water solubles and apparently independent of the lipid fraction. However, first-order phase transitions of lipids may occur over the same temperature range as those of ice and water, which may also affect physical characteristics of foods containing lipids.

2. Boiling temperature elevation

Boiling temperatures of liquids are significantly affected by pressure. The phase diagram of water (Figure 4.1) showed pressure and temperature dependence of the equilibrium values of the liquid and gaseous states of water. It should be noticed that a change in ambient pressure causes a change in the boiling temperature and that boiling occurs when ambient pressure is equal to or lower than the vapor pressure.

Raoult's law may be applied to predict vapor pressure and boiling temperature elevation of dilute solutions. Figure 4.5 shows the effect of solute, e.g., sodium chloride, on the boiling temperature of water at atmospheric pressure. A fairly high amount of salt is required to cause a significant boiling temperature elevation. However, low molecular weight solutes in foods may significantly elevate the boiling temperature in evaporation processes as the concentration of solutes increases. The boiling temperature elevation together with flow properties of liquids must be taken into account in evaporator design and operation (Hartel, 1992).

Since boiling temperature elevation of solutions is a result of vapor pressure depression of the solvent, equations used to evaluate relationships between solute concentration and boiling temperature are similar to those used in the calculation of freezing temperatures. The mole fraction of water for a solution boiling at temperature, $T$, is obtained from equation (4.9), where

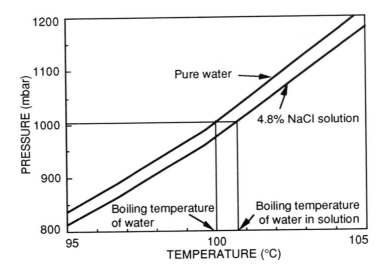

**Figure 4.5** Vapor pressure of a 4.8% NaCl solution in comparison with that of pure water at atmospheric pressure. Depression of the vapor pressure caused by the solute according to Raoult's law increases the boiling temperature of the solution.

$\Delta H_0$ is the molar latent heat of vaporization (40.606 kJ/mol at 100°C) and $T_0$ is boiling temperature of pure water. For dilute solutions the boiling temperature may be calculated with equation (4.10), where $x_s$ is solute molality (mol/kg of water), and $L_v$ is the mass latent heat of vaporization (2254 kJ/kg).

$$\ln X_w = \frac{\Delta H_0}{R}\left(\frac{1}{T_0} - \frac{1}{T}\right) \tag{4.9}$$

$$\Delta T = \frac{RT^2 M_w x_s}{1000 L_v} \tag{4.10}$$

Boiling temperature elevations have been tabulated for several foods and they can be obtained, e.g., for sugar solutions, from Hugot (1986). Boiling temperatures of solutions with various concentrations may also be obtained from Dühring's rule. Dühring's rule states that the boiling temperature of a solution is a linear function of that of water at the same pressure. Plots of boiling temperatures of solutions against the boiling temperature of water show linearity over practical and industrially applicable temperature ranges.

## 3. Eutectic solutions

Removal of solvent from a solution increases the remaining solute concentration and at some stage the solubility of the solute becomes less than the actual concentration. The phenomenon is applied in crystallization processes, but it occurs also in evaporation and freezing. The term *eutectic* refers to crystallization of both solvent and solute, which, in biological materials, may occur at freezing temperatures. Binary solutions of food components such as salts and sugars often form eutectic solutions, e.g., both the solute and water crystallize at some temperature below the initial freezing temperature of water in the solution. Below the eutectic temperature both the solvent and solute exist in the crystalline state.

Various inorganic salts form eutectic solutions with water. One of the most common of such salts is sodium chloride, which crystallizes with water from binary solution at -21.1°C. The process in a dilute solution occurs according to the phase diagram. As shown in Figure 4.6 ice formation in the solution initiates at a concentration-dependent temperature, $T_m$, that is defined by the freezing temperature depression. Ice formation proceeds with decreasing temperature to an extent that is defined by the $T_m$ curve. At the eutectic

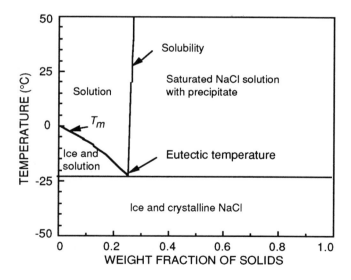

**Figure 4.6** Phase diagram for NaCl. The solubility of the salt in water decreases with decreasing temperature. Ice formation occurs below the melting temperature, $T_m$, which causes freeze-concentration of the remaining unfrozen solution. At temperatures below the eutectic temperature (-21.1°C) both water and the salt exist in the crystalline state.

temperature the solution becomes supersaturated and both water and the solute exist in the crystalline state below the eutectic temperature. The eutectic temperature in more complicated multicomponent solutions depends on the solutes. Eutectic temperatures for mixtures of inorganic salts may be determined (Fennema, 1973). The eutectic behavior of organic compounds, e.g., sugars, is more complicated and their supersaturated solutions often solidify into the glassy state before eutectic crystallization occurs. In sugar mixtures the eutectic behavior and also crystallization behavior are affected by the component compounds. Sucrose hydrates developed during frozen storage have been found to have various eutectic temperatures (Young and Jones, 1949) and a method for predicting the eutectic behavior of sugar mixtures such as fruit juices was reported by Chandrasekaran and King (1971).

The eutectic behavior of various systems was also discussed by Fennema (1973). One of the lowest eutectic temperatures is that of $CaCl_2$ at -55°C. Eutectic temperatures of sugars are fairly high. Eutectic temperatures for glucose, lactose, and sucrose are at -5, -0.7, and -9.5°C, respectively (Fennema, 1973). Eutectic crystallization in foods is probably rare due to the complex nature of the materials. However, crystallization of solutes due to supersaturation may occur in some foods. One example is lactose crystallization in ice cream during frozen storage. Ice formation in aqueous solutions can be significantly decreased by the addition of various solutes at appropriate concentrations. The phase behavior of sugar solutions has been of great interest due to the importance of sugars as food components and also as cryoprotectants of biological materials.

## III. Water in Foods

Water sorption characteristics as well as most other interactions of food solids with water are defined by the composition of nonfat food solids. Sorption properties are mainly due to carbohydrates and proteins, which often represent most of the nonfat fraction of food solids. A large number of sorption isotherms have been determined and are readily available (Iglesias and Chirife, 1982). However, sorption properties may be affected by time-dependent phenomena due to structural transformations and phase transitions. Most structural transitions and phase transitions of foods are significantly affected by water. Such transitions occur and may affect rates of deteriorative changes in low-moisture and frozen foods. Therefore, knowledge of water sorption properties is extremely important in predicting the physical state of foods at various conditions.

## A. Sorption Behavior

Water sorption may be either adsorption or desorption. Adsorption occurs when food solids are exposed to conditions where the vapor pressure of water is higher than the vapor pressure of water within the solids. At inverse conditions a lower vapor pressure of water in the surroundings of food solids is the driving force for desorption. It should be remembered that both adsorption and desorption are dependent on temperature and that the amount of water adsorption and desorption of the same food may differ at equal vapor pressure, which is referred to as sorption hysteresis.

## 1. Sorption isotherms

Sorption isotherms show the amount of water adsorbed as a function of steady state relative vapor pressure (RVP) at a constant temperature (Figure 4.7). A true equilibrium between a food and its surroundings is unlikely to exist (Peleg, 1988), but it may be assumed that at a steady state the ratio of the vapor pressure in the food, $p$, and that of pure water, $p_0$ equals water activity, $a_w$. The relative vapor pressure of water in a food and in the

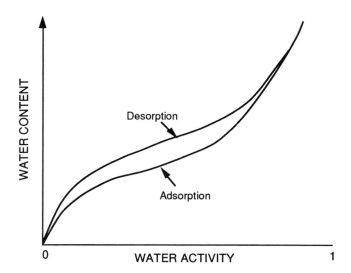

**Figure 4.7** A schematic representation of a sorption isotherm with a typical hysteresis between the adsorption and desorption isotherms of food materials. The water content is often given in g/g or g/100 g of solids.

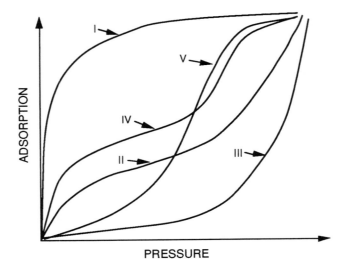

**Figure 4.8** Brunauer's five types of adsorption isotherms.

surrounding atmosphere are also equal and therefore it may be stated that water activity is obtained from relative humidity (RH) of air (RVP = $p/p_0$ = $a_w$ = RH/100). Typical sorption isotherms of food materials are sigmoid curves and they exhibit hysteresis between the adsorption and desorption isotherms as shown in Figure 4.7.

## 2. Sorption models

The shapes of isotherms showing sorption of gases by solid materials may have various forms. According to the classification of Brunauer adsorption isotherms may have five basic forms (Brunauer *et al.*, 1938), which are shown in Figure 4.8. Water adsorption isotherms of biological and food materials often follow the shape of the sigmoid, type II isotherm. Some crystalline materials, e.g., sugars, may have a fairly low adsorption of water until water activity becomes sufficient for solubilization and the adsorption increases. Such sorption follows the type III isotherm.

Water adsorption in foods is one of the main factors that affects their physical state and stability. Therefore, prediction of water adsorption is needed to establish water activity and water content relationships for food materials. Such prediction is often based on the determination of sufficient experimental data and fitting adsorption models to the data. Several empirical

**Table 4.3** Water activities of saturated salt solutions used in the determination of water sorption at temperatures typical of food storage.

| Salt | Water activity at | | | |
|------|-------|--------|--------|--------|
|      | 5°C   | 15°C   | 25°C   | 35°C   |
| LiBr | 0.074 | 0.069 | 0.064 | 0.597 |
| LiCl | 0.113 | 0.113 | 0.113 | 0.113 |
| KCH$_3$CO$_2$ | 0.291 | 0.234 | 0.225 | 0.216 |
| MgCl$_2$ | 0.336 | 0.333 | 0.328 | 0.321 |
| K$_2$CO$_3$ | 0.431 | 0.432 | 0.432 | 0.436 |
| Mg(NO$_3$)$_2$ | 0.589 | 0.559 | 0.529 | 0.499 |
| NaNO$_2$ | 0.732 | 0.693 | 0.654 | 0.628 |
| SrCl$_2$ | 0.771 | 0.741 | 0.709 | - |
| NaCl | 0.757 | 0.756 | 0.753 | 0.749 |
| (NH$_4$)$_2$SO$_4$ | 0.824 | 0.817 | 0.803 | 0.803 |
| KCl | 0.877 | 0.859 | 0.843 | 0.830 |
| BaCl$_2$ | - | 0.910 | 0.903 | 0.895 |
| K$_2$SO$_4$ | 0.985 | 0.979 | 0.973 | 0.967 |

*Source*: Greenspan (1977), Labuza *et al.* (1985), and Resnik and Chirife (1988)

and theoretical adsorption models are available (e.g., Chirife and Iglesias, 1978; van den Berg and Bruin, 1981). These models have proved to be useful in predicting water adsorption, since experimental data are usually obtained only at a few relative vapor pressures. A common method is equilibration of samples over saturated salt solutions in evacuated desiccators at a constant temperature and gravimetric measurement of the steady state water content. Relative humidities for various salt solutions and temperatures may be obtained, e.g., from Greenspan (1977) and Labuza *et al.* (1985). Those commonly used are listed in Table 4.3 with data for selected temperatures over the range from 5 to 35°C. Wolf *et al.* (1985) proposed a method for preparing the saturated salt solutions to be used in water sorption studies. Experimental data and sorption isotherms for a wide range of food materials are also available (Iglesias and Chirife, 1982).

*a. BET Model.* The Brunauer-Emmett-Teller (BET) adsorption model by Brunauer *et al.* (1938) is given by equation (4.11).

$$\frac{m}{m_m} = \frac{a_w}{(1 - a_w)[1 + (K-1)a_w]}$$

(4.11)

where $m$ is water content (g/100 g of solids), $m_m$ is the monolayer value, and $K$ is a constant.

The BET model is often used in modeling water adsorption of foods and particularly to obtain the monolayer value. The monolayer value expresses the amount of water that is sufficient to form a layer of water molecules of the thickness of one molecule on the adsorbing surface. The BET model can also be written into the form of a straight line, which is given by equation (4.12).

$$\frac{a_w}{m(1-a_w)} = \frac{1}{m_m K} + \frac{K-1}{m_m K} a_w \qquad (4.12)$$

Equation (4.12) may be given in the form of equation (4.13), where the constants $b$ and $c$ are defined by $b = (K-1)/(m_m K)$ and $c = 1/(m_m K)$, respectively.

$$\frac{a_w}{m(1-a_w)} = b a_w + c \qquad (4.13)$$

Experimental sorption data can be plotted as a function of $a_w$ as suggested by equation (4.13), which allows the use of linear regression in model fitting. Values for $K$ and $m_m$ can be obtained from $K = (b+c)/c$ and $m_m = 1/(b+c)$, respectively (Roos, 1993a). The applicability of the BET equation is often limited, since it has proved to fit water adsorption data of most food materials only over the narrow $a_w$ range from 0.1 to 0.5 (Labuza, 1968). However, the BET monolayer value has often been found to be an optimal water content for stability of low-moisture foods (Labuza *et al.*, 1970; Labuza, 1980).

*b. GAB Model.* The Guggenheim-Andersson-DeBoer (GAB) adsorption model was introduced by van den Berg (1981; van den Berg *et al.*, 1975; van den Berg and Bruin, 1981). The GAB model that is given in equation (4.14) is similar to the BET equation with the exception that it has an additional parameter, $C$.

$$\frac{m}{m_m} = \frac{K'C a_w}{(1-Ca_w)[1+(K'-1)Ca_w]} \qquad (4.14)$$

For data-fitting purposes the GAB equation can be written into the form of a second-order polynomial. The GAB equation in the form of the second-order polynomial is given by equation (4.15), where values for the parameters are defined by $\alpha = C/\{m_m[(1/K') - 1]\}$, $\beta = 1/\{m_m[1 - (2/K')]\}$, and $\gamma = 1/(CK'm_m)$ Equation (4.15) suggests that a plot showing experimental data on $m$ against $a_w$ has the shape of a parabola. Therefore, regression analysis can be used to solve the constants $\alpha$, $\beta$, and $\gamma$. The parameters, $C$, $K'$, and $m_m$ are defined by equations (4.16), (4.17), and (4.18), respectively. Thus, values obtained for $\alpha$, $\beta$, and $\gamma$ can be used to derive numerical values for $C$, $K'$, and $m_m$.

$$\frac{a_w}{m} = \alpha(a_w)^2 + \beta a_w + \gamma \qquad (4.15)$$

$$C = \frac{\beta - (1/m_m)}{-2\gamma} \qquad (4.16)$$

$$K' = \frac{1}{m_m C \gamma} \qquad (4.17)$$

$$m_m = \sqrt{-\frac{1}{4\alpha\gamma - \beta^2}} \qquad (4.18)$$

The GAB equation has been shown to fit experimental adsorption data over almost the whole $a_w$ range (van den Berg *et al.*, 1975; van den Berg, 1981; van den Berg and Bruin, 1981). The model is applicable to predict water sorption of most foods and it can also be used to calculate the monolayer value. A comparison of fitting the BET and GAB models to experimental data and sorption isotherms predicted with the BET and GAB models is shown in Figure 4.9. Water sorption data are often needed in establishing relationships between phase transition temperatures and water content. The GAB isotherm is a particularly useful sorption model, since (1) it can be applied over a wide $a_w$ range; (2) it can be used to fit experimental data from various temperatures; (3) it provides an estimate for the monolayer water content; and (4) it can be used to fit sorption data of most foods. However, it should be noticed that the monolayer water content is not a well-defined concept (Peleg, 1993).

## B. Water Plasticization

Water is the most important nonnutrient component, solvent, and plasticizer of food solids, food components, and almost all biological materials. The effect of water on the physical state of food materials is often observed from the

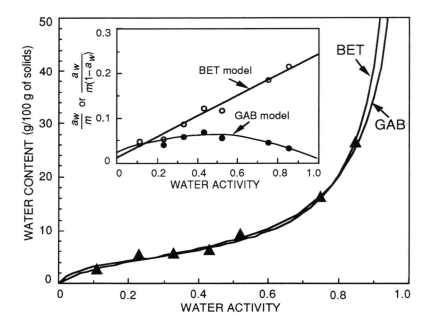

**Figure 4.9** Fitting of BET and GAB equations to experimental data in modeling water sorption of maltodextrin Maltrin M200. The inset figure shows experimental adsorption data linearized for fitting the BET model and in the form of a second-order polynomial for fitting the GAB model. Data from Roos and Karel (1991b).

lowering of the melting temperature of water-soluble food solids and structural changes that are dependent on water activity and water content. In amorphous foods water plasticization causes a dramatic decrease of the glass transition temperature. Although such plasticization may be due to extremely small amounts of water, the plasticization is observed from a significant depression of the observed $T_g$.

### 1. Melting temperature depression of crystalline components

Melting temperatures of synthetic polymers are depressed in the presence of diluents (Flory, 1953). Since water in foods is the ubiquitous diluent, the melting temperatures of food components are often affected by water. In a similar manner with the freezing temperature depression of water by the vapor pressure-lowering effect of various solutes, at low concentrations water as a diluent depresses melting temperatures of food solids. According to Flory (1953) the melting temperature of a diluted polymer may be regarded as the temperature at which the composition is that of a saturated solution, which presumably applies to melting temperatures of crystalline food components in the presence of water.

Although food solids are complex mixtures of various solids the effect of water on the equilibrium melting temperatures of single food components can be important. One example is extrusion, in which at least partially crystalline food solids are melted with small amounts of water to form a homogeneous mass that can be extruded through the dye, which is followed by rapid evaporation-induced expansion, dehydration, and cooling of the melt to the glassy state. A number of extruded foods contain starch as the main component. Lelievre (1976) applied the Flory-Huggins relationship that is given in its simplified form by equation (4.19) to predict the effect of sugars on the melting temperatures of starch-water systems.

$$\frac{1}{T_m} = \frac{RV_u}{\Delta H_u V_1}(1 - \chi_1)v_1 + \frac{1}{T_m^0} \tag{4.19}$$

Here, $T_m$ is the melting temperature, $\Delta H_u$ is the latent heat of melting per repeating polymer unit, $V_u$ and $V_1$ are the molar volumes of the polymer

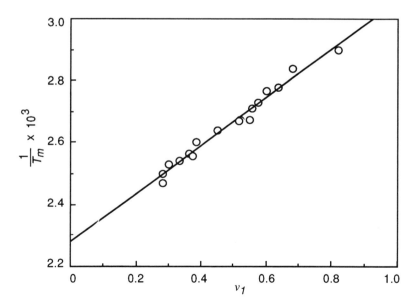

**Figure 4.10** Melting temperature, $T_m$, of starch as a function of volume fraction of water, $v_1$, shown in the form of a Flory-Huggins plot. Data from Donovan (1979).

repeating unit and the diluent, respectively, $v_1$ is the volume fraction of the diluent, and $\chi_1$ is the polymer-diluent interaction parameter.

Lelievre (1976) found that the polymer melting theory could be used to predict the decrease of melting temperature of starch with increasing water content and also the increase of gelatinization temperature with increasing sugar concentration. However, Lelievre (1976) stressed that the use of the idealized theory may not be feasible for modeling melting of such complex systems as starch. Gelatinization of starch when observed with DSC produces two endotherms at intermediate water contents. Donovan (1979) used the Flory-Huggins relationship to model the increase of the higher-temperature endotherm of potato starch with decreasing water content. The assumption that the Flory-Huggins polymer-diluent interaction parameter, $\chi_1$, was zero allowed plotting of $1/T_m$ against the diluent volume fraction, $v_1$, to produce a straight line as shown in Figure 4.10. Donovan (1979) predicted that the equilibrium melting temperature of the crystals in starch was 168°C. The melting temperature obtained was probably too low. The low value may reflect the fact that melting of crystalline regions within starch is not equilibrium melting and application of the relationship to model melting temperature depression in starch when diluted with water is probably not feasible as was suggested by Slade and Levine (1989). However, Whittam *et al.* (1990) fitted the Flory-Huggins equation to experimental crystallization data of amylose with a constant molecular weight. The equation fitted well to the data, but the $\chi_1$ value was found to be 0.5. The Flory-Huggins prediction for the melting temperature of amylose was 257°C.

Biliaderis *et al.* (1986) showed that the melting temperature of rice starches increased with decreasing water content. At low and intermediate water contents the results followed the Flory-Huggins relationship, but when $v_1 > 0.7$ a significant deviation from the theoretical depression of $T_m$ was observed. Biliaderis *et al.* (1986) concluded that at high water contents the $T_m$ could not be depressed to below the $T_g$ of the amorphous regions of starch as was proposed by Maurice *et al.* (1985). Therefore, the $T_m$ leveled off at a constant value. They also pointed out that the extrapolated values for the anhydrous melting temperatures were lower than those determined experimentally and that the $T_m$ values determined from DSC curves were probably not those of the most perfect crystals, as would be required by the theory (Flory, 1953). Although the applicability of the Flory-Huggins equation for starch-water systems has been questioned, Flory plots showing $1/T_m$ against $v_1$ can be used to compare differences between various starches and to obtain estimates for $T_m$ at various water contents (Biliaderis, 1991).

Melting temperature depression of component compounds in food solids probably occurs in most foods due to water plasticization. However, experimental data on melting temperature depressions at low water contents of food components are scarce. It should also be noticed that melting temperature data for anhydrous proteins and polysaccharides are often inaccessible due to thermal decomposition of the materials below $T_m$. Therefore, predictions of melting temperatures, e.g., with the Flory-Huggins equation, are extremely useful. Melting temperatures of polymeric food components decrease with increasing water content, which allows the formation of viscoelastic melts in such processes as extrusion and starch gelatinization.

## 2. Plasticization of food solids

In polymers low molecular weight diluents in the vicinity of the polymeric chains lower the local viscosity and enhance molecular motion, which may result in the reduction of relaxation times (Ferry, 1980). The molecular weight of water is significantly lower than that of the component compounds of food solids. Plasticization of food solids by water is observed from the depression of the glass transition temperature with increasing water content and from the dramatic change in the mechanical properties of foods that occurs as the $T_g$ is depressed to below ambient temperature.

Ferry (1980) pointed out that addition of low molecular weight diluents to polymers depresses $T_g$ sharply and almost linearly. Water is an extremely strong plasticizer even in synthetic polymers. Chan *et al.* (1986) stated that plasticization of organic molecular glasses by water may affect their relaxation characteristic similarly to plasticization in synthetic polymers, which is extremely important to the characterization of food behavior in processing and storage (Levine and Slade, 1988a,b; Slade and Levine, 1991). Ellis (1988) studied water plasticization of polyamides, which had different chemical compositions. They found that an increase in water content by 1% may induce a 15 to 20°C reduction of $T_g$ while comparable concentrations of organic diluents decrease the $T_g$ only about 5°C. Such strong plasticization by water has been found to be typical of food solids (e.g., Maurice *et al.*, 1985; Biliaderis *et al.*, 1986; Cocero and Kokini, 1991; Roos and Karel, 1991a,b,c; Slade and Levine, 1991).

According to Slade and Levine (1991) the key elements of water plasticization in foods include its role as an ubiquitous plasticizer of all amorphous food ingredients and products, and the effect of water as a plasticizer on glass transition temperature and the resulting non-Arrhenius, diffusion-limited behavior of food materials at temperatures above $T_g$. The effect of water on the $T_g$ of food components with various molecular weights is shown in Figure 4.11. The $T_g$ values of food polymers such as starch and glutenin are fairly high at low water contents and water plasticization is significant. The $T_g$

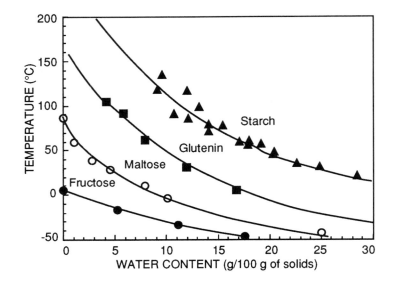

**Figure 4.11** Water plasticization observed from experimental glass transition, $T_g$, data for amorphous fructose (Roos and Karel, 1991d), maltose (Roos and Karel, 1991e), glutenin (Cocero and Kokini, 1991), and starch (Laine and Roos, 1994). The $T_g$ of biopolymers and amorphous foods decreases dramatically with increasing water content, especially at low water contents.

values of anhydrous food components generally decrease with decreasing molecular weight. Some monosaccharides, e.g., fructose, have a low anhydrous $T_g$, and even small amounts of water may depress the $T_g$ to below 0°C. Consequently, foods with high sugar contents are very sensitive to water at room temperature.

Plasticization is applied in polymer technology to increase workability, flexibility, or extensibility of synthetic polymers (Ferry, 1980). Ferry (1980) stated that the depression of $T_g$ in polymers results from an introduction of additional free volume by the diluent. Plasticization of amorphous food materials occurs accordingly. It is obvious that the fractional free volume of water, $f_1$, exceeds that of food solids, $f_2$. According to Ferry (1980) the free volumes are additive and almost always $f_1 > f_2$ and $T_{g1} < T_{g2}$ (1 and 2 refer to diluent and polymer, respectively). In foods the free volume of water is probably significantly higher than that of the amorphous solids. Water is also one of the lowest molecular weight plasticizers, with an extremely low $T_g$. The assumption that there is a linear additivity of the free volumes may be used in modeling plasticization and in particular the effect of composition on the $T_g$ (Couchman, 1978; Ferry, 1980). Several food materials are partially crystalline and it should be noticed that water plasticization occurs only in the amorphous regions (Slade and Levine, 1991).

In addition to water other low molecular weight food components may act as plasticizers within food solids. Buera *et al.* (1992) studied the effect of water and other plasticizers on the glass transition temperature of poly(vinylpyrrolidone) (PVP), which is often used as a model compound of biopolymers. A typical $T_g$ increase of polymers with increasing molecular weight was observed. The $T_g$ of PVP of various molecular weights decreased with increasing water content, but other compounds, including D-glucose, D-xylose, and L-lysine, which have been used as reactants in food models to study the rate of nonenzymatic browning (Karmas *et al.*, 1992), were also found to plasticize PVP. It may be concluded that several compounds in foods may act as plasticizers and contribute to the physical state. The most important plasticizer, however, is water, which governs the workability and behavior of food materials in various processes and contributes to the shelf life of foods.

## C. Ice Formation and Freeze-Concentration

Freezing of water in foods below 0°C follows temperature depression and supercooling at temperatures below the initial, composition-dependent freezing temperature. Ice formation continues until the equilibrium amount of ice at the freezing temperature has been formed. The concomitant freeze-concentration increases viscosity of the remaining freeze-concentrated matrix that contains solutes and unfrozen water. Such freeze-concentrated materials may be considered to be composed of ice and solutes that are plasticized by the unfrozen water.

### 1. Equilibrium freezing

Equilibrium ice formation may occur in solutions and in food materials below the initial melting temperature, provided that a driving force to the equilibrium state can be maintained without kinetic restrictions. Equilibrium freezing is typical of solutions which contain eutectic solutes, but it may proceed to some extent in supersaturated solutions.

*a. Eutectic Solutions.* Equilibrium ice formation may occur in solutions that show eutectic behavior. In such solutions the amount of ice formed below the equilibrium melting temperature is defined by the $T_m$ curve that is

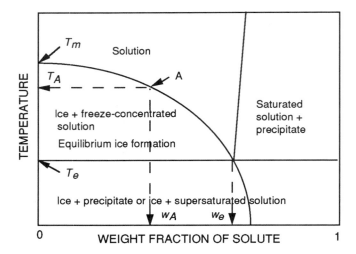

**Figure 4.12**  Ice formation in solutions below the freezing temperature of water.

shown in Figure 4.12. The maximum amount of ice is formed at the eutectic temperature, $T_e$, i.e., all water freezes provided that the solute crystallizes into an anhydrous form. At higher temperatures the weight fraction of the solute in the freeze-concentrated unfrozen phase is obtained from the $T_m$ curve. The amount of ice is the difference between the initial water content of the solution and the amount of water in the freeze-concentrated phase. According to Figure 4.12 a solution with an initial solute weight fraction, $w_i$, being less than the solute weight fraction, $w_A$, at temperature, $T_A$, may become freeze-concentrated at $T_A$. At $T_A$ an equilibrium amount of ice is formed and the weight fraction of the solute in the freeze-concentrated solute phase equals $w_A$ as defined by the point A on the $T_m$ curve. Solutions with solute concentrations of $w_e$ remain unfrozen at temperatures above $T_e$ and crystallize at temperatures below $T_e$.

*b. Supersaturated Solutions.*  Many food components, e.g., sugars, may become supersaturated in freeze-concentrated solutions. Supersaturation occurs also in foods, which contain complex mixtures of solutes. As shown in Figure 4.12 ice formation at temperatures above $T_e$ occurs only to the extent defined by the freezing temperature. However, at temperatures below $T_e$ if the solutes remain in the unfrozen solution ice formation may proceed according to the extended $T_m$ curve. Such solutions contain ice and a supersaturated unfrozen phase. Foods with supersaturated solutes may show solute crystallization during frozen storage such as lactose crystallization in ice cream.

## 2. Nonequilibrium freezing

Nonequilibrium ice formation is a typical phenomenon of rapidly cooled carbohydrate solutions and probably the most common form of ice formation in foods and other biological materials at low temperatures. One of the first studies reporting nonequilibrium freezing phenomena in frozen foods was that of Troy and Sharp (1930). They found that rapid freezing of ice cream caused concentration of lactose solution and at sufficiently low temperatures lactose crystallization was avoided.

Rey (1960) used thermal analysis to study thermal behavior of frozen solutions. Various solutes, e.g., glycerol, in water were found to remain noncrystalline at low temperatures. Rey (1960) concluded that freeze-concentration and lowering of temperature increased viscosity of the unfrozen phase until the unfrozen, concentrated solution became a glass. He also pointed out that rapidly cooled materials can be rewarmed to a devitrification temperature to allow ice formation. The existence of dissolved lactose within a freeze-concentrated glass in ice cream was also suggested by White and Cakebread (1966).

Luyet and Rasmussen (1968) found that glass transition temperatures of rapidly cooled solutions of glycerol, ethylene glycol, glucose, and sucrose increased with increasing solute concentration, but no eutectic freezing or melting was observed. They established phase diagrams that showed thermal transitions occurring in such solutions at low temperatures. The phase diagrams contained data on glass transition, devitrification (ice formation during warming of a frozen solution), and ice melting temperatures as a function of concentration. The glass transition temperatures of slowly frozen samples were higher than those of rapidly cooled samples and they occurred at an initial concentration-independent temperature (Rasmussen and Luyet, 1969). This transition was considered to be the glass transition of the freeze-concentrated solutes. Rasmussen and Luyet (1969) reported temperature values for two other initial concentration-independent transitions, which occurred above the glass transition and which were referred to as $T_{am}$ (ante-melting) and $T_{im}$ (incipient melting). Several sugars form freeze-concentrated amorphous matrices (Bellows and King, 1973) and the transitions reported by Rasmussen and Luyet (1969) are typical of frozen biological and food materials (Simatos and Turc, 1975). Franks *et al.* (1977) referred

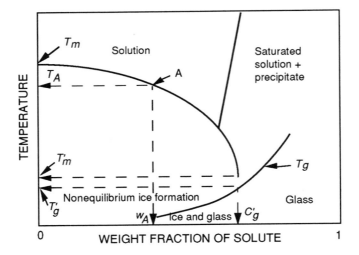

**Figure 4.13** Schematic representation of nonequilibrium ice formation in frozen solutions and foods. Ice formation above the melting temperature of ice in contact with the maximally freeze-concentrated solution, $T'_m$, proceeds to an equilibrium extent determined by the equilibrium melting temperature, $T_m$, curve. Solutions cooled rapidly to temperatures lower than the glass transition temperature of the maximally freeze-concentrated solute matrix, $T'_g$, show nonequilibrium ice formation. The amount of ice formed is defined by the glass transition, $T_g$, curve, since ice formation ceases at temperatures below the $T_g$ of the unfrozen solute matrix. Maximum ice formation, i.e., formation of an unfrozen solute phase with solute concentration equal to $C'_g$, occurs at $T'_g < T < T'_m$.

to the glass transition of freeze-concentrated solutes with the symbol $T'_g$. Levine and Slade (1986) defined $T_{am}$ and $T_{im}$ to be the onset and completion, respectively, of a single thermal event that was referred to as the glass transition of the freeze-concentrated solute matrix surrounding the ice crystals in a maximally frozen solution, $T'_g$. Roos and Karel (1991a) suggested that formation of such maximally frozen solutions with solute concentrations in the unfrozen matrix equal to $C'_g$, as shown in Figure 4.13, requires annealing slightly below the initial ice melting temperature within the maximally frozen solution. Indeed, such annealed solutions show a glass transition that corresponds to that reported by Rasmussen and Luyet (1969) and onset for ice melting at $T_{am}$. Roos and Karel (1991a) considered these transitions to be the glass transition of the maximally freeze-concentrated matrix, $T'_g$, and onset of ice melting, $T'_m$.

A schematic description of ice formation in noneutectic solutions is shown in Figure 4.13. At temperatures above $T'_m$ the equilibrium amount of ice is formed. A solution that is cooled rapidly with no ice formation to a temperature below $T_g$ becomes a glass. The high viscosity of the glassy material does not allow sufficient molecular motion of water required for crystallization. At temperatures above $T_g$ but below $T_g'$ a nonequilibrium amount of ice is formed as defined by the $T_g$ curve, since ice formation ceases as the concentration of the freeze-concentrated solute matrix approaches that of a solution with $T_g$ at the same temperature. Consequently the maximum amount of ice may form only at temperatures higher than $T_g'$ but lower than $T_m'$. It should be noticed that the glass transition occurs over a temperature range with a concomitant dramatic decrease of viscosity. The viscosity at the onset of glass transition indicated by $T_g'$ in Figure 4.13 is probably close to that of a glass ($10^{12}$ Pa s). The viscosity above $T_g'$ may be assumed to decrease according to the WLF relationship (Levine and Slade, 1986). Luyet and Rasmussen (1967; Rasmussen and Luyet, 1969) observed that ice formation in rapidly cooled solutions occurred at about a constant temperature above $T_g$ of the solution. They suggested that the viscosity limit for ice formation was $10^8$ Pa s. Roos and Karel (1991a) suggested that the viscosity of maximally freeze-concentrated sucrose solutions at $T_m'$ was $10^{7.4}$ Pa s. Therefore, $T_m'$ occurs in the vicinity of $T_g'$ and the temperatures may by defined to be equal according to the interpretation of Levine and Slade (1986), as was also suggested by Ablett *et al.* (1992a, 1993). However, the definition of $T_g'$ to be the onset of the glass transition and locating $T_m'$ at the onset temperature of ice melting allows separation of the two phenomena from each other. Such definition also allows evaluation of time-dependent phenomena within the glass transition but below $T_m'$, e.g., formation of the maximally freeze-concentrated solute matrix. Considering that ice melting occurs above $T_m'$ is extremely important in the analysis of rates of various kinetic phenomena above $T_m'$ (Kerr and Reid, 1994).

### 3. State diagrams

State diagrams are simplified phase diagrams that describe the concentration dependence of the glass transition temperature of solutes or food solids and relationships between ice formation and solute concentration at low temperatures. State diagrams are useful in the characterization of the physical state of food solids at various temperatures and water contents.

Rasmussen and Luyet (1969) used phase diagrams to show the concentration-dependence of the thermal transitions of frozen solutions. Such phase diagrams, which were similar to the schematic phase diagram showing nonequilibrium ice formation in Figure 4.13 characterize rate-controlled phenomena. Therefore, Franks *et al.* (1977) stressed that the term *state*

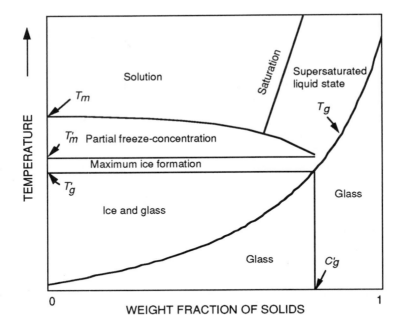

**Figure 4.14** A schematic state diagram typical of food solids. Glass transition temperature, $T_g$, decreases with an increasing water content due to plasticization. Ice formation occurs in solutions with initial solute concentrations that are lower than the solute concentration in the maximally freeze-concentrated solute matrix, $C'_g$. The maximally freeze-concentrated solute matrix has the onset of glass transition at $T'_g$ and ice melting occurs at temperatures above $T'_m$.

should be used instead of *phase* due to the nonequilibrium nature of the physical state of freeze-concentrated biological materials. State diagrams have been used by Levine and Slade (1988c) to characterize effects of frozen storage temperature on food quality. According to Roos and Karel (1991f) state diagrams are useful in characterization of physical properties and stability of food materials both in processing and storage.

State diagrams have been established for a number of food components, e.g., sucrose (Blanshard and Franks, 1987), various sugars and starch (van den Berg, 1986; Roos and Karel, 1991e,f; Roos, 1993b), cereal proteins (de Graaf *et al.*, 1993), and milk powders (Jouppila and Roos, 1994). A schematic state diagram typical of food components and foods is shown in Figure 4.14. The solute concentration of the maximally freeze-concentrated solute matrix may be obtained with various methods. Levine and Slade (1986) have suggested the determination of the latent heat of ice melting for 20% (w/w) solutions with DSC. The method allows calculation of the amount of ice in the sample based on the assumption that the latent heat of ice melting in the food is equal to that of pure water. $C'_g$ is then obtained from the difference between total water content and the amount of ice in the sample, provided that no solute crystallization has occurred. The unfrozen water content may also be estimated by the determination of latent heat for ice melting in samples having various initial water contents. The latent heat of melting divided by the weight of solids is plotted against the initial water content and the straight line obtained is extrapolated to zero latent heat, which gives the amount of unfrozen water (Simatos *et al.*, 1975; Blond, 1989; Roos, 1992). Ablett *et al.* (1992) showed that the true value of $C'_g$ may be obtained with DSC only when the temperature dependence of the latent heat of melting is considered together with a proper heat capacity correction for the samples. The most precise $C'_g$ values can be derived from state diagrams established with experimental $T_g$ data for a number of solute concentrations (Roos and Karel, 1991a). The solute concentration of maximally freeze-concentrated solute matrices has been found to be about 80% (w/w) for most food components (Ablett *et al.*, 1992; Roos, 1993b).

State diagrams are important tools in establishing proper processing and storage conditions for dehydrated and frozen foods. They can be also applied in evaluating amounts of ice in biological materials at various temperatures, which is extremely important in cryopreservation. State diagrams can also be used to define proper freeze-drying conditions and to define proper processing and storage conditions for dehydrated foods.

# References

Ablett, A., Izzard, M.J. and Lillford, P.J. 1992. Differential scanning calorimetric study of frozen sucrose and glycerol solutions. *J. Chem. Soc., Faraday Trans. 88*: 789-794.

Ablett, S., Izzard, M.J., Lillford, P.J., Arvanitoyannis, I. and Blanshard, J.M.V. 1993. Calorimetric study of the glass transition occurring in fructose solutions. *Carbohydr. Res. 246*: 13-22.

Angell, C.A. 1983. Supercooled water. *Annu. Rev. Phys. Chem. 34*: 593-630.

Angell, C.A. and Tucker, J.C. 1980. Heat capacity changes in glass-forming aqueous solutions and the glass transition in vitreous water. *J. Phys. Chem. 84*: 268-272.

Angell, C.A., Bressel, R.D., Green, J.L., Kanno, H., Oguni, M. and Sare, E.J. 1994. Liquid fragility and the glass transition in water and aqueous solutions. *J. Food Eng. 22*: 115-142.

Bellows, R.J. and King, C.J. 1973. Product collapse during freeze drying of liquid foods. *AIChE Symp. Ser. 69(132)*: 33-41.

Biliaderis, C.G. 1991. The structure and interactions of starch with food constituents. *Can. J. Physiol. Pharmacol. 69*: 60-78.

Biliaderis, C.G., Page, C.M., Maurice, T.J. and Juliano, B.O. 1986. Thermal characterization of rice starches: A polymeric approach to phase transitions of granular starch. *J. Agric. Food Chem. 34*: 6-14.

Blanshard, J.M.V. and Franks, F. 1987. Ice crystallization and its control in frozen-food systems. Chpt. 4 in *Food Structure and Behaviour*, ed. J.M.V. Blanshard and P. Lillford. Academic Press, London, pp. 51-65.

Blond, G. 1989. Water-galactose system: Supplemented state diagram and unfrozen water. *Cryo-Lett. 10*: 299-308.

Brunauer, S., Emmett, P.H. and Teller, E. 1938. Adsorption of gases in multimolecular layers. *J. Am. Chem. Soc. 60*: 309-319.

Buera, M.P., Levi, G. and Karel, M. 1992. Glass transition in poly(vinylpyrrolidone): Effect of molecular weight and diluents. *Biotechnol. Prog. 8*: 144-148.

Chan, R.K., Pathmanathan, K. and Johari, G.P. 1986. Dielectric relaxations in the liquid and glassy states of glucose and its water mixtures. *J. Phys. Chem. 90*: 6358-6362.

Chandrasekaran, S.K. and King, C.J. 1971. Solid-liquid phase equilibria in multicomponent aqueous sugar solutions. *J. Food Sci. 36*: 699-704.

Chen, C.S. 1985. Thermodynamic analysis of the freezing and thawing of foods: Enthalpy and apparent specific heat. *J. Food Sci. 50*: 1158-1162.

Chirife, J. and Iglesias, H.A. 1978. Equations for fitting water sorption isotherms of foods: Part 1 - A review. *J. Food Technol. 13*: 159-174.

Cocero, A.M. and Kokini, J.L. 1991. The study of the glass transition of glutenin using small amplitude oscillatory rheological measurements and differential scanning calorimetry. *J. Rheol. 35*: 257-270.

Couchman, P.R. 1978. Compositional variation of glass-transition temperatures. 2. Application of the thermodynamic theory to compatible polymer blends. *Macromolecules 11*: 1156-1161.

de Graaf, E.F., Madeka, H., Cocero, A.M. and Kokini, J.L. 1993. Determination of the effect of moisture on gliadin glass transition using mechanical spectrometry and differential scanning calorimetry. *Biotechnol. Prog. 9*: 210-213.

Donovan, J.W. 1979. Phase transitions of the starch-water system. *Biopolymers 18*: 263-275.

Ellis, T.S. 1988. Moisture-induced plasticization of amorphous polyamides and their blends. *J. Appl. Polym. Sci. 36*: 451-466.

Fennema, O.R. 1973. Solid-liquid equilibria. Chpt. 3 in *Low-Temperature Preservation of Foods and Living Matter*, ed. O.R. Fennema, W.D. Powrie and E.H. Marth. Marcel Dekker, New York, pp. 101-149.

Fennema, O.R. 1985. Water and ice. Chpt. 2 in *Food Chemistry*, ed. O.R. Fennema, 2nd ed. Marcel Dekker, New York, pp. 23-67.

Ferry, J.D. 1980. *Viscoelastic Properties of Polymers*. 3rd ed. John Wiley & Sons, New York.

Flory, P.J. 1953. *Principles of Polymer Chemistry*. Cornell University Press, Ithaca, N Y.

Franks, F., Asquith, M.H., Hammond, C.C., Skaer, H.B. and Echlin, P. 1977. Polymeric cryoprotectants in the preservation of biological ultrastructure I. Low temperature states of aqueous solutions of hydrophilic polymers. *J. Microsc. (Oxford) 110*: 223-238.

Greenspan, L. 1977. Humidity fixed points of binary saturated aqueous solutions. *J. Res. Natl. Bur. Stand., Sect. A 81A*: 89-96.

Hallbrucker, A., Mayer, E., and Johari, G.P. 1989a. Glass-liquid transition and the enthalpy of devitrification of annealed vapor-deposited amorphous solid water. A comparison with hyperquenched glassy water. *J. Phys. Chem. 93*: 4986-4990.

Hallbrucker, A., Mayer, E. and Johari, G.P. 1989b. Glass transition in pressure-amorphized hexagonal ice. A comparison with amorphous forms made from the vapor and liquid. *J. Phys. Chem. 93*: 7751-7752.

Hartel, R.W. 1992. Evaporation and freeze concentration. Chpt. 8 in *Handbook of Food Engineering*, ed. D.R. Heldman and D.B. Lund. Marcel Dekker, New York, pp. 341-392.

Heldman, D. R. 1992. Food freezing. Chpt. 6 in *Handbook of Food Engineering*, ed. D.R. Heldman and D.B. Lund. Marcel Dekker, New York, pp. 277-315.

Heldman, D.R. and Singh, R.P. 1981. *Food Process Engineering*. 2nd ed. AVI Publishing Co., Westport, CT.

Hugot, E. 1986. *Handbook of Cane Sugar Engineering*, 3rd ed. Elsevier, Amsterdam.

Iglesias, H.A. and Chirife, J. 1982. *Handbook of Food Isotherms*. Academic Press, New York.

Jenniskens, P. and Blake, D.F. 1994. Structural transitions in amorphous water ice and astrophysical implications. *Science 265*: 753-756.

Johari, G.P., Hallbrucker, A. and Mayer, E. 1987. The glass-liquid transition of hyperquenched water. *Nature 330*: 552-553.

Johari, G.P., Hallbrucker, A. and Mayer, E. 1990. Calorimetric study of pressure-amorphized cubic ice. *J. Phys. Chem. 94*: 1212-1214.

Jouppila, K. and Roos, Y.H. 1994. Glass transitions and crystallization in milk powders. *J. Dairy Sci. 77*: 2907-2915.

Karmas, R., Buera, M.P. and Karel, M. 1992. Effect of glass transition on rates of nonenzymatic browning in food systems. *J. Agric. Food Chem. 40*: 873-879.

Kerr, W.L. and Reid, D.S. 1994. Temperature dependence of the viscosity of sugar and maltodextrin solutions in coexistence with ice. *Lebensm.-Wiss. u. -Technol. 27*: 225-231.

Labuza, T.P. 1968. Sorption phenomena in foods. *Food Technol. 22*: 263-265, 268, 270, 272.

Labuza, T.P. 1980. The effect of water activity on reaction kinetics of food deterioration. *Food Technol. 34(4)*: 36-41, 59.

Labuza, T.P., Tannenbaum, S.R. and Karel, M. 1970. Water content and stability of low-moisture and intermediate-moisture foods. *Food Technol. 24*: 543-544, 546-548, 550.

Labuza, T.P., Kaanane, A. and Chen, J.Y. 1985. Effect of temperature on the moisture sorption isotherms and water activity shift of two dehydrated foods. *J. Food Sci. 50*: 385-391.

Laine, M.J.K. and Roos, Y. 1994. Water plasticization and recrystallization of starch in relation to glass transition. In *Proceedings of the Poster Session, International Symposium on the Properties of Water, Practicum II*, ed. A. Argaiz, A. López-Malo, E. Palou and P. Corte. Universidad de las Américas-Puebla, pp. 109-112.

Larsen, B.L., Rasmussen, P. and Fredenslund, A. 1987. A modified UNIFAC group-contribution model for prediction of phase equilibria and heats of mixing. *Ind. Eng. Chem. Res. 26*: 2274-2286.

Lelievre, J. 1976. Theory of gelatinization in a starch-water-solute system. *Polymer 17*: 854-858.

Levine, H. and Slade, L. 1986. A polymer physico-chemical approach to the study of commercial starch hydrolysis products (SHPs). *Carbohydr. Polym. 6*: 213-244.

Levine, H. and Slade, L. 1988a. Principles of "cryostabilization" technology from structure/property relationships of carbohydrate/water systems. *Cryo-Lett. 9*: 21-63.

Levine, H. and Slade, L. 1988b. 'Collapse' phenomena - A unifying concept for interpreting the behaviour of low moisture foods. Chpt. 9 in *Food Structure - Its Creation and Evaluation*, ed. J.M.V. Blanshard and J.R. Mitchell. Butterworths, London, pp. 149-180.

Levine, H. and Slade, L. 1988c. Thermomechanical properties of small-carbohydrate-glasses and 'rubbers.' *J. Chem. Soc., Faraday Trans. 1, 84*: 2619-2633.

Luyet, B. and Rasmussen, D. 1967. Study by differential thermal analysis of the temperatures of instability in rapidly cooled solutions of polyvinylpyrrolidone. *Biodynamica 10(209)*: 137-147.

Luyet, B. and Rasmussen, D. 1968. Study by differential thermal analysis of the temperatures of instability of rapidly cooled solutions of glycerol, ethylene glycol, sucrose and glucose. *Biodynamica 10(211)*: 167-191.

Maurice, T.J., Slade, L., Sirett, R.R. and Page, C.M. 1985. Polysaccharide-water interactions - thermal behavior of rice starch. In *Properties of Water in Foods*, ed. D. Simatos and J.L. Multon. Martinus Nijhoff Publishers, Dordrecht, the Netherlands, pp. 211-227.

Peleg, M. 1988. An empirical model for the description of moisture sorption curves. *J. Food Sci. 53*: 1216-1217, 1219.

Peleg, M. 1993. Assessment of a semi-empirical four parameter general model for sigmoid moisture sorption isotherms. *J. Food Process Eng. 16*: 21-37.

Perry, R.H. and Green, D. 1984. *Perry's Chemical Engineers' Handbook*, 6th ed. McGraw-Hill, New York.

Pham, Q.T. 1987. Calculation of bound water in frozen food. *J. Food Sci. 52*: 210-212.

Pongsawatmanit, R. and Miyawaki, O. 1993. Measurement of temperature-dependent ice fraction in frozen foods. *Biosci. Biotechnol. Biochem. 57*: 1650-1654.

Rasmussen, D. and Luyet, B. 1969. Complementary study of some nonequilibrium phase transitions in frozen solutions of glycerol, ethylene glycol, glucose and sucrose. *Biodynamica 10(220)*: 319-331.

Rasmussen, D.H. and MacKenzie, A.P. 1971. The glass transition in amorphous water. Application of the measurements to problems arising in cryobiology. *J. Phys. Chem. 75*: 967-973.

Resnik, S.L. and Chirife, J. 1988. Proposed theoretical water activity values at various temperatures for selected solutions to be used as reference sources in the range of microbial growth. *J. Food Prot. 51*: 419-423.

Rey, L.R. 1960. Thermal analysis of eutectics in freezing solutions. *Ann. N.Y. Acad. Sci. 85*: 510-534.

Roos, Y.H. 1992. Phase transitions and transformations in food systems. Chpt. 3 in *Handbook of Food Engineering*, ed. D.R. Heldman and D.B. Lund. Marcel Dekker, New York, pp. 145-197.

Roos, Y. H. 1993a. Water activity and physical state effects on amorphous food stability. *J. Food Process. Preserv. 16*: 433-447.

Roos, Y. 1993b. Melting and glass transitions of low molecular weight carbohydrates. *Carbohydr. Res. 238*: 39-48.

Roos, Y. and Karel, M. 1991a. Amorphous state and delayed ice formation in sucrose solutions. *Int. J. Food Sci. Technol. 26*: 553-566.

Roos, Y. and Karel, M. 1991b. Phase transitions of mixtures of amorphous polysaccharides and sugars. *Biotechnol. Prog. 7*: 49-53.

Roos, Y. and Karel, M. 1991c. Plasticizing effect of water on thermal behavior and crystallization of amorphous food models. *J. Food Sci. 56*: 38-43.

Roos, Y. and Karel, M. 1991d. Nonequilibrium ice formation in carbohydrate solutions. *Cryo-Lett. 12*: 367-376.

Roos, Y. and Karel, M. 1991e. Water and molecular weight effects on glass transitions in amorphous carbohydrates and carbohydrate solutions. *J. Food Sci. 56*: 1676-1681.

Roos, Y. and Karel, M. 1991f. Applying state diagrams to food processing and development. *Food Technol. 45(12)*: 66, 68-71, 107.

Schwartzberg, H.G. 1976. Effective heat capacities for the freezing and thawing of food. *J. Food Sci. 41*: 152-156.

Simatos, D. and Blond, G. 1993. Some aspects of the glass transition in frozen foods systems. Chpt. 19 in *The Glassy State in Foods*, ed. J.M.V. Blanshard and P.J. Lillford. Nottingham University Press, Loughborough, pp. 395-415.

Simatos, D. and Turc, J.M. 1975. Fundamentals of freezing in biological systems. Chpt. 2 in *Freeze Drying and Advanced Food Technology*, ed. S.A. Goldblith, L. Rey and W.W. Rothmayr. Academic Press, New York, pp. 17-28.

Simatos, D., Faure, M., Bonjour, E. and Couach, M. 1975. The physical state of water at low temperatures in plasma with different water contents as studied by differential thermal analysis and differential scanning calorimetry. *Cryobiology 12*: 202-208.

Slade, L. and Levine, H. 1989. A food polymer science approach to selected aspects of starch gelatinization and retrogradation. In *Frontiers in Carbohydrate Research - 1: Food Applications*, ed. R.P. Millane, J.N. BeMiller and R. Chandrasekaran. Elsevier, London, pp. 215-270.

Slade, L. and Levine, H. 1991. Beyond water activity: Recent advances based on an alternative approach to the assessment of food quality and safety. *Crit. Rev. Food Sci. Nutr. 30*: 115-360.

Sugisaki, M., Suga, H. and Seki, S. 1968. Calorimetric study of the glassy state. IV. Heat capacities of glassy water and cubic ice. *Bull. Chem. Soc. Jpn. 41*: 2591-2599.

Troy, H.C. and Sharp, P.F. 1930. $\alpha$ and $\beta$ lactose in some milk products. *J. Dairy Sci. 13*: 140-157.

van den Berg, C. 1981. Vapour sorption equilibria and other water-starch interactions; a physico-chemical approach. Ph.D thesis, Agricultural University Wageningen, the Netherlands.

van den Berg, C. 1986. Water activity. In *Concentration and Drying of Foods*, ed. D. MacCarthy. Elsevier, Amsterdam, pp. 11-36.

van den Berg, C. and Bruin, S. 1981. Water activity and its estimation in food systems: Theoretical aspects. In *Water Activity: Influences on Food Quality*, ed. L.B. Rockland and G.F. Stewart. Academic Press, New York, pp. 1-61.

van den Berg, C., Kaper, F.S., Weldring, J.A.G. and Wolters, I. 1975. Water binding by potato starch. *J. Food Technol. 10*: 589-602.

Weast, R.C. 1986. *CRC Handbook of Chemistry and Physics*, 67th ed. CRC Press, Boca Raton, FL.

White, G.W. and Cakebread, S.H. 1966. The glassy state in certain sugar-containing food products. *J. Food Technol. 1*: 73-82.

Whittam, M.A., Noel, T.R. and Ring, S.G. 1990. Melting behaviour of A- and B-type crystalline starch. *Int. J. Biol. Macromol. 12*: 359-362.

Wolf, W., Spiess, W.E.L. and Jung, G. 1985. Standardisation of isotherm measurements (COST-project 90 and 90 bis). *Properties of Water in Foods*, ed. D. Simatos and J.L. Multon. Martinus Nijhoff Publishers, Dordrecht, the Netherlands, pp. 661-679.

Young, F.E. and Jones, F.T. 1949. Sucrose hydrates. The sucrose-water phase diagram. *J. Phys. Colloid Chem. 53*: 1334-1350.

# CHAPTER 5

# *Food Components and Polymers*

## I. Introduction

Phase transitions of food components are likely to alter the physical properties of food materials. Foods are complicated systems, but their physical state is usually governed by phase transitions of the main components, i.e., carbohydrates, lipids, proteins, and water. The phase transition behavior of food solids has similarities with that of synthetic polymers. However, in foods water is probably the most significant compound and diluent, which may significantly affect the physical state and properties of other component compounds.

Various first- and second-order phase transitions in food materials may occur during a number of processes, storage, and distribution. Food materials that are rich in water obviously become solid at freezing temperatures. The physical state of oils and spreads at various temperatures is dependent on the location of melting temperatures, which is an extremely important quality attribute. The relationships between food properties and the physical state are complicated. The physical state of foods may be difficult to detect and it is often extremely sensitive to temperature, time, and water. The main constituents of foods may exist in the liquid state or in the solid crystalline or amorphous noncrystalline state. Many of the component compounds, e.g., sugars, fats, and water, in their chemically pure form, crystallize below the equilibrium melting temperature. However, the large number of chemical compounds within food solids does not always allow the formation of such highly ordered equilibrium states.

Phase transitions of food components, which often occur during various processes and storage, are emphasized in this chapter. Foods can seldom be considered as being equilibrium systems, which makes their phase behavior complicated and time-dependent.

## II. Carbohydrates

Carbohydrates are present in almost all foods. Phase transitions of low molecular weight sugars as well as those of carbohydrate polymers are extremely important to food properties. Chemically pure low molecular weight sugars may exist in the crystalline state, but the formation of equilibrium structures in foods is often restricted. However, carbohydrates may exist as crystalline compounds, semicrystalline compounds, partially crystalline compounds, or totally amorphous compounds. First-order phase transitions of carbohydrates in foods include melting and crystallization, and gelatinization of starch. Transitions between the nonequilibrium and equilibrium states are particularly important to the stability of carbohydrate foods with low water contents.

### A. Sugars

The phase behavior of sugars is important in manufacturing of various sweet foods. Phase transitions of sugars in such foods may affect their behavior in the manufacturing processes and quality changes during storage. Numerous examples of sweet foods include candies, confectionery, bakery products, and almost all food materials that contain sugar. Phase transitions of sugars are also important in freezing processes and to frozen food stability. Moreover, phase transitions of sugars are related to their ability to act as cryoprotectants or cryopreservatives. The physical properties of sugars in the frozen state are particularly important to the stability of ice cream and frozen desserts.

## 1. Melting and crystallization

Melting and crystallization are opposite phenomena, which depend on several factors. Melting and crystallization temperatures of sugars are sensitive to water, which affects the crystallization and melting transition temperatures of sugars in most foods. Sugars may also crystallize as hydrates, which may show phase behavior significantly different from that of the anhydrides. In addition physicochemical properties of sugar anomers may differ.

Melting of crystalline sugars occurs when they are heated to above their melting temperature, although some sugars may caramelize and brown concomitantly with the melting process. Caramelization occurs especially in sugars that have melting temperatures above 150°C. Carbohydrates with melting temperatures well above 200°C tend to decompose before melting occurs. Such decomposition may be observed from thermogravimetric measurements or DSC curves that suggest chemical changes over the temperature range from 200 to 400°C (e.g., Pavlath and Gregorski, 1985). Sugars do not have exact melting temperatures and their melting proceeds over a temperature range. Therefore, literature values for melting temperatures of sugars may slightly differ. Melting temperatures, $T_m$, and heats of fusion, $\Delta H_f$, for sugars are given in Table 5.1. The melting temperatures and heats of fusion may be determined with DSC. Melting endotherms of sugars occur over the melting temperature range and they are fairly broad. Integration of the melting endotherm can be used to obtain $\Delta H_f$. It is important to notice that both $T_m$ and $\Delta H_f$ may slightly differ between sugar anomers. The melting temperatures and heats of fusion are significantly decreased by water in hydrates. Generally melting temperatures of sugars increase with increasing molecular weight in the order monosaccharides < disaccharides < oligosaccharides.

Crystallization of sugars may occur at temperatures below their melting temperature. However, a sufficiently rapid cooling of a sugar melt results in the formation of amorphous sugar melts that solidify below the glass transition temperature. The presence of water may enhance crystallization of sugar melts and especially crystallization of sugars that form hydrates. Crystallization of sugars occurs usually from supersaturated solutions. According to Hartel and Shastry (1991) crystallization occurs only when supersaturation exceeds some critical value. Such behavior is related to the energy needed for the formation of critical nuclei and formation of the solid phase. The solubility of sugars in water increases with increasing temperature

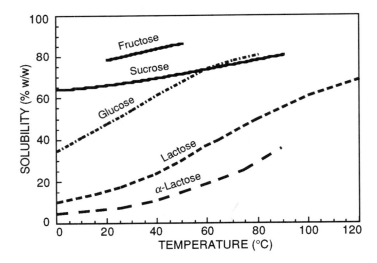

**Figure 5.1** Solubilities for common food component sugars as a function of temperature. Data from Pancoast and Junk (1980).

as shown in Figure 5.1 for common sugars. Most sugars are fairly soluble in water, but solubility may decrease in solutions that contain other sugars or compounds. One example is lactose, which exhibits decreased solubility in the presence of sucrose although sucrose solubility is not greatly affected by lactose (Nickerson and Moore, 1972; Nickerson and Patel, 1972; Hartel and Shastry, 1991). The solubility of sugars in alcohols has been found to be lower than in water and some sugars are totally insoluble in alcohols (Hartel and Shastry, 1991).

## 2. Glass transitions

Glass transition temperatures of anhydrous sugars vary from the very low temperatures of alditols to those of mono-, di-, and oligosaccharides at higher temperatures. Glass transition temperatures of sugars govern the physical state of a wide range of foods. The noncrystalline state of sugars is typical of sugar candies, confectionery, and even ice cream, but manipulation of the physical state has been a skillful art rather than a science.

The amorphous state of glucose and sucrose is fairly well characterized, but information on glass transition temperatures of several sugars and especially on transitions between the amorphous and crystalline states is scarce. All sugars exhibit significant water plasticization, which affects the

**Table 5.1** Melting temperatures, $T_m$, for various sugars and sugar alcohols with latent heat of fusion, $\Delta H_f$, determined with DSC.

| Compound | $T_m$ (°C)[a] | | | $\Delta H_f$ (J/g) |
|---|---|---|---|---|
| | Onset | Peak | Other[b] | |
| Pentoses | | | | |
| D-Arabinose | 150 | 160 | 155.5-156.5 | 238 |
| L-Arabinose | - | - | 159-160 | - |
| D-Ribose | 70 | 86 | 95 | 146 |
| D-Xylose | 143 | 157 | 145 | 211 |
| Hexoses | | | | |
| D-Fructose | 108 | 127 | 103-105 | 169 |
| $\alpha$-D-Fucose | 133 | 145 | - | 186 |
| D-Galactose | 163 | 170 | - | 243 |
| D-Glucose | 143 | 158 | 146 | 179 |
| $\alpha$-D-Glucose x $H_2O$ | | | 86 | - |
| D-Mannose | 120 | 134 | 132 | 137 |
| $\alpha$-L-Rhamnose x $H_2O$ | 86 | 99 | 92 | 191 |
| L-Sorbose | 153 | 163 | 165 | 245 |
| Disaccharides | | | | |
| $\alpha$-Lactose x $H_2O$ | - | - | 201-202 | - |
| $\alpha$-Lactose | - | - | 222.8 | - |
| $\beta$-Lactose | - | - | 253 | - |
| Lactulose[c] | 162 | 169 | - | 136 |
| Maltose | - | - | 160-165 | - |
| Maltose x $H_2O$ | 104 | 123 | 102-103 | 126 |
| $\alpha$-Melibiose x 0.5 $H_2O$ | 138 | - | - | - |
| Sucrose | 173 | 190 | 185-186 | 118 |
| $\alpha,\alpha$-Trehalose | - | - | 214-216 | - |
| $\alpha,\alpha$-Trehalose x 2 $H_2O$ | 91 | 97 | 103 | 127 |
| Oligosaccharides[c] | | | | |
| Raffinose x $5H_2O$ | 76 | 80 | 80 | 138 |
| Sugar alcohols | | | | |
| Maltitol | 139 | 149 | - | 150 |
| D-Mannitol[c] | 163 | 167 | 168 | 251 |
| D-Sorbitol | 85 | 99 | 110-112 | 150 |
| Xylitol | 89 | 95 | 93-94 | 250 |

[a] Onset of melting endotherm; peak of melting endotherm
[b] Weast (1986)
[c] Y.H. Roos, unpublished data (1992)
*Source*: Roos (1993)

stability of a number of foods. White and Cakebread (1966) realized the importance of the glassy state of sugars to the stability of such food products as boiled sweets, freeze-dried foods, ice cream, and milk powders.

Heating of crystalline sugars to above melting temperature followed by rapid cooling often results in the formation of a solid, transparent, glassy material, which was realized quite early (Parks and Thomas, 1934; Kauzmann, 1948). Parks and co-workers conducted a number of studies on the physical properties of glucose glasses and they were well aware of the time-dependent nature of the amorphous state (Parks and Reagh, 1937). The glass transition temperature of sugars generally depends on the molecular weight. Monosaccharides have lower $T_g$ values than disaccharides, which have lower values than oligosaccharides (Slade and Levine, 1991, 1995; Roos, 1993) Several studies have reported transition temperatures for common sugars. The differences between the reported $T_g$ values for the same sugars are probably due to residual water in samples, differences in sample handling techniques, and differences in techniques used to measure $T_g$. Most studies have used DSC, but the use of other techniques such as DMTA (Schenz *et al.*, 1991; MacInnes, 1993) and TMA (Maurice *et al.*, 1991; Le Meste and Huang, 1992) has been reported. Chan *et al.* (1986) studied glass transition temperatures of glucose using dielectric and calorimetric techniques. The dielectric relaxation features of glucose and water-plasticized glucose were found to be remarkably similar to those of amorphous polymers. $T_g$ may also be derived from data for spin-lattice relaxation times (Karger and Lüdemann, 1991). Observed glass transition temperatures for sugars are given in Table 5.2.

The glass transition temperature of amorphous sugars is extremely sensitive to water. State diagrams are particularly useful in the characterization of the physical state of sugars and the water content dependence of transition temperatures. In addition state diagrams describe conditions needed for ice formation. The first phase diagrams that showed the glass transition dependence of water content for glucose and sucrose were established by Luyet and Rasmussen (1968). Other state diagrams have been reported for a number of sugars by various authors. These include those of sucrose (Blanshard and Franks, 1987; Hatley *et al.*, 1991; Izzard *et al.*, 1991; Roos and Karel, 1991a), galactose (Blond, 1989), fructose (Roos and Karel, 1991b; Ablett *et al.*, 1993a), maltose (Roos and Karel, 1991d), and maltotriose (Ablett *et al.*, 1993b). The state diagram for sucrose with experimental data for various temperature- and concentration-dependent properties is shown in Figure 5.2. State diagrams can be established with experi mental data for freezing temperature depression, maximum freeze-concentration, solubility, and water plasticization, i.e., concentration dependence of the $T_g$.

**Table 5.2** Glass transition temperatures, $T_g$, change of heat capacity occurring over the glass transition, $\Delta C_p$, and the ratio of melting temperature, $T_m$, and $T_g$.

| Compound | $T_g$ (°C)[a] Onset | Midpoint | Other | $\Delta C_p$ (J/g°C)[d] I | II | $T_m / T_g$[e] I | II |
|---|---|---|---|---|---|---|---|
| Pentoses | | | | | | | |
| Arabinose | -2 | 3 | 4[c] | 0.66 | 0.80 | 1.56 | - |
| Ribose | -20 | -13 | -10[b], -11[c] | 0.67 | 0.94 | 1.36 | 1.37 |
| Xylose | 6 | 14 | 9.5[b], 13[c] | 0.66 | 0.95 | 1.49 | 1.51 |
| Hexoses | | | | | | | |
| Fructose | 5 | 10 | 100[b], 7[c] | 0.75 | 0.84 | 1.37 | 1.06 |
| Fucose | 26 | 31 | - | - | - | 1.36 | - |
| Galactose | 30 | 38 | 110[b], 32[c] | 0.50 | 0.85 | 1.44 | 1.16 |
| Glucose | 31 | 36 | 31[b], 38[c] | 0.63 | 0.88 | 1.37 | 1.42 |
| Mannose | 25 | 31 | 30[b], 36[c] | 0.72 | 0.97 | 1.32 | 1.36 |
| Rhamnose | -7 | 0 | 27[c] | 0.69 | 1.00 | - | - |
| Sorbose | 19 | 27 | - | 0.69 | - | 1.46 | - |
| Disaccharides | | | | | | | |
| Lactose | 101 | - | - | - | - | - | - |
| Lactulose[f] | 79 | 88 | - | 0.45 | - | - | - |
| Maltose | 87 | 92 | 43[b], 95[c] | 0.61 | 0.79 | - | 1.27 |
| Melibiose | 85 | 91 | 95[c] | 0.58 | 0.65 | - | - |
| Sucrose | 62 | 67 | 52[b], 70[c] | 0.60 | 0.77 | 1.33 | 1.43 |
| Trehalose | 100 | 107 | 79[b] | 0.55 | - | - | 1.35 |
| Oligosaccharides | | | | | | | |
| Raffinose[f] | 70 | 77 | - | 0.45 | - | - | - |
| Sugar alcohols | | | | | | | |
| Maltitol | 39 | 44 | - | 0.56 | - | 1.32 | - |
| Sorbitol | -9 | -4 | -2[b] | 0.96 | - | 1.36 | 1.42 |
| Xylitol | -29 | -23 | -18.5[b], -19[c] | 1.02 | 1.52 | 1.48 | 1.44 |

[a] Onset and midpoint of the glass transition temperature range
[b] Slade and Levine (1991)
[c] Orford et al. (1990)
[d] I Roos (1993), II Orford et al. (1990)
[e] I Roos (1993), II Slade and Levine (1991)
[f] Y.H. Roos, unpublished data (1992)
Source: Roos (1993)

The Gordon and Taylor equation has proved to be a reliable predictor of glass transition temperatures of sugars at various water contents. Application of the equation requires that the anhydrous $T_g$ (Table 5.1) and numerical value for the constant $k$ are known. One of the experimental $T_g$

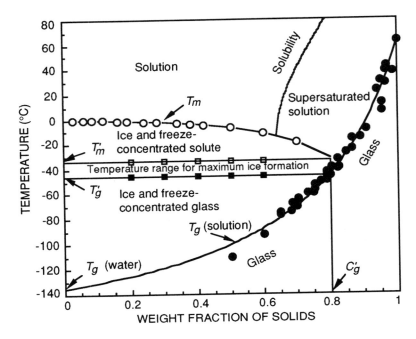

**Figure 5.2** State diagram of sucrose showing the physical state of sucrose solutions as a function of solute weight fraction and temperature. The glass transition temperature, $T_g$, of solutions decreases with increasing water content according to the $T_g$ curve. Solutions with an initial solute concentration lower than the solute concentration in a maximally freeze-concentrated solute matrix, $C'_g$, may form ice according to the equilibrium melting temperature, $T_m$, curve. Maximally freeze-concentrated solutions show glass transition of the freeze-concentrated solute matrix, $T'_g$, and maximum ice formation within the temperature range $T'_g < T < T'_m$. $T'_m$ is the onset temperature of ice melting in maximally freeze-concentrated solutions. Experimental data shown with symbols are from Young and Jones (1949), Luyet and Rasmussen (1968), Pancoast and Junk (1980), Weast (1986), Hatley *et al.* (1991), Izzard *et al.* (1991), and Roos and Karel (1991a).

values for amorphous water (Table 4.2) may be used in calculations. Roos (1993) noticed that the $k$ value for amorphous sugars increased linearly with increasing anhydrous $T_g$. Such $k$ values derived with equation (5.1) and data for $T'_g$, $T'_m$, and $C'_g$ for sugars are given in Table 5.3.

$$k = 0.0293T_g + 3.61 \qquad (5.1)$$

Glass transition temperatures of sugars are responsible for the thermal behavior of a number of food products with various water contents during processing and storage. Characterization of the physical state can be based on established state diagrams, which allow determination of the physical state at low water contents as well as in rapidly cooled or freeze-concentrated solutions. The physical state governs ice formation and the thermal behavior of frozen sugar solutions, which is particularly important in cryopreservation and cryostabilization.

## 3. Mixtures of sugars

Sugars in foods are usually present as mixtures. The most natural sugar composition is a mixture of fructose, glucose, and sucrose, which is typical of fruits and vegetables. Prepared foods may also contain mixtures of lactose and sucrose and maltose and sucrose. The number of possible compositions is large, but the physical properties may often be due to the principal sugar. Compositional effects are also important to the physical state of corn syrups and other blends of sweeteners.

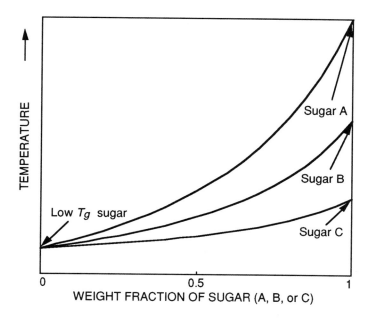

**Figure 5.3** A schematic representation of the effect of sugars with low glass transition temperatures, $T_g$, on the $T_g$ of binary sugar mixtures. Sugars with the lower $T_g$ values may significantly depress the $T_g$ of the mixture in comparison with the effect of the higher $T_g$ sugar components.

**Table 5.3** Glass transition temperature of the maximally freeze-concentrated solute matrix, $T'_g$, onset temperature of ice melting in the maximally freeze-concentrated solution, $T'_m$, estimated constant, $k$, for the Gordon and Taylor equation for predicting water plasticization, and the estimated solute concentration in the maximally freeze-concentrated solute matrix, $C'_g$, for sugars.

| Compound | $T'_g$ (°C)[a] Onset | Midpoint | Other[b] | $T'_m$ (°C) | $k$ | $C'_g$ (% w/w) |
|---|---|---|---|---|---|---|
| Pentoses | | | | | | |
| Arabinose | -66 | -61 | -47.5 | -53 | 3.55 | 79.3 |
| Ribose | -67 | -62 | -47 | -53 | 3.02 | 81.4 |
| Xylose | -65 | -60 | -48 | -53 | 3.78 | 78.9 |
| Hexoses | | | | | | |
| Fructose | -57 | -53 | -42 | -46 | 3.76 | 82.5 |
| Fucose | -62 | -57 | -43 | -48 | 4.37 | 78.4 |
| Galactos | -56 | -51 | -41.5 | -45 | 4.49 | 80.5 |
| Glucose | -57 | -53 | -43 | -46 | 4.52 | 80.0 |
| Mannose | -58 | -53 | -41 | -45 | 4.34 | 80.1 |
| Rhamnose | -60 | -55 | -43 | -47 | 3.40 | 82.8 |
| Sorbose | -57 | -52 | -41 | -44 | 4.17 | 81.0 |
| Disaccharides | | | | | | |
| Lactose | -41 | -36 | -28 | -30 | 6.56 | 81.3 |
| Lactulose[c] | -42 | -37 | -30 | -32 | 5.92 | 82.0 |
| Maltose | -42 | -37 | -29.5 | -32 | 6.15 | 81.6 |
| Melibiose | -42 | -37 | -30.5 | -32 | 6.10 | 81.7 |
| Sucrose | -46 | -41 | -32 | -34 | 5.42 | 81.7 |
| Trehalose | -40 | -35 | -29.5 | -30 | 6.54 | 81.6 |
| Oligisaccharides | | | | | | |
| Raffinose[c] | -36 | -32 | -26.5 | -28 | 5.66 | 84.1 |
| Sugar alcohols | | | | | | |
| Maltitol | -47 | -42 | -34.5 | -37 | 4.75 | 82.9 |
| Sorbitol | -63 | -57 | -43.5 | -49 | 3.35 | 81.7 |
| Xylitol | -72 | -67 | -46.5 | -57 | 2.76 | 80.2 |

[a] Onset and midpoint of the glass transition temperature range
[b] Slade and Levine (1991)
[c] Y.H. Roos, unpublished data (1992)
*Source*: Roos (1993)

Several factors that affect various physical properties of sugar blends can be found in the work of Pancoast and Junk (1980). The thermal behavior of sugar mixtures has been investigated in some detail. Finegold *et al.* (1989) studied glass transitions of binary sugar blends using comelted quenched samples of glucose-fructose and fructose-sucrose as models. The $T_g$ was found to be a function of composition. The $T_g$ of both sugar blends decreased with increasing fructose content due to its lower glass transition temperature. In such mixtures the lower $T_g$ component can be considered to act as a plasticizer. The $T_g$ values of amorphous sugar mixtures can probably be predicted with the Gordon and Taylor or other equations used to model composition dependence of the $T_g$ of polymer blends. The effect of composition on the $T_g$ of binary sugar mixtures is shown in Figure 5.3. In foods sugar blends also become plasticized by water, which has to be taken into account in the evaluation of the phase behavior of foods that contain several sugars.

Roos and Karel (1991c) found almost equal $T_g$ values for sucrose and a mixture of sucrose and fructose (7:1) at the same water activities, but crystallization of sucrose was significantly delayed in the sugar mixture. Arvanitoyannis *et al.* (1993) studied glass transitions of amorphous mixtures of fructose, glucose, and water. Glucose was found to be plasticized by fructose in anhydrous glasses, which were further plasticized by water. State diagrams were established for mixtures with various fructose-glucose compositions. Arvanitoyannis and Blanshard (1994) studied the physical state of lactose-sucrose mixtures. The glass transition temperature was a function of composition. As expected the $T_g$ was highest for lactose and it decreased linearly with increasing sucrose content towards that of sucrose. Physical properties of amorphous sugar mixtures affect their applicability as food ingredients, e.g., in powdered food products or as encapsulating agents of flavor compounds.

## B. Starch

Starch is the most common carbohydrate polymer in foods. Starch is obtained from cereal and legume seed endosperm, potato tuber, and other plant reserve organs, where it exists in the form of granules. The size of starch granules varies depending of the origin over the range of 1 to 100 $\mu$m. Starch granules are insoluble in cold water. The granules are composed of two glucose homopolysaccharides. Starch usually contains linear amylose molecules as a minor component and highly branched amylopectin molecules as the major component. The amounts of amylose and amylopectin differ significantly in various starches. Phase transitions of starch are extremely important in food processing. According to Whittam *et al.* (1991) the melting and glass transition temperatures are the most important parameters that characterize physical properties of starch polymers over a wide temperature range.

## 1. Physical state of native starches

Starch components, amylose and amylopectin, may exist in crystalline, partially crystalline, and amorphous states. In native granular starches amylose exists in the amorphous noncrystalline state and amylopectin exhibits partial crystallinity. According to Biliaderis (1991a) long-range ordering, i.e., crystallinity in a granular starch, is attributed to clusters of short chains of amylopectin, which form ordered arrays of hexagonally packed double helices. In some starches a part of the amylose fraction may have co-crystallized with amylopectin (Biliaderis, 1991b). The amorphous phase of granular starch is heterogeneous and consists of amorphous amylose and partially crystalline regions of amylopectin (Biliaderis, 1992). The physical properties of starch depend on the chemical structure and crystallinity, which have been reviewed by Zobel (1992).

Phase transitions of native granular starches are observed when they are heated in excess water. The transitions observed include transitions in both the amorphous and crystalline portions of starch. Biliaderis *et al.* (1986a)

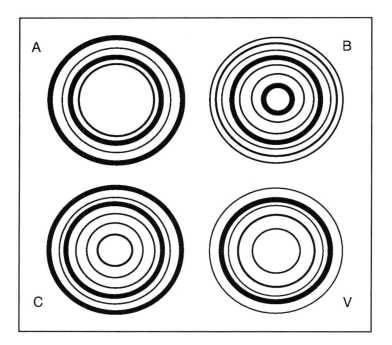

**Figure 5.4** Schematic representation of X-ray diffraction patterns of various starches used for their classification into the A-, B-, C-, and V-type starches. After Zobel (1988a).

reported that during heating of granular starch with water a glass transition and a subsequent melting transition were observed. These transitions occurred over the temperature range that is often referred to as starch gelatinization temperature. The gelatinization of starch occurs in baking, extrusion, and thickening of starch-containing sauces, pie fillings, etc.

Native starches are classified according to their wide-angle X-ray diffraction patterns as shown in Figure 5.4 to A-, B-, or C-type starches (e.g., Zobel, 1988a,b; Biliaderis, 1991a,b). An additional diffraction pattern known as V-type crystallinity corresponds to structures of helical inclusion complexes of amylose. Such complexes may contain a variety of polar and nonpolar compounds and it is typical of amylose-lipid complexes (Biliadersis, 1991a,b). The A-type pattern is exhibited by cereal starches (rice, wheat, and corn); the B-type pattern is shown by tuber, fruit, and high-amylose corn starches and by retrograded starch. C-type pattern starch is an intermediate between A- and B- type starches and it is typical of legume seed starches (e.g., pea and bean starches). The V-type structure has not been found in native cereal starches, but it may form due to heating of lipid-containing starches (Biliaderis, 1991a,b).

Hizukuri *et al.* (1983) found that A-type starches have a shorter chain length than B-type starches and that amylodextrins with short chain lengths tend to crystallize into the A-form. Gidley (1987) and Gidley and Bulpin (1987) reported that the A-type crystal is thermodynamically the most stable form and that the B-type is the kinetically favored polymorphic form. The A-type structure is preferred over the B-type at higher crystallization temperatures, higher polymer concentrations, and shorter chain lengths. This conclusion agreed with the findings of Hizukuri *et al.* (1983) that amylopectins of B-type starches have longer chain lengths than those of A-type starches and those of Whittam *et al.* (1990) showing that A-type starch spherulites have approximately a 20°C higher melting temperature than B-type starch spherulites with various water contents. X-ray diffraction patterns typical of starch samples are shown in Figure 5.4. Maize has an A-type structure with 27% amylose. V-type structures are formed from collapsed amylose helices inside which chemical adjuncts are trapped. Potato starch has an A-type structure and 22% amylose. C-form is typical of root and tuber starches (Zobel *et al.*, 1988).

Phase transitions that are associated with gelatinization and loss of native structure of granular starches define and explain differences in physical properties of starches and their behavior in food products. Phase transitions and partial crystallinity of starch are similar to those found in partially crystalline polymers (Levine and Slade, 1990). The main importance of starch behavior is due to its dominance as a food component and the significance of its physical properties to food stability and texture. Phase transitions of starch are particularly important to the physical state of cereal foods.

## 2. Physical state of starch and starch components

The polymer science approach (Lelievre, 1976; Donovan, 1979; van den Berg, 1981; Maurice *et al.*, 1985; Levine and Slade, 1986, 1990) has significantly contributed to the understanding of the physicochemical principles and phase behavior of starch and starch-water systems. Biliaderis (1992) pointed out that glass transition of starch controls texture- and stability-related phenomena, e.g., annealing, gelatinization, and retrogradation, which may proceed only over the temperature range $T_g < T < T_m$.

*a. Starch and Starch Components.* The physical state of starch is often characterized by the X-ray diffraction patterns obtained from starches of various origins, by the amylose content, and by gelatinization behavior. In native starch granules amylose and amylopectin are considered to exist in the amorphous and partially crystalline states, respectively. The extent of crystallinity in starch granules is generally 15 to 35%. However, understanding of various phenomena that are associated with heating of starch-water suspensions requires knowledge of changes in the physical state that may occur as a function of temperature and water content. State diagrams for starch and starch components have been established in some studies. The first attempt to estimate $T_g$ for starch and its dependence on water content was reported by van den Berg (1981). The reported anhydrous $T_g$ for starch at 151°C was used by Marsh and Blanshard (1988) in the prediction of the $T_g$ of starch at various water contents. Establishing reliable state diagrams requires experimental values for $T_g$ at several water contents. Zeleznak and Hoseney (1987) used DSC for the determination of $T_g$ of native and pregelatinized wheat starch at several water contents. $T_g$ values for native wheat starch were obtained at water contents ranging from 13 to 22%. The $T_g$ of a commercially pregelatinized amorphous starch was found to occur at a lower temperature probably due to the lower amount of crystalline regions in the pregelatinized product. It has been well established that increasing crystallinity in partially crystalline polymers increases the $T_g$ of amorphous regions (Jin *et al.*, 1984). Consequently, a decreasing crystallinity in starch should decrease the observed $T_g$ at a corresponding water content.

The glass transition temperature of anhydrous starch has not been determined experimentally due to probable thermal decomposition and other difficulties in detecting glass transition-induced changes in physical properties at low water contents. Probably the first and the highest $T_g$ of 330°C for anhydrous amylose was extrapolated with experimental $T_g$ data by Nakamura and Tobolsky (1967). Orford *et al.* (1989) studied the effect of the degree of polymerization, DP, on the $T_g$ of malto-oligomers. The results showed that the $T_g$ increased with increasing DP. Orford *et al.* (1989) fitted equation (5.2), where $T_g^0$ is the $T_g$ limit for high molecular weight polymers and $a$ is a constant, to experimental data and the equation was used to predict the $T_g$ of anhydrous starch.

$$\frac{1}{T_g} = \frac{1}{T_g^0} + \frac{a}{DP} \qquad\qquad (5.2)$$

Orford *et al.* (1989) predicted that amorphous glucose polymers with high DP such as amylose and amylopectin have the anhydrous $T_g$ at 227 ± 10°C. Whittam *et al.* (1990) reported extrapolated values for the $T_g$ and melting temperature, $T_m$, values for anhydrous starch, which were 197 and 257°C, respectively. Roos and Karel (1991d) used the data of Orford *et al.*

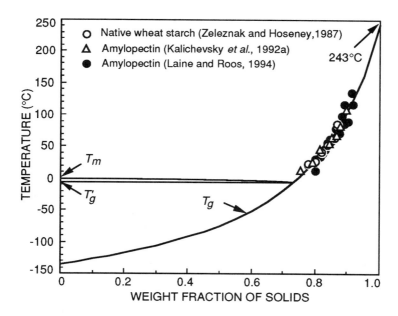

**Figure 5.5** State diagram of starch with experimental data from various studies. The predicted $T_g$ curve was obtained with the Gordon and Taylor equation using the anhydrous $T_g$ of 243°C and the constant $k = 5.2$ according to Roos and Karel (1991d).

(1989) and experimental $T_g$ data of maltodextrins (starch hydrolysis products) to derive the $T_g$ for high molecular weight glucose polymers. The $T_g$ estimate of 243°C was obtained with the Fox and Flory equation.

Levine and Slade (1986) observed that the glass transition temperature of maximally freeze-concentrated solute matrices, $T_g'$, of starch hydrolysis products increased linearly with decreasing dextrose equivalent, DE, or inverse value of the number average molecular weight, $M_n$. The relationship predicted that the $T_g'$ of starch was at -5.9°C, in good agreement with experimental values found at -5 to -7°C (Levine and Slade, 1989; Liu and Lelievre, 1992). Roos and Karel (1991d) reported that the experimental $T_g'$ value for waxy corn starch was -6°C. They also found that the $T_g'$ and $T_m'$ coincided for high molecular weight carbohydrates, suggesting that the glass transition and onset of ice melting in gelatinized starch occurred at the same temperature. Roos and Karel (1991d) concluded that ice formation in such materials to the extent of the maximally freeze-concentrated state cannot be completed due to the nonequilibrium state at a coinciding first-order (ice melting) and second-order (glass transition) transition temperatures. Glass transitions for amylopectin have been determined with various techniques and it has been shown to become plasticized by water, in agreement with other biopolymers (Kalichevsky and Blanshard, 1992; Kalichevsky *et al.*, 1992a). The state diagram established for starch with experimental $T_g$ data and the predicted anhydrous $T_g$ of 243°C is shown in Figure 5.5.

The state diagram of starch can be used in predicting the physical state of a number of starch-containing food materials at various water contents and temperatures. Examples of such products include various cereal products with a crispy, glassy structure and bakery products (Slade and Levine, 1994, 1995).

*b. Effect of Composition.* Effects of sugars and proteins on the physical state of starch at various water contents have been studied due to their importance as component compounds of most starch-based foods. von Krüsi and Neukom (1984) found that maltose and malto-oligosaccharides acted as plasticizers in starch gels. Kalichevsky and Blanshard (1992) studied the physical state of mixtures of amylopectin, casein, and gluten. Amylopectin-gluten mixtures (1:1) were found to become plasticized by water and to show miscibility after heating with sufficient water. However, the $T_g$ of such materials was not reported at various water contents. Sugars are fairly low molecular weight compounds in comparison with starch and its components. Sugars may be considered as plasticizers for starch and to depress the $T_g$ as shown in Figure 5.3 for sugar mixtures, which was reported by Roos and Karel (1991e) to apply for starch hydrolysis products. The expected plasticization of amylopectin by fructose was reported by Kalichevsky and

Blanshard (1993) to be significant at low concentrations. Results obtained for mixtures with fructose contents above 20% showed that the mechanical properties of the mixture became more dependent on the sugar component. Similar plasticization of amylopectin by glucose, sucrose, and xylose has also been reported (Kalichevsky *et al.*, 1993). It should also be noticed that the $T'_g$ and $T'_m$ values of starch-sugar mixtures are dependent on the relative amounts of the solid components and lower than those of pure starch.

3. Gelatinization and melting

Gelatinization of starch occurs during heating of native starch with a sufficient amount of water. Gelatinization is observed from several changes in the starch fraction. According to Olkku and Rha (1978) gelatinization includes (1) loss of birefringence of the granules; (2) diffusion from ruptured granules and dissolving of linear molecules; (3) increasing clarity of the starch-water mixture; (4) hydration and swelling of granules to several times their original size; (5) a marked increase of consistency to a maximum; and (6) formation of a paste-like mass or gel.

Lelievre (1974) related gelatinization of starch to the melting of homogeneous polymers. Marchant and Blanshard (1978) pointed out that at least three processes occur in gelatinization. These include diffusion of water into the unswollen starch granule, disappearance of the birefringence due to hydration-facilitated melting, and swelling of the granule. Melting of crystallites during gelatinization is supported by X-ray diffraction studies, which show loss of semicrystalline order and calorimetric measurements, which indicate loss of ordered structure (Evans and Haisman, 1982). Atwell *et al.* (1988) defined gelatinization as a phenomenon during which molecular orders within starch granules collapse. Factors that govern gelatinization onset temperature and the temperature range over which gelatinization occurs included starch concentration, method of observation, granule type, and heterogeneities within the granule population under observation. Atwell *et al.* (1988) defined *pasting* to be the phenomenon that follows gelatinization. Pasting is observed as dissolution of starch, which involves granular swelling, exudation of molecular components from the granule, and eventually a total disruption of the granules. Gelatinization often occurs over a temperature range of 10 to 15°C. However, a single starch granule may become gelatinized within 1 to 2°C (Evans and Haisman, 1982).

The major determination techniques of gelatinization temperatures of various starches have included polarizing microscopic hot stage examination of birefringence, detection of changes in viscosity with a viscoamylograph detection of changes in X-ray diffraction, and determination of thermal properties with DSC (Lund, 1984).

*a. Birefringence.* The loss of birefringence of starch-water dispersions with 0.1 to 0.2% starch can be used to determine the gelatinization temperature range with a polarized light microscope (Eliasson and Larsson, 1993). The temperature range for the loss of birefringence observed during heating is dependent on water content. Burt and Russell (1983) showed that the loss of birefringence was associated with the completion of the melting transition observed with DSC and it occurred even at high temperatures and low water contents. Wheat starch loses birefringence completely when heated above 65°C with excess water, but when the water content is reduced to 50% the temperature needed is 75°C. At a water content of 30% a heat treatment at 132°C is insufficient to cause the loss of birefringence in wheat starch (Eliasson and Larsson, 1993).

*b. X-Ray Diffraction.* According to Zobel *et al.* (1988) the use of the term *melting* to refer to starch gelatinization originates from X-ray and synthetic polymer technologies. X-ray diffraction studies of crystalline specimens yield reflections from crystalline planes. Melting causes the disappearance of the crystals due to the formation of the amorphous liquid state (Zobel *et al.*, 1988). Synthetic polymers are known to have a high viscosity even in the liquid state, which obviously applies to biopolymers and water-plasticized biopolymers such as starch.

The loss of crystallinity during heating of starch with a sufficient water content occurs as the temperature is increased to above the gelatinization temperature. The extent of crystallinity can be observed from decreasing intensity of the X-ray diffraction pattern. Zobel *et al.* (1988) compared results of X-ray diffraction and DSC studies of starch gelatinization. At high and intermediate water contents the loss of crystallinity that was observed from the intensity of the X-ray diffraction pattern for potato starch agreed with the melting temperature determined with DSC. However, starch with 22% (w/w) water showed a residual pattern even at 148°C. Samples with higher water contents showed a steady loss of crystallinity throughout the temperature range of melting. Zobel *et al.* (1988) found that the two endotherms, M1 and M2, typical of starch melting DSC scans of samples with intermediate water contents both indicated starch melting and were concluded to describe the same thermal event. They also stated that the X-ray and DSC data agreed with the theory that gelatinization and granule disruption are dependent on starch crystallites with various degrees of perfection. The crystallites become disordered at temperatures that depend on the amount of water available for

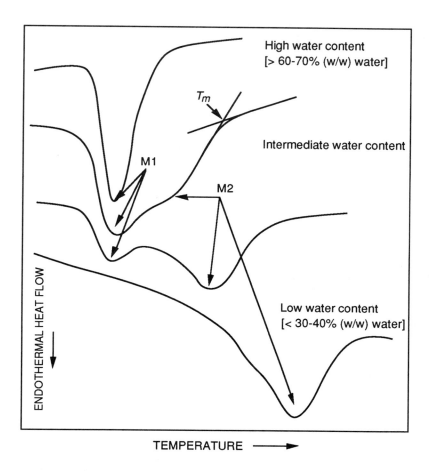

**Figure 5.6** A schematic representation of DSC thermograms typical of starch when heated with various amounts of water. Heating at high water contents over the gelatinization temperature range produces a single endotherm, M1. At intermediate water contents another endotherm, M2, is observed as a shoulder of the M1 endotherm or as a separate endotherm. At low water contents only M2 is obtained. The melting temperature, $T_m$, increases with increasing water content.

melting. Zobel *et al.* (1988) concluded that at a water content of 30% or higher starches were analogous to other natural and synthetic polymers in which the degree of the order of crystallinity and plasticizer interactions are major factors in the determination of the melting behavior and other properties. The results of the study did not confirm occurrence of the increasing melting temperature of starch with less than 20% (w/w) water as was reported for potato starch by Donovan (1979). Cameron and Donald (1993a,b) used small-angle X-ray scattering to study the gelatinization of wheat starch. When excess water was available water was absorbed to the amorphous phase at temperatures below 55°C. At higher temperatures the amorphous phase became saturated and a drop in the crystalline order became evident due to cooperative melting. At limiting water contents the loss of the crystalline order occurred at higher temperatures.

*c. Differential Scanning Calorimetry.* DSC thermograms of starch with low and intermediate water contents often show multiple melting endotherms, which have been interpreted differently in a number of studies (Donovan, 1979; von Eberstein *et al.*, 1980; Biliaderis *et al.*, 1980, 1986a; Evans and Haisman, 1982; Maurice *et al.*, 1985). Thermal transitions suggesting gelatinization and crystallite melting during heating of starch are shown in Figure 5.6. These transitions may produce two endotherms, M1 and M2, which reflect the water- and heat-induced disorganization of crystallites (e.g., Donovan, 1979; Biliaderis *et al.*, 1986a; Biliaderis, 1991a). Transitions may also be observed at higher temperatures, which have been attributed to order-disorder processes of amylose-lipid complexes (Biliaderis *et al.*, 1986b; Biliaderis, 1991a,b).

Biliaderis *et al.* (1986a) reported that rice starch-water mixtures containing more than 60% (w/w) water show a single, symmetrical melting endotherm that occurs over the gelatinization temperature range. At lower water contents the size of the melting endotherm decreased with decreasing water content and confirmed the results reported by Donovan (1979) for potato starch. Evans and Haisman (1982) suggested that the M1 endotherm resulted from the melting of the least stable crystallites in accordance with the polymer melting theory of Flory (1953) and less water remained available for the melting of the remaining crystallites, thus increasing the melting temperature. Biliaderis *et al.* (1986a) proposed that the multiple melting endotherms resulted from crystalline melting and reorganization processes that occurred simultaneously during a dynamic DSC run. The theories of Donovan (1979) and Evans and Haisman (1982) postulated that melting of starch was a solvent-facilitated process and they did not take into account that starch as a semicrystalline polymer may also undergo reorganization during heating (Biliaderis *et al.*, 1986a).

Biliaderis *et al.* (1986a) studied gelatinization of rice starch with 50% (w/w) water using DSC at various heating rates. They assumed that at low heating rates, immediately after the onset of the M1 endotherm, molecular mobility increased and there was a greater opportunity for chain rearrangements. Therefore, the granular structure approached a new equilibrium and

**Table 5.4** Gelatinization temperature range, $T_{gel}$, of various starches as determined with microscopy and DSC. The heats of gelatinization, $\Delta H_{gel}$, were obtained with DSC.

| Starch | Amylose content (%) | Crystallinity (%) | $T_{gel}$ (°C) | $\Delta H_{gel}$ (J/g) |
|---|---|---|---|---|
| Barley | 22 | - | 51-60 | - |
| Corn | 23-28 | 40 | 62-76 | 13.8-20.5 |
| High-amylose corn | 52 | 15-22 | 67-86 | 28.0 |
| Waxy corn | 1 | 40 | 63-80 | 16.7-20.1 |
| Oat | 23-24 | 33 | 52-64 | 9.2 |
| Pea | 29 | - | 62 | 12.5 |
| Potato | 19-23 | 28 | 58-71 | 17.6-18.8 |
| Rice | 17-21 | 38 | 68-82 | 13.0-16.3 |
| Waxy rice | - | 37 | 69.5 | - |
| Rye | 27 | 34 | 49-70 | 10.0 |
| Sorghum | 25 | 37 | 68-78 | - |
| Tapioca | 17-18 | 38 | 63-80 | 15.1-16.7 |
| Triticale | 23-24 | - | 55-62 | - |
| Wheat | 23-26 | 36 | 52-66 | 9.7-12.0 |

*Source*: von Eberstein *et al.* (1980), Lineback (1986), Ring *et al.* (1988), and Zobel (1988a,b)

less crystallites melted at M1, which resulted in melting at a higher temperature and the occurrence of the M2 shoulder. An increase in the heating rate decreased such annealing and apparently a higher amount of crystallites melted at M1. When the heating rate was 30°C/min only a single endotherm was obtained.

*d. Effects of Water on Gelatinization and Melting.* The need of water for complete gelatinization of starches has been well established. Evans and Haisman (1982) discussed the effect of water activity of starch on the gelatinization temperature. Their results on potato starch suggested that the water activity approached 1 at 0.54 g of water/g of dry starch. Both the initial and final gelatinization temperatures increased when less water was present. At water contents up to 2 g/g of dry starch the final gelatinization temperature decreased with increasing water content. The minimum water content for complete gelatinization is often about 60% (w/w) (Lund, 1984). Gelatinization temperatures for various starches are given in Table 5.4.

Although gelatinization can be considered to be a solvent- and heat-induced melting of cystallites it involves kinetic limitations due to the nonequilibrium state of the native starch granules (Biliaderis, 1991a,b). The kinetic constraints include at least (1) diffusion of water into the granule; (2) melting of crystallites; and (3) swelling due to the hydration of the disordered polymer chains as was suggested by Marchant and Blanshard (1978). Maurice *et al.* (1984) suggested that the gelatinization endotherm obtained with DSC is affected by crystalline and amorphous transitions and that water as a plasticizer alters the thermal behavior of starch granules. Maurice *et al.* (1985) pointed out that a glass transition of the amorphous regions precedes the melting of crystallites and gelatinization is controlled by the molecular mobility in the amorphous phase surrounding crystallites. This theory was based on the fact that crystallite melting in partially crystalline synthetic polymers occurs above the glass transition of the amorphous regions. Experimental evidence of the glass transition followed by the melting endotherm was reported by Biliaderis *et al.* (1986a) for rice starch, which could not be confirmed by Liu and Lelievre (1991), who observed the $T_g$ at much lower temperatures. However, the change in the heat capacity at such $T_g$ measured with DSC for aqueous starch dispersions is often too small and detection of the $T_g$ may be difficult (Biliaderis, 1991a,b). Biliaderis *et al.* (1986a) found that the minimum $T_g$ of granular rice starch was constant at 68°C for water contents higher than 30% (w/w). They concluded that the minimum requirement for water to fully exert its plasticizing effect on the material was 30% (w/w). Therefore, higher water contents formed a separate solvent phase outside the granules. Levine and Slade (1990) have used the fringed-micelle model to describe the semicrystalline structure of starch and emphasized that for such materials a glass transition of the amorphous phase and melting of the crystalline phase may be observed.

*e. Effects of Solutes.* Sugars often increase the gelatinization temperature of starches and may or may not affect the heat of gelatinization. This has been considered to be due to (1) the ability of sugars to reduce water availability (Derby *et al.*, 1975); (2) starch-sugar interactions that stabilize the amorphous regions (Lelievre, 1976; Spies and Hoseney, 1982; Hansen *et al.*, 1989); and (3) the increase in free volume and antiplasticization effect of sucrose in comparison with pure water (Levine and Slade, 1990).

Wootton and Bamunuarachchi (1980) used DSC to observe effects of sucrose and sodium chloride on the gelatinization behavior of wheat starch. Sucrose concentrations in the aqueous phase ranging from 0 to 45% (w/w) caused a decrease in the heat of gelatinization, $\Delta H_{gel}$, from 19.7 to 9.6 J/g. The onset of gelatinization temperature range (50°C) was not affected by sucrose, but the peak temperature of the gelatinization endotherm increased with increasing sucrose concentration from 68 to 75°C. Wootton and Bamunuarachchi (1980) reported that sodium chloride also decreased $\Delta H_{gel}$ to 9.6 J/g when the concentration was 6% (w/w) in the aqueous phase.

Higher amounts of sodium chloride increased $\Delta H_{gel}$ and it was 13.8 J/g for starch gelatinized in a 30% (w/w) sodium chloride solution. The onset temperature of gelatinization increased with increasing sodium chloride concentration up to 9% and reached 68°C. Higher sodium chloride concentrations decreased the onset temperature and it was 59°C for the 30% solution. The highest peak temperature (80°C) was observed when the sodium chloride concentration was 21%. The ratio of starch and water was 1:2 in all experiments. Eliasson (1992) reported that the $\Delta H_{gel}$ was not affected by sucrose, but at limited water contents the shoulder observed in DSC endotherms disappeared as sucrose increased the gelatinization temperature.

Chungcharoen and Lund (1987) reported that sucrose and salt shifted the gelatinization temperature of rice starch to higher temperatures and decreased heat of gelatinization. They agreed with Maurice *et al.* (1985) and Biliaderis *et al.* (1986a) that granular starch consists of three phases, which are crystallites, a bulk amorphous phase consisting mainly of amylose, and intracrystalline amorphous regions consisting of dense branches of amylopectin. The results of these studies among others suggested that sucrose increased the weight-average molecular weight of the plasticizing solution, which increased the gelatinization temperature as was emphasized by Levine and Slade (1990). Results of NMR measurements have shown that mobility of starch increases during gelatinization (Lelievre and Mitchell, 1975). However, the mobility of water and its ability to act as a plasticizer have been observed to decrease in the presence of solutes such as sugars and salts (Chinachoti *et al.*, 1991a). Sodium chloride at a 1 to 2% concentration may cause a much higher increase in the gelatinization temperature than sucrose, probably because of ionic interactions between sodium and the hydroxyl groups of starch (Chinachoti *et al.*, 1991b).

In addition to sugars and salts other food components may also affect the gelatinization behavior of starch. Evans and Haisman (1982) reported that sugar and other hydroxylated compounds increased the gelatinization temperature with increasing solute concentration, but the heat of gelatinization and the shape of the DSC endotherm were unchanged. They showed that each solute had a specific effect on the gelatinization temperature. A plot showing gelatinization temperature as a function of water activity had a different curve for each solute. However, when glycerol was used as solute the gelatinization temperature increased linearly with decreasing water activity. Other polymers, such as gluten in bread, may affect the gelatinization behavior of starch. Eliasson (1983) reported that gluten decreased the heat of gelatinization and increased the gelatinization temperature of wheat starch.

4. Amylose-lipid complexes

Amylose forms complexes with iodine, alcohols, and fatty acids. The ability of starch to bind iodine is commonly used as a measure of the amylose content of starch. Several baked products contain emulsifiers due to their ability to retard firming and retrogradation of starch. The exact mechanism of their function is unclear, but it is often related to the formation of the amylose-lipid complexes (Biliaderis, 1991a,b).

Amylose-lipid complexes give a V-type X-ray diffraction pattern, which has been used in their chemical characterization. Kugimiya *et al.* (1980) used X-ray data to show that a DSC endotherm observed over the temperature range of 110 to 120°C was due to the melting of crystalline V-type amylose-lipid complexes. As shown in Figure 5.7 DSC thermograms for starch containing lipids often show one endotherm at high water contents,

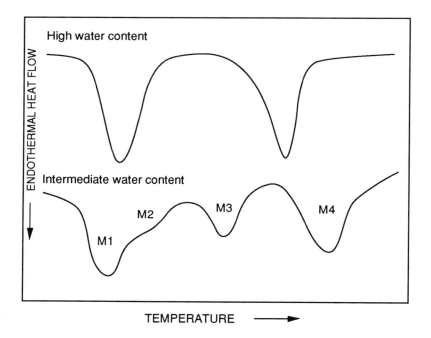

**Figure 5.7** Thermal transitions observed with DSC during heating of starch in the presence of lipids. At intermediate water contents the endotherms M1 and M2 occur during gelatinization. The M3 and M4 transitions are due to melting of amylose-lipid complexes. An exotherm between the M3 and M4 transitions indicates recrystallization of the amylose-lipid complex melted at M3. At high water contents two separate endotherms are obtained due to starch gelatinization and melting of the amylose-lipid complexes, respectively.

which indicates cooperative melting, but two melting endotherms that are separated by an exotherm may occur at intermediate and low water contents (Biliaderis *et al.*, 1985, 1986b). Biliaderis *et al.* (1985) assumed that the behavior was indicative of nonequilibrium melting of a metastable solid material. The first endotherm at a lower temperature was considered to appear due to melting of the original crystallites. The exotherm appeared to be probably due to recrystallization of the amylose-lipid complex and melting of the recrystallized material produced the higher temperature endotherm. Biliaderis *et al.* (1985) pointed out that water was essential to both the melting and crystallization phenomena. They assumed that water was needed for plasticization of the amorphous chain segments, which facilitated chain mobility and allowed melting at a lower temperature. Crystallization at high water contents produced metastable, less perfect crystalline complexes. It should be noticed that the formation of complexes is thermoreversible and DSC cooling curves show an exotherm, which indicates crystallization of the melted amylose-lipid complex during cooling (Biliaderis *et al.*, 1985).

The amylose-lipid complexes may form during gelatinization. Kugimiya and Donovan (1981) found that the heat of gelatinization of various starches was lower when gelatinization occurred with lipids. They suggested that the apparent heat of gelatinization was lower due to the concurrent exothermal crystallization of amylose-lipid complexes. The gelatinized materials showed only one melting endotherm when they were reheated in the DSC. Eliasson (1986) also found that the observed heat of gelatinization decreased in the presence of surface-active agents, probably due to the simultaneous release of heat from the formation of amylose-lipid complexes. Zobel *et al.* (1988) reported that the V-type structure in maize starch with 45 to 50% water showed development after annealing at 120 and 142°C. After storage of 18 hrs at room temperature a well-formed V-type structure as well as B-type structure were evident in the X-ray diffraction pattern. The V-type structure was typical of the amylose-lipid complex. The B-type structure was reported to be due to crystallization of mainly amylopectin from the starch melt. Zobel *et al.* (1988) concluded that simultaneous crystallization and melting phenomenon may occur in starch-water systems.

## III. Proteins

Phase transitions of proteins affect the physical state and textural characteristics of various foods. The most important transition that occurs in proteins is denaturation. Native proteins denature when they are heated in the presence of water. Denaturation may be considered to be an irreversible phase transition that includes an endothermal heat of denaturation. Another important state transition of proteins is glass transition. Proteins are biopolymers that are plasticized by water (e.g., Yannas, 1972; Slade and Levine, 1995). In enzymes such as lysozyme water plasticization increases the flexibility of the enzyme structure. Bone and Pethig (1985) reported that the effect of plasticization of proteins at room temperature became significant with a water content of 20% (w/w), which corresponded to the water content at which enzyme activity begins. They reported similar plasticization for cytochrome-c, collagen, and elastin. Cereal proteins have also been shown to be water-plasticizible amorphous polymers (e.g., Slade and Levine, 1995).

## A. Denaturation

Kauzmann (1959) defined *protein denaturation* to be a process or a sequence of processes in which the spatial arrangement of the polypeptide chain within the protein molecule is changed from that typical of the native protein to a more disordered arrangement. Foegeding (1988) defined the denaturation temperature, $T_d$, to be the temperature at which the concentration of the native protein became equal to the concentration of the denatured protein. Most proteins denature at temperatures from 50 to 80°C. Denaturation temperatures and heats of denaturation, $\Delta H_d$, are often obtained with DSC.

Protein denaturation occurs over a temperature range that is specific for each protein and also dependent on other compounds and pH. The DSC thermogram for a single protein shows a fairly sharp endotherm at $T_d$. The size and location of the endotherm may depend on pH, e.g., denaturation of lysozyme occurs at a lower temperature at pH 2 than at pH 4.5 (Privalov and Khechinashvili, 1974). The thermograms also show that the heat capacity of the native protein is lower than that of the denatured protein (Privalov and Khechinashvili, 1974). According to Privalov (1979) the heat of denaturation includes the heat of the conformational transition, the heat of ionization of the protein, and the heat of ionization of the buffer compound if denaturation occurs in a buffer solution. It should be noticed that the aggregation of protein molecules and breakup of hydrophobic interactions are exothermic processes that are associated with the total denaturation process (Arntfield and Murray, 1981).

The protein fraction of food materials is often composed of several proteins, e.g., muscle proteins include myosin, sarcoplasmic proteins and collagen, and actin. Therefore, thermograms of a particular food material

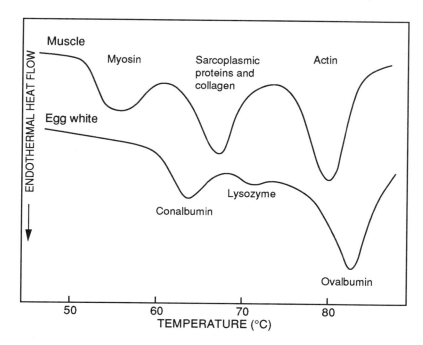

**Figure 5.8** A schematic representation of DSC thermograms typical of muscle proteins and egg white. Both materials show three main endotherms due to the denaturation of the various protein fractions.

may exhibit several denaturation endotherms. Meat products show at least three separate endotherms. Egg white has three main endotherms, which are due to the denaturation of conalbumin, lysozyme, and ovalbumin fractions (Donovan *et al.*, 1975). According to Donovan and Mapes (1976) ovalbumin is converted to S-ovalbumin during storage of eggs, which is observed from the higher-denaturation endotherm of the S-ovalbumin. A schematic representation of DSC thermograms typical of egg white and muscle proteins is given in Figure 5.8. Salts and sugars as well as other food components have been found to affect denaturation temperatures and heats of denaturation of food proteins (Arntfield *et al.*, 1990). Extremely high and low pH values tend to decrease both $T_d$ and $\Delta H_d$. Denaturation behavior of proteins is also dependent on ionic strength, heating rate, and processing history of the protein.

De Wit and Swinkels (1980) studied denaturation and denaturation kinetics of $\beta$-lactoglobulin. The protein in a buffer solution at pH 6.7 showed a large endotherm between 60 and 90°C. The denaturation temperature was

**Table 5.5** Denaturation temperatures, $T_d$, and heats of denaturation, $\Delta H_d$, for proteins.

| Protein | $T_d$ (°C) | $\Delta H_d$ (J/g) | Reference |
|---|---|---|---|
| Ceral proteins | | | |
|   Oat | 112 | 18.8 | Arntfield and Murray (1981) |
|   Wheat gluten | 88.4 and 101.4 | 0.25 and 0.13 | Eliasson and Hegg (1980) |
| Egg white proteins | | | Donovan *et al*. (1975) |
|   Conalbumin | 61 | 15.2 | |
|   Globulins | 92.5 | 11.8 | |
|   Lysozyme | 75 | 28.2 | |
|   Ovomucoid | 79 | 21.9 | |
|   Ovalbumin | 84 | 15.2 | |
| Legume proteins | | | Arntfield and Murray (1981) |
|   Fababean | 88 | 18.4 | |
|   Field pea | 86 | 15.6 | |
|   Soybean | 93 | 14.6 | |
| Muscle proteins | | | Wright *et al*. (1977) |
|   Actin | 83.5 | 14.5 | |
|   Myofibrils | 59.5 and 74.5 | 22.6 | |
|   Myosin | 55 | 13.9 | |
|   Sarcoplasmic proteins | 63, 67, and 75 | 16.5 | |
| Whey proteins | | | De Wit and Klarenbeek (1984) |
|   $\alpha$-Lactalbumin | 62 | 17.8 | |
|   $\beta$-Lactoglobulin | 78 | 16.9 | |
|   Immunoglobulin | 72 | 13.9 | |
|   Bovine serum albumin | 64 | 12.2 | |
|   Whey protein concentrate | 62/78 | 11.5 | |

dependent on the heating rate, but extrapolation of the data to zero heating rate gave a denaturation temperature of $70.5 \pm 0.5°C$. They concluded that 70°C is a critical temperature for $\beta$-lactoglobulin. At higher temperatures aggregation and denaturation occurred. A kinetic analysis showed that the denaturation process followed first-order kinetics over the temperature range from 65 to 72°C with an apparent activation energy of 343 kJ/mol. Denaturation temperatures and heats of denaturation for various food proteins are given in Table 5.5. The thermal behavior of various proteins and the effect of pH, processing, and other compounds on denaturation behavior have been reviewed by Arntfield *et al*. (1990), Findlay and Barbut (1990), Barbut and Findlay (1990), and Ma and Harwalkar (1991).

## B. Glass Transition

Proteins in foods are water-plasticizable amorphous polymers. The amorphous state is particularly important to the structural properties of baked products. Levine and Slade (1990) have emphasized the thermosetting properties of gluten and also its role as a cryostabilizer in frozen bread dough. Levine and Slade (1990) have stated that a number of examples in the animal kingdom emphasize the importance of proteins, rather than polysaccharides, as structural biopolymers. They have pointed out that such materials as silk, wool, and hair contain crystallites of fibrous proteins that are embedded in a relatively amorphous protein matrix. The physical state of proteins is also important to the physicochemical properties of various protein films that may be used as edible coatings.

### 1. Physical state of proteins

The physical state of proteins in accordance with other biopolymers is significantly affected by water. Information on the crystallinity in food proteins is almost nonexistent and only a few studies have reported changes in the physical state of amorphous proteins. A number of proteins are not water soluble, but they become plasticized by water in the amorphous state.

Kakivaya and Hoeve (1975) reported that elastin showed glass transition behavior typical of amorphous polymers. They used DSC to study the change in the heat capacity of the protein as a function of water content. The anhydrous $T_g$ approached temperatures above 200°C and it could not be determined due to possible decomposition of the protein. However, small increases in water content depressed the $T_g$ and it was observed to decrease to below 20°C as the water content was higher than 20% (w/w). The depression of the $T_g$ to below 20°C was also observed from changes in mechanical properties and electrical resistance. Kakivaya and Hoeve (1975) suggested that the water content-dependent state of elastin was responsible for the hardening of arteries. The finding of Bone and Pethig (1982) that water plasticization of lysozyme increases the vibrational freedom of the protein structure, which, perhaps, after sufficient plasticization results in enzyme activity is of great importance to food applications. Foods with low water contents may contain several natural enzymes, which become active due to water plasticization. Also the stability of dehydrated enzyme preparations may be due to their existence in the glassy state. Lillie and Gosline (1993) pointed out that the $T_g$ of proteins especially at low water contents may occur over a wide temperature range. The main glassy proteins in foods are those present in various cereals and cereal foods.

*a. Cereal Proteins.*    One of the main important properties of cereal proteins is their ability to form polymer networks during baking. Schofield *et al.* (1983) showed that free sulfhydryl groups in glutenin were involved in rheological changes that occurred at temperatures between 55 and 75°C. They postulated that glutenin proteins were unfolded on heating, which facilitated sulfhydryl/disulfide interchange between exposed groups. Similar phenomena occurred in gliadin proteins, which with glutenins belong to the low molecular weight (LMW) subunits of wheat gluten proteins (Shewry *et al.*, 1986), after heating at 100°C. Hoseney *et al.* (1986) reported glass transition temperatures for freeze-dried wheat gluten as a function of water content. Their results indicated that wheat gluten was similarly plasticized by water as elastin in the study of Kakivaya and Hoeve (1975). Slade *et al.* (1988) suggested that heating of gluten in the presence of water to above the glass transition temperature allows sufficient mobility, due to thermal and water plasticization, for the molecules to form a thermoset network via disulfide cross-linking. Such thermosetting, which is analogous to chemical curing and vulcanization of rubber, may occur only at temperatures above $T_g$ (Levine and Slade, 1990). Thermosetting of gluten is beneficial and extremely

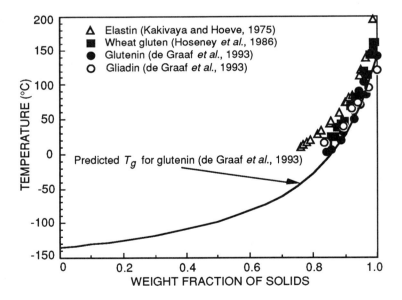

**Figure 5.9**  Glass transition temperatures, $T_g$, of elastin and cereal proteins. The $T_g$ is depressed with decreasing solids concentration due to water plasticization. The Gordon and Taylor equation was used with constant $k = 6.3$ to predict the $T_g$ curve for glutenin according to de Graaf *et al.* (1993).

important to the quality and structure of baked bread, but deleterious to the quality of such products as sugar-snap cookies (Levine and Slade, 1993; Slade and Levine, 1994). Levine and Slade (1993) stated that the presence of sucrose with high-quality soft wheat flours prevents thermosetting of glutenins and $\alpha$-gliadins. Therefore, the thermoplastic material may cool into the hard, glassy state.

Wheat gluten proteins are divided into gliadin and glutenin fractions. Gliadin is composed of proteins that are solubilized in 70% ethanol-water solution, while the glutenin proteins remain insoluble. Glutenins are higher molecular weight, cross-linked, and elastic polymers. Cocero and Kokini (1991) studied the thermal and mechanical behaviors of the glutenin fraction using DSC and mechanical spectroscopy. The glass transition of glutenin and its depression due to water plasticization were observed from both measurements. The results suggested correlation between the endset temperature of the change in heat capacity determined with DSC and the temperature at the loss modulus, $G''$, maximum. Cocero and Kokini (1991) reported softening of glutenin above $T_g$, but the storage modulus, $G'$, did not decrease to the typical value of the rubbery region. Cocero and Kokini (1991) assumed that the high $G'$ value above $T_g$ was due to thermosetting of the protein via formation of the disulfide cross-linking as was suggested by Slade *et al.* (1988). In a subsequent study de Graaf *et al.* (1993) showed that the gliadin fraction of the gluten proteins was also a water-plasticizable material. However, the glass transition temperature of gliadin was lower than that of gluten at water contents below 8% as shown in Figure 5.9. The lower $T_g$ would be expected due to the lower molecular weight of the gliadin proteins. Madeka and Kokini (1994) studied the mechanical behavior of gliadin at 25% (w/w) water content. A large increase in $G'$ occurred above 70°C, which was considered to reflect increasing elasticity due to formation of a cross-linked protein network structure. Cross-linking was also suggested by a significantly higher $G'$ in comparison with $G''$, which is typical of cross-linked polymers. A reduction of $G'$ and a high value of $G''$ were observed at 135°C, which suggested softening of the structure in accordance with the finding of Attenburrow *et al.* (1992) that the $T_g$ of gluten thermoset at 90°C was higher than the $T_g$ of gluten thermoset at 150°C.

de Graaf *et al.* (1993) and Kokini *et al.* (1993) have applied the Gordon and Taylor equation to predict water plasticization of cereal proteins. The equation was proved to be useful in predicting water plasticization of both gliadin and glutenin as well as the water plasticization of zein, which is the hydrophobic protein in corn. Kalichevsky *et al.* (1992a,b, 1993) have studied effects of composition on the glass transition and water plasticization behavior of mixtures of proteins and sugars and proteins and lipids. Kalichevsky *et al.* (1992a) used DSC, pulsed NMR, and a three-point bend test to observe the $T_g$ of gluten and gluten-sugar mixtures. The results on the water plasticization of gluten agreed well with those reported by Hoseney *et al.* (1986). The anhydrous $T_g$ was found to be 162°C, which was higher than the 121.25 and 141.5°C reported for gliadin (de Graaf *et al.*, 1993) and glutenin (Cocero, 1993), respectively. Fructose, glucose, and sucrose were found to have only little effect on the glass transition, when they were added at a gluten-sugar ratio of 10:1. At a higher fructose content of 2:1 some phase separation was evident, but the $T_g$ of the gluten fraction was depressed by 20 to 40°C. Kalichevsky *et al.* (1992a) reported that the Gordon and Taylor equation was applicable to predict water plasticization of gluten, but suggested the use of the Couchman and Karasz equation to predict the $T_g$ of the three-component mixtures of gluten, sugar, and water. Kalichevsky *et al.* (1992a) estimated a $\Delta C_p$ of 0.39 J/g°C for anhydrous gluten. However, the authors did not consider the possibility of applying the Gordon and Taylor equation to predict the $T_g$ curves for binary mixtures of gluten-sugar solids and water.

In addition to enzymes, elastin, and cereal proteins, glass transitions and water plasticization have been found to occur in milk proteins. Kalichevsky *et al.* (1993) used the Gordon and Taylor equation to fit the $T_g$ data of casein and sodium caseinate at various water contents. The anhydrous $T_g$ values for the materials were 144 and 130°C, respectively. The depression of the $T_g$ due to water plasticization was observed from Young's modulus, which was determined by a three-point bend test. Glucose and lactose in mixtures of 1:10 with casein had little effect on the $T_g$, although at water contents above 12% the transition was broader in comparison with casein alone. Casein was not found to become plasticized by fructose, glucose, and sucrose, which suggested incompatibility of casein and sugars, but high sugar contents decreased the Young's modulus substantially. A comparison of casein and sodium caseinate with other food polymers showed that they were less plasticized by water than such polymers as amylopectin and gluten. The low plasticization of these materials as food components by water was considered to be advantageous when they are added in foods to increase $T_g$ and to reduce hygroscopicity.

*b. State Diagrams.* State diagrams similar to those of carbohydrates can be established for proteins. State diagrams of proteins may show the glass curve as a function of concentration with the $T_g'$ and $C_g'$ values for freeze-concentrated protein matrices. According to Slade *et al.* (1988) the $T_g'$ values of commercial vital wheat glutens are between -10 and -6.5°C. Kokini *et al.* (1993) reported that the $T_g'$ and $C_g'$ values for glutenin were found to be

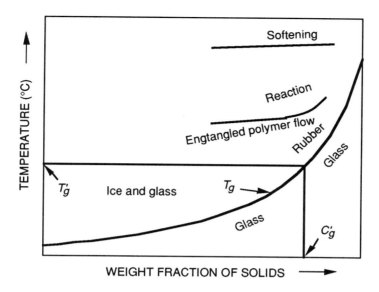

**Figure 5.10** A schematic state diagram for thermosetting cereal proteins showing various transition "zones" and their dependence on water content as suggested by Kokini *et al.* (1993, 1994).

-12°C and 80%, respectively. Slade *et al.* (1988) found an unexpectedly large variation between the $C'_g$ values of commercial vital wheat glutens between 70.9 and 93.5%, but most materials studied had $T'_g$ between -8.5 and -6.5°C and $C'_g$ from about 71 to 77% (respective water contents, $W'_g$, were 40 to 30 g/100 g of solids). Although $T_g$ data are available only for a few proteins, state diagrams showing various transition zones for thermosetting proteins have become important tools in the characterization of their physical state (Kokini *et al.*, 1993, 1994; Madeka and Kokini, 1994).

Cocero (1993) and Kokini *et al.* (1993, 1994) have established state diagrams for cereal proteins. The proteins included wheat glutenin, wheat gliadin, and zein. These proteins were found to become plasticized by water and show thermosetting behavior during heating. Other common properties of the proteins were that they became rubbers above $T_g$, showed entangled polymer flow at sufficient $T - T_g$, and exhibited softening of the thermoset structure at high temperatures. However, these properties were slightly different for each material depending on the water content. Most changes in the protein structure and rheological properties were derived from data obtained with mechanical spectroscopy using dynamic measurements at constant frequency or temperature. High-temperature data were obtained with a pressure rheometer. A schematic state diagram with the various transition zones typical of thermosetting proteins is shown in Figure 5.10.

Application of the gluten state diagram has been proved to be useful by Slade *et al.* (1988). Slade *et al.* (1988) reported that frozen bread dough that is yeast leavened before frozen storage can be cryostabilized with vital wheat gluten to ensure stability at -18°C for at least three months. The frozen product is baked directly from the frozen state. Phase behavior of amorphous proteins has not been thoroughly studied. It is obvious that further studies may provide additional information on their functional properties in foods. The cryostabilizing and thermosetting properties of proteins may provide new important applications for proteins, including their use as edible and biodegradable films. There is also an increasing interest in establishing relationships between water plasticization of enzymes and enzyme activity.

## IV.  Lipids

Lipids together with carbohydrates and proteins are the main food solids. However, lipids as food components differ significantly from carbohydrates and proteins. Lipids are fairly insoluble in water and they exhibit phase transitions that are not significantly affected by the presence of water. The main phase transitions of lipids in foods that affect their behavior in processing, storage, and consumption are transitions between the solid and liquid states of fats and oils. Such transitions include crystallization into various polymorphic forms and effects of fat composition on food properties in various applications.

### A.  Polymorphic Forms

The physical state of food lipids is essential to the quality of fat spreads and confectionery products and to their utility and stability. Most triglycerides exist at least in three crystalline forms, $\alpha$, $\beta'$, and $\beta$, that can be separated according to their X-ray diffraction patterns. Fatty acids and triglycerides are in most cases monotropic, i.e., polymorphic transformations occur only in one direction and transformation of the material from a stable form to a less stable form must occur through the liquid state. The crystallization of triglycerides from melt occurs after sufficient supercooling. Crystallization takes place first to the least stable $\alpha$-form (Bailey, 1950). Crystallization to the other forms occurs in inverse order of their stability. Increasing the temperature slightly above the melting temperature of the $\alpha$-form may allow melting and crystallization to the $\beta'$-form. Similarly, the $\beta$-form may crystallize via melting of the $\beta'$-form. The stability, melting temperature, and heat of fusion of the polymorphic forms increase in the order $\alpha > \beta' > \beta$. According to Hoffmann (1989) crystallization of fats is a slow process that can be made faster with mixing and rapid cooling. The modern analytical methods applied in studies of the crystalline forms of lipids include DSC, IR spectroscopy, NMR, and X-ray diffraction.

## 1. Calorimetric studies

DSC is one of the main techniques in the determination of phase transitions of edible fats and oils. It can be used to study the solidification behavior of lipids as well as melting and transitions between the various polymorphic forms.

A DSC scan of solid triglycerides often shows an endotherm for the melting of the $\alpha$-form followed by an exotherm. The exotherm results from the release of heat as the $\alpha$-form recrystallizes into the $\beta'$-form. Melting and recrystallization phenomena occur also at a higher temperature as the $\beta'$-form melts and recrystallizes to the $\beta$-form, and finally the $\beta$-form melts to the liquid state as shown in Figure 5.11. Various types of DSC curves obtained for lipids were reported by Hagemann (1988). The transitions are not always detected due to different kinetics of the experiment and the concomitant recrystallization and melting phenomena.

The crystalline forms of fats are important in determining physical characteristics of food products. Garti (1988) stated that the stable $\alpha$-forms in food applications often have too high melting temperatures, large crystals, and unpleasant texture and the preferred crystalline form in such products as butter and margarine is the $\beta'$-form (Garti, 1988; Precht, 1988). In addition to butter and margarine cocoa butter is one of the most studied natural fats. Cocoa butter has six polymorphic forms that have been observed with DSC and X-ray techniques (Schlichter-Aronhime and Garti, 1988). Although DSC may be used to observe the formation of the various crystalline forms of fats and their melting behavior, concomitant melting and crystallization phenomena may cause difficulties in the interpretation of thermograms and determination of the latent heats involved in transformations. It should also be noticed that in food materials other thermal phenomena, e.g., melting of ice, may occur simultaneously with fat melting and crystallization phenomena.

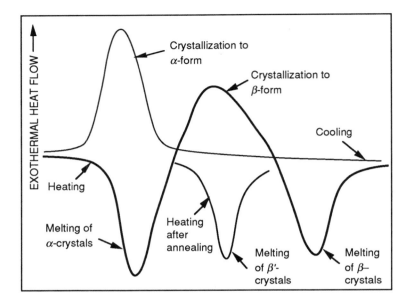

**Figure 5.11** A schematic representation of DSC thermograms typical of fat crystallization and melting. The cooling curve shows an exotherm, which occurs due to crystallization of fat to the least stable $\alpha$-form. The heating curve shows an endotherm indicating melting of the $\alpha$-crystals followed by an exotherm due to the formation of the $\beta$-crystals, which produce another endotherm at their melting temperature. Crystallization to the $\beta'$-form may occur during annealing and the melting of the $\beta'$-crystals causes the appearance of an intermediate endotherm.

## 2. Other techniques

The various crystalline forms of lipids have been identified with X-ray diffraction techniques. Complementary techniques such as dilatometry, DSC, IR and Raman spectroscopy, and NMR have given additional evidence of physical properties of the various polymorphic forms and confirmed their existence. Most published data on the crystalline forms available have been obtained for pure triglycerides such as monoacid triglycerides.

Most natural triglycerides and those in food materials are mixed saturated and unsaturated triglycerides with palmitic, stearic, and oleic acids as the main components. The polymorphism of natural triglycerides may differ from that of saturated monoacid triglycerides. However, the basic characteristics of the crystalline forms are comparable. The X-ray diffraction pattern for the $\alpha$-form shows a strong line at 0.415 nm. The $\beta'$- and $\alpha$-crystals show strong short spacings, which occur for the $\beta'$-form at 0.38 and 0.42 nm and for the

$\beta$-form at 0.37, 0.39, and 0.46 nm (Hagemann, 1988). Moreover, IR spectroscopy has been successfully applied in the characterization of the crystalline states of triglycerides. IR spectra for the $\alpha$-form of tristearin show a singlet at 720 cm$^{-1}$ and for the $\beta$-form a singlet at 717 cm$^{-1}$. The IR spectra for the $\beta'$-forms include a doublet at 719 and 726 cm$^{-1}$ for $\beta'_1$-crystals and a singlet at 719 cm$^{-1}$ for $\beta'_2$-crystals (Hagemann, 1988).

The crystalline form obtained by cooling of fat melts is dependent on the cooling rate and annealing. X-ray diffraction studies of monoacid triglycerides have shown that the crystalline form obtained by cooling from melt is preferably the $\alpha$-form. The $\alpha$-form is transformed to an enantiotropic sub-$\alpha$-form during further cooling (Hagemann, 1988). Heating of the $\alpha$-form to above the melting temperature and annealing can be applied to crystallize the $\beta$- and $\beta'$-forms, respectively. The crystalline form is particularly important to the structure and properties of fat spreads.

### B. Melting of Fats and Oils

Many of the most important characteristics of edible fats and oils are related to their phase behavior, e.g., solid-liquid and liquid-solid transitions during melting and solidification. The melting temperature is a characteristic property of each fatty compound and it can be used in the identification of fats as well as in the evaluation of their technical properties. An important feature of fat melting temperature is that it increases with an increase in chain length and decreases with increasing degree of unsaturation. Food fats are mixtures of triglycerides with melting and crystallization properties that are governed by composition. Natural lipids are also subject to compositional variations that may originate from differences in seasonal, geographical, and processing conditions.

### 1. Melting behavior of fats and oils

Melting behavior of triglycerides includes several general features that are related to their fatty acid composition. Melting temperatures, $T_m$, and heats of fusion, $\Delta H_f$, of saturated monoacid triglycerides increase with increasing fatty acid chain length or molecular weight. The melting temperature and heat of fusion, density, and dilation for each crystalline form increase in the order $\alpha < \beta' < \beta$ as shown in Figure 5.12. The melting temperatures of triglycerides containing unsaturated fatty acids are lower than the melting temperatures of triglycerides, which have a corresponding saturated composition.

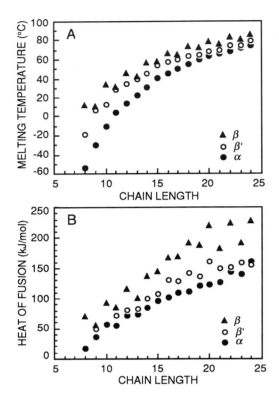

**Figure 5.12** Effect of fatty acid chain length on the melting temperature and heat of fusion of saturated monoacid triglycerides. The difference between the melting temperature of various polymorphic forms decreases with increasing chain length, but the difference in the heat of fusion between the polymorphic forms increases with increasing chain length of the fatty acid. Data from Hagemann (1988).

Most natural triglycerides exist in at least one of the polymorphic forms depending on composition, temperature, and time. Crystallization to the $\beta$-form is common in such fats as coconut, corn, olive, peanut, safflower, and soybean oils as well as in cocoa butter and lard. Cottonseed, palm, and rapeseed oils, milk fat tallow, modified lard, and most natural fats and fat blends, including butter and margarine, have the $\beta'$-form as the preferred crystalline state (Nawar, 1985; Birker and Padley, 1987). Typical melting temperatures of natural fats and oils are given in Table 5.6. Fats with high unsaturated fatty acid contents are usually in the liquid state at room temperature, e.g., marine oils an several plant oils. Solid fats contain

**Table 5.6** Typical melting temperatures, $T_m$, of natural fats and oils and the amount of palmitic, stearic, oleic, and linoleic acids of total fatty acids.

| Fat or oil | $T_m$ (°C) | Saturated acids (%, w/w) | | Unsaturated acids (%, w/w) | |
|---|---|---|---|---|---|
| | | Palmitic | Stearic | Oleic | Linoleic |
| Butterfat | 32.2 | 22.5 | 8.4 | 21.1 | 3.5 |
| Lard oil | 30.5[a] | 22.1 | 10.6 | 32.2 | 5.7 |
| Cocoa butter | 34.1 | 19.6 | 26.1 | 27.6 | 2.1 |
| Coconut oil | 25.1 | 9.5 | 2.2 | 7.0 | - |
| Corn oil | -20.0[a] | 9.3 | 2.9 | 33.2 | 25.5 |
| Cottonseed oil | -1.0[a] | 19.0 | 1.1 | 18.6 | 32.3 |
| Olive oil | -6.0[a] | 6.5 | 2.2 | 45.8 | 4.4 |
| Palm oil | 35.0 | 28.6 | 5.2 | 29.9 | 9.3 |
| Palmkernel oil | 24.1 | 8.1 | 1.3 | 15.6 | 0.7 |
| Peanut oil | 3.0[a] | 7.7 | 3.0 | 35.9 | 20.6 |
| Rapeseed oil | -10[a] | 1.0 | - | 24.2 | 13.0 |
| Sesame oil | -6.0[a] | 8.3 | 4.1 | 31.2 | 28.8 |
| Soybean oil | -16.0[a] | 8.9 | 2.3 | 22.4 | 33.6 |
| Sunflowerseed oil | -17.0[a] | 5.3 | 2.2 | 20.1 | 39.8 |

[a] Solidification temperature
*Source*: Weast (1986)

fairly high amounts of palmitic and stearic acids. According to Hoffmann (1989) triglycerides can be classified into four groups according to their fatty acid composition. These groups are SSS with melting temperatures of 54 to 65°C, SSU with melting temperatures of 27 to 42°C, SUU with melting temperatures of 1 to 23°C, and UUU with melting temperatures of -14 to 1°C (S refers to saturated and U, to unsaturated fatty acid component).

Partial solidification or melt crystallization of natural oils is the basis of fractional crystallization and separation of triglycerides on the basis of their melting temperatures. A single-step fractionation of milk fat yields a hard stearin fraction and a soft or liquid olin fraction (e.g., Deffence, 1993). Milk fat is one of the most complicated fatty materials and contains over 40 different fatty acids, with about 70% being saturated. The saturated fraction contains about 25% short-chain acids and 45% long-chain acids. The unsaturated fraction contains 27% unsaturated and 3% polyunsaturated acids. The melting behavior of such natural fatty materials is a complicated result of the physical state, processing history, and composition.

## 2. Solid fat content

Most commercial fat products are blends of several triglycerides. The solid fat content measured by pulsed NMR or DSC and solid fat index based on dilatation are important measures of the physical properties of fats at various temperatures. However, the solid fat content depends on the thermal history of the material (Walstra, 1987).

The solid fat content can be measured at several temperatures for a particular composition. Various diagrams can be established to show the effect of composition of fat blends on the melting temperature or solid fat content. Fats are considered to be plastic or elastic solids or viscoelastic liquids. Proper blending of the various fat groups is essential in developing fat products with desired characteristics. Pulsed NMR can be used to determine differences in molecular mobility, which, as a function of temperature, is used to obtain the solid fat content (Templeman *et al.*, 1977). The solid fat content can be used to establish isosolids diagrams, which show the solids content as a function of temperature and concentration as shown in Figure 5.13.

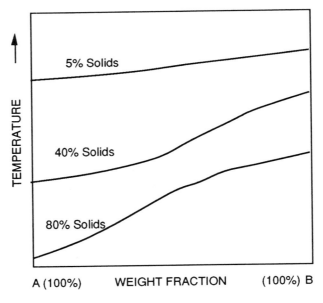

**Figure 5.13** A schematic isosolids diagram for a binary fat blend. The solid fat content is determined for blends with various weight fractions of components A and B as a function of temperature. The isosolids lines define temperatures for constant solid fat contents.

Crystallization behavior and extent of crystallization to the various polymorphic forms of fatty materials and fat blends are dependent on composition. Plotting the solid fat content with the amount of various crystalline forms against temperature can be used to evaluate changes that may occur in the solid fat content and crystalline state during processing and storage due to changes in temperature (Hoffmann, 1989). In rapidly cooled products crystallization occurs to the $\alpha$-form, which, at sufficient conditions, crystallizes to the higher-melting $\beta'$-form and increases the solid fat content. According to Hoffmann (1989) the solid fat content due to $\alpha$-crystals after rapid cooling and the increase of solid fat content due to the formation of $\beta'$-crystals is roughly proportional to the maximum hardness of a fatty product, which is applied in margarine production.

Mechanical properties of fats and spreads are related to temperature and the relative amount of the solid and liquid phases in the product. The difficulty in the prediction of mechanical properties is the large number of raw materials and fats from various sources that make each blend original. The main methods that can be used to predict the physical state of fats and spreads at various temperatures are making the product based on the raw material characteristics and establishing diagrams that show isodilation values or isosolid concentrations as a function of temperature.

# References

Ablett, S., Izzard, M.J., Lillford, P.J., Arvanitoyannis, I. and Blanshard, J.M.V. 1993a. Calorimetric study of the glass transition occurring in fructose solutions. *Carbohydr. Res. 246*: 13-22.

Ablett, S., Darke, A.H., Izzard, M.J. and Lillford, P.J. 1993b. Studies of the glass transition in malto-oligomers. Chpt. 9, in *The Glassy State in Foods*, ed. J.M.V. Blanshard and P.J. Lillford. Nottingham University Press, Loughborough, pp. 189-206.

Arntfield, S.D. and Murray, E.D. 1981. The influence of processing parameters on food protein functionality. I. Differential scanning calorimetry as an indicator of protein denaturation. *Can. Inst. Food Sci. Technol. 14*: 289-294.

Arntfield, S.D., Ismond, M.A.H. and Murray, E.D. 1990. Thermal analysis of food proteins in relation to processing effects. Chpt. 3 in *Thermal Analysis of Foods*, ed. V.R. Harwalkar and C.-Y. Ma. Elsevier, London, pp. 51-91.

Arvanitoyannis, I. and Blanshard, J.M.V. 1994. Rates of crystallization of dried lactose-sucrose mixtures. *J. Food Sci. 59*: 197-205.

Arvanitoyannis, I., Blanshard, J.M.V., Ablett, S., Izzard, M.J., and Lillford, P.J. 1993. Calorimetric study of the glass transition occurring in aqueous glucose:fructose solutions. *J. Sci. Food Agric. 63*: 177-188.

Attenburrow, G.E., Davies, A.P., Goodband, R.M. and Ingman, S.J. 1992. The fracture behaviour of starch and gluten in the glassy state. *J. Cereal Sci. 16*: 1-12.

Atwell, W.A., Hood, L.F., Lineback, D.R., Varriano-Marston, E. and Zobel, H.F. 1988. The terminology and methodology associated with basic starch phenomena. *Cereal Foods World 33*: 306-311.

Bailey, A.E. 1950. *Melting and Solidification of Fats*. Interscience Publishers, New York.

Barbut, S. and Findlay, C.J. 1990. Thermal analysis of egg proteins. Chpt. 5 in *Thermal Analysis of Foods*, ed. V.R. Harwalkar and C.-Y. Ma. Elsevier, London, pp. 126-148.

Biliaderis, C.G. 1991a. The structure and interactions of starch with food constituents. *Can. J. Physiol. Pharmacol. 69*: 60-78.

Biliaderis, C.G. 1991b. Non-equilibrium phase transitions of aqueous starch systems. In *Water Relationships in Food*, ed. H. Levine and L. Slade. Plenum Press, New York, pp. 251-273.

Biliaderis, C.G. 1992. Structures and phase transitions of starch in food systems. *Food Technol. 46*(6): 98-100, 102, 104, 106, 108-109.

Biliaderis, C.G., Maurice, T.J. and Vose, J.R. 1980. Starch gelatinization phenomena studied by differential scanning calorimetry. *J. Food Sci. 45*: 1669-1674, 1680.

Biliaderis, C.G., Page, C.M., Slade, L. and Sirett, R.R. 1985. Thermal behavior of amylose-lipid complexes. *Carbohydr. Polym. 5*: 367-389.

Biliaderis, C.G., Page, C.M., Maurice, T.J. and Juliano, B.O. 1986a. Thermal characterization of rice starches: A polymeric approach to phase transitions of granular starch. *J. Agric. Food Chem. 34*: 6-14.

Biliaderis, C.G., Page, C.M. and Maurice, T.J. 1986b. On the multiple melting transitions of starch/monoglyceride systems. *Food Chem. 22*: 279-295.

Birker, P.J.M.W.L. and Padley, F.B. 1987. Physical properties of fats and oils. In *Recent Advances in Chemistry and Technology of Fats and Oils*, ed. R.J. Hamilton and A. Bhati. Elsevier, London, pp. 1-11.

Blanshard, J.M.V. and Franks, F. 1987. Ice crystallization and its control in frozen-food systems. Chpt. 4, in *Food Structure and Behaviour*, ed. J.M.V. Blanshard and P. Lillford. Academic Press, Orlando, FL, pp. 51-65.

Blond, G. 1989. Water-galactose system: Supplemented state diagram and unfrozen water. *Cryo-Lett. 10*: 299-308.

Bone, S. and Pethig, R. 1982. Dielectric studies of the binding of water to lysozyme. *J. Mol. Biol. 157*: 571-575.

Bone, S. and Pethig, R. 1985. Dielectric studies of protein hydration and hydration-induced flexibility. *J. Mol. Biol. 181*: 323-326.

Burt, D.J. and Russell, P.L. 1983. Gelatinization of low water content wheat starch-water mixtures. A combined study by differential scanning calorimetry and light microscopy. *Starch 35*: 354-360.

Cameron, R.E. and Donald, A.M. 1993a. A small-angle X-ray scattering study of starch gelatinization in excess and limiting water. *J. Polym. Sci.; Part B: Polym. Phys. 31*: 1197-1203.

Cameron, R.E. and Donald, A.M. 1993b. A small-angle X-ray scattering study of the absorption of water into the starch granule. *Carbohydr. Res. 244*: 225-236.

Chan, R.K., Pathmanathan, K. and Johari, G.P. 1986. Dielectric relaxations in the liquid and glassy states of glucose and its water mixtures. *J. Phys. Chem. 90*: 6358-6362.

Chinachoti, P., White, V.A., Lo, L. and Stengle, T.R. 1991a. Application of high resolution carbon-13, oxygen-17, and sodium-23 nuclear magnetic resonance to study the influence of water, sucrose and sodium chloride on starch gelatinization. *Cereal Chem.* *68*: 238-244.

Chinachoti, P., Kim-Shin, M.-S., Mari, F. and Lo, L. 1991b. Gelatinization of wheat starch in the presence of sucrose and sodium chloride: Correlation between gelatinization temperature and water mobility as determined by oxygen-17 nuclear magnetic resonance. *Cereal Chem.* *68*: 245-248.

Chungcharoen, A. and Lund, D.B. 1987. Influence of solutes and water on rice starch gelatinization. *Cereal Chem.* *64*: 240-243.

Cocero, A.M. 1993. The rheology and phase transitions of wheat glutenin. Ph.D. thesis, Rutgers-The State University of New Jersey, New Brunswick, NJ.

Cocero, A.M. and Kokini, J.L. 1991. The study of the glass transition of glutenin using small amplitude oscillatory rheological measurements and differential scanning calorimetry. *J. Rheol.* *35*: 257-270.

Deffence, E. 1993. Milk fat fractionation today. *J. Am. Oil Chem. Soc. 70*: 1193-1201.

de Graaf, E.M., Madeka, H., Cocero, A.M. and Kokini, J.L. 1993. Determination of the effect of moisture on gliadin glass transition using mechanical spectrometry and differential scanning calorimetry. *Biotechnol. Prog. 9*: 210-213.

Derby, R.I., Miller, B.S., Miller, B.F. and Trimbo, H.B. 1975. Visual observation of wheat-starch gelatinization in limited water systems. *Cereal Chem. 52*: 702-713.

De Wit, J.N. and Klarenbeek, G. 1984. Effects of various heat treatments on structure and solubility of whey proteins. *J. Dairy Sci. 67*: 2701-2710.

De Wit, J.N. and Swinkels, G.A.M. 1980. A differential scanning calorimetric study of the thermal denaturation of bovine $\beta$-lactoglobulin. Thermal behaviour at temperatures up to 100°C. *Biochim. Biophys. Acta 624*: 40-50.

Donovan, J.W. 1979. Phase transitions of the starch-water system. *Biopolymers 18*: 263-275.

Donovan, J.W. and Mapes, C.J. 1976. A differential scanning calorimetric study of conversion of ovalbumin to S-ovalbumin in eggs. *J. Sci. Food Agric. 27*: 197-204.

Donovan, J.W., Mapes, C.J., Davis, J.G. and Garibaldi, J.A. 1975. A differential scanning calorimetric study of the stability of egg white to heat denaturation. *J. Sci. Food Agric. 26*: 73-83.

Eliasson, A.-C. 1983. Differential scanning calorimetry studies on wheat starch-gluten mixtures. I. Effect of gluten on the gelatinization of wheat starch. *J. Cereal. Sci. 1*: 199-205.

Eliasson, A.-C. 1986. On the effects of surface active agents on the gelatinization of starch - A calorimetric investigation. *Carbohydr. Polym. 6*: 463-476.

Eliasson, A.-C. 1992. A calorimetric investigation of the influence of sucrose on the gelatinization of starch. *Carbohydr. Polym. 18*: 131-138.

Eliasson, A.-C. and Hegg, P.-O. 1980. Thermal stability of wheat gluten. *Cereal Chem. 57*: 436-437.

Eliasson, A.-C. and Larsson, K. 1993. *Cereals in Breadmaking*. Marcel Dekker, New York.

Evans, I.D. and Haisman, D.R. 1982. The effect of solutes on the gelatinization temperature range of potato starch. *Starch 34*: 224-231.

Findlay, C.J. and Barbut, S. 1990. Thermal analysis of meat. Chpt. 4 in *Thermal Analysis of Foods*, ed. V.R. Harwalkar and C.-Y. Ma. Elsevier, London, pp. 92-125.

Finegold, L. Franks, F. and Hatley, R.H.M. 1989. Glass/rubber transitions and heat capacities of binary sugar blends. *J. Chem. Soc., Faraday Trans. 1 85*: 2945-2951.

Flory, P.J. 1953. *Principles of Polymer Chemistry*. Cornell University Press, Ithaca, NY.

Foegeding, E.A. 1988. Thermally induced changes in muscle proteins. *Food Technol. 42(6)*: 58, 60-62, 64.

Garti, N. 1988. Effects of surfactants on crystallization and polymorphic transformation of fats and fatty acids. Chpt. 7 in *Crystallization and Polymorphism of Fats and Fatty Acids*, ed. N. Garti and K. Sato. Marcel Dekker, New York, pp. 267-303.

Gidley, M.J. 1987. Factors affecting the crystalline type (A-C) of native starches and model compounds. A rationalisation of observed effects in terms of polymorphic structures. *Carbohydr. Res. 161*: 301-304.

Gidley, M.J. and Bulpin, P.V. 1987. Crystallization of malto-oligosaccharides as models of the crystalline forms of starch: Minimum chain-length requirement for the formation of double helices. *Carbohydr. Res. 161*: 291-300.

Hagemann, J.W. 1988. Thermal behavior and polymorphism of acylglycerides. Chpt. 2 in *Crystallization and Polymorphism of Fats and Fatty Acids*, ed. N. Garti and K. Sato. Marcel Dekker, New York, pp. 9-95.

Hansen, L.M., Setser, C.S. and Paukstelis, J.V. 1989. Investigations of sugar-starch interactions using carbon-13 nuclear magnetic resonance. I. Sucrose. *Cereal Chem. 66*: 411-415.

Hartel, R.W. and Shastry, A.V. 1991. Sugar crystallization in food products. *Crit. Rev. Food Sci. Nutr. 30*: 49-112.

Hatley, R.H.M., van den Berg, C. and Franks, F. 1991. The unfrozen water content of maximally freeze concentrated carbohydrate solutions: Validity of the methods used for its determination. *Cryo-Lett. 12*: 113-124.

Hizukuri, S., Kaneko, T. and Takeda, Y. 1983. Measurement of the chain length of amylopectin and its relevance to the origin of crystalline polymorphism of starch granules. *Biochim. Biophys. Acta 760*: 188-191.

Hoffmann, G. 1989. *The Chemistry and Technology of Edible Oils and Fats and Their High Fat Products*. Academic Press, San Diego, CA.

Hoseney, R.C., Zeleznak, K. and Lai, C.S. 1986. Wheat gluten: A glassy polymer. *Cereal Chem. 63*: 285-286.

Izzard, M.J., Ablett, S. and Lillford, P.J. 1991. Calorimetric study of the glass transition occurring in sucrose solutions. Chpt. 23 in *Food Polymers, Gels, and Colloids*, ed. E. Dickinson. The Royal Society of Chemistry, Cambridge, pp. 289-300.

Jin, X., Ellis, T.S. and Karasz, F.E. 1984. The effect of crystallinity and crosslinking on the depression of the glass transition temperature in nylon 6 by water. *J. Polym. Sci.: Polym. Phys. Ed. 22*: 1701-1717.

Kakivaya, S.R. and Hoeve, C.A.J. 1975. The glass point of elastin. *Proc. Nalt. Acad. Sci. U.S.A. 72*: 3505-3507.

Kalichevsky, M.T. and Blanshard, J.M.V. 1992. A study of the effect of water on the glass transition of 1:1 mixtures of amylopectin, casein and gluten using DSC and DMTA. *Carbohydr. Polym. 19*: 271-278.

Kalichevsky, M.T. and Blanshard, J.M.V. 1993. The effect of fructose and water on the glass transition of amylopectin. *Carbohydr. Polym. 20*: 107-113.

Kalichevsky, M.T., Jaroszkiewicz, E.M., Ablett, S., Blanshard, J.M.V. and Lillford, P.J. 1992a. The glass transition of amylopectin measured by DSC, DMTA and NMR. *Carbohydr. Polym. 18*: 77-88.

Kalichevsky, M.T., Jaroszkiewicz, E.M. and Blanshard, J.M.V. 1992b. Glass transition of gluten. 1: Gluten and gluten-sugar mixtures. *Int. J. Biol. Macromol. 14*: 257-267.

Kalichevsky, M.T., Jaroszkiewicz, E.M. and Blanshard, J.M.V. 1993. A study of the glass transition of amylopectin-sugar mixtures. *Polymer 34*: 346-358.

Karger, N. and Lüdemann, H.-D. 1991. Temperature dependence of the rotational mobility of the sugar and water molecules in concentrated aqueous trehalose and sucrose solutions. *Z. Naturforsch. C: Biosci. 46C*: 313-317.

Kauzmann, W. 1948. The nature of the glassy state and the behavior of liquids at low temperatures. *Chem. Rev. 43*: 219-255.

Kauzmann, W. 1959. Some factors in the interpretation of protein denaturation. *Adv. Protein Chem. 14*: 1-63.

Kokini, J.L., Cocero, A.M., Madeka, H. and de Graaf, E. 1993. Order-disorder transitions and complexing reactions in cereal proteins and their effect on rheology. In *Proceedings of International Symposium Advances in Structural and Heterogeneous Continua*, Moscow, August 1993. In press.

Kokini, J.L., Cocero, A.M., Madeka, H. and de Graaf, E. 1994. The development of state diagrams for cereal proteins. *Trends Food Sci. Technol. 5*: 281-288.

Kugimiya, M. and Donovan, J.W. 1981. Calorimetric determination of the amylose content of starches based on formation and melting of the amylose-lysolecithin complex. *J. Food Sci. 46*: 765-777.

Kugimiya, M., Donovan, J.W. and Wong, R.Y. 1980. Phase transitions of amylose-lipid complexes in starches: A calorimetric study. *Starch 32*: 265-270.

Laine, M.J.K. and Roos, Y. 1994. Water plasticization and recrystallization of starch in relation to glass transition. In *Proceedings of the Poster Session, International Symposium on the Properties of Water, Practicum II*, ed. A. Argaiz, A. López-Malo, E. Palou and P. Corte. Universidad de las Américas-Puebla, pp. 109-112.

Lelievre, J. 1974. Starch gelatinization. *J. Appl. Polym. Sci. 18*: 293-296.

Lelievre, J. 1976. Theory of gelatinization in a starch-water-solute system. *Polymer 17*: 854-858.

Lelievre, J. and Mitchell, J. 1975. A pulsed NMR study of some aspects of starch gelatinization. *Starch 27*: 113-115.

Le Meste, M. and Huang, V. 1992. Thermomechanical properties of frozen sucrose solutions. *J. Food Sci. 57*: 1230-1233.

Levine, H. and Slade, L. 1986. A polymer physico-chemical approach to the study of commercial starch hydrolysis products (SHPs). *Carbohydr. Polym. 6*: 213-244.

Levine, H. and Slade, L. 1989. A food polymer science approach to the practice of cryostabilization technology. *Comments Agric. Food Chem. 1*: 315-396.

Levine, H. and Slade, L. 1990. Influences of the glassy and rubbery states on the thermal, mechanical, and structural properties of doughs and baked products. Chpt. 5 in *Dough Rheology and Baked Product Texture*, ed. H. Faridi and J.M. Faubion. AVI Publishing Co., New York, pp. 157-330.

Levine, H. and Slade, L. 1993. The glassy state in applications for the food industry, with an emphasis on cookie and cracker production. Chpt. 17 in *The Glassy State in Foods*, ed. J.M.V. Blanshard and P.J. Lillford. Nottingham University Press, Loughborough, pp. 333-373.

Lillie, M.A. and Gosline, J.M. 1993. The effects of swelling solvents on the glass transition in elastin and other proteins. Chpt. 14 in *The Glassy State in Foods*, ed. J.M.V. Blanshard and P.J. Lillford. Nottingham University Press, Loughborough, pp. 281-301.

Lineback, D.R. 1986. Current concepts of starch structure and its impact on properties. *J. Jpn. Soc. Starch Sci. 33*: 80-88.

Liu, H. and Lelievre, J. 1991. A differential scanning calorimetry study of glass and melting transitions in starch suspensions and gels. *Carbohydr. Res. 219*: 23-32.

Liu, H. and Lelievre, J. 1992. Transitions in frozen gelatinized starch systems studied by differential scanning calorimetry. *Carbohydr. Polym. 19*: 179-183.

Lund, D. 1984. Influence of time, temperature, moisture, ingredients, and processing conditions on starch gelatinization. *CRC Crit. Rev. Food Sci. Nutr. 20*: 249-273.

Luyet, B. and Rasmussen, D. 1968. Study by differential thermal analysis of the temperatures of instability of rapidly cooled solutions of glycerol, ethylene glycol, sucrose and glucose. *Biodynamica 10(211)*: 167-191.

Ma, C.-Y. and Harwalkar, V.R. 1991. Thermal analysis of food proteins. *Adv. Food Nutr. Res. 35*: 317-366.

MacInnes, W.M. 1993. Dynamic mechanical thermal analysis of sucrose solutions. Chpt. 11 in *The Glassy State in Foods*, ed. J.M.V. Blanshard and P.J. Lillford. Nottingham University Press, Loughborough, pp. 223-248.

Madeka, H. and Kokini, J.L. 1994. Changes in rheological properties of gliadin as a function of temperature and moisture: Development of a state diagram. *J. Food Eng. 22*: 241-252.

Marchant, J.L. and Blanshard, J.M.V. 1978. Studies of the dynamics of the gelatinization of starch granules employing a small angle light scattering system. *Starch 30*: 257-264.

Marsh, R.D.L. and Blanshard, J.M.V. 1988. The application of polymer crystal growth theory to the kinetics of formation of the $\beta$-amylose polymorph in a 50% wheat-starch gel. *Carbohydr. Polym. 9*: 301-317.

Maurice, T.J., Page, C.M. and Slade, L. 1984. Applications of thermal analysis to starch research. *Cereal Foods World 29*: 504.

Maurice, T.J., Slade, L., Sirett, R.R. and Page, C.M. 1985. Polysaccharide-water interactions - Thermal behavior of rice starch. In *Properties of Water in Foods*, ed. D. Simatos and J.L. Multon. Martinus Nijhoff Publishers, Dordrecht, the Netherlands, pp. 211-227.

Maurice, T.J., Asher, Y.J. and Thomson, S. 1991. Thermomechanical analysis of frozen aqueous systems. In *Water Relationships in Foods*, ed. H. Levine and L. Slade. Plenum Press, New York, pp. 215-223.

Nakamura, S. and Tobolsky, A.V. 1967. Viscoelastic properties of plasticized amylose films. *J. Appl. Polym. Sci. 11*: 1371-1386.

Nawar, W.W. 1985. Lipids. Chpt. 4 in *Food Chemistry*, ed. O.R. Fennema, 2nd ed. Marcel Dekker, Inc., New York, pp. 139-244.

Nickerson, T.A. and Moore, E.E. 1972. Solubility interrelations of lactose and sucrose. *J. Food Sci. 37*: 60-61.

Nickerson, T.A. and Patel, K.N. 1972. Crystallization in solutions supersaturated with sucrose and lactose. *J. Food Sci. 37*: 693-697.

Olkku, J. and Rha, C.K. 1978. Gelatinisation of starch and wheat flour starch. *Food Chem. 3*: 293-317.

Orford, P.D., Parker, R., Ring, S.G. and Smith, A.C. 1989. Effect of water as a diluent on the glass transition behaviour of malto-oligosaccharides, amylose and amylopectin. *Int. J. Biol. Macromol. 11*: 91-96.

Orford, P.D., Parker, R. and Ring, S.G. 1990. Aspects of the glass transition behaviour of mixtures of carbohydrates of low molecular weight. *Carbohydr. Res. 196*: 11-18.

Pancoast, H.M. and Junk, W.R. 1980. *Handbook of Sugars*. 2nd ed. AVI Publishing Co. Westport, CT.

Parks, G.S. and Reagh, J.D. 1937. Studies on glass. XV. The viscosity and rigidity of glucose glass. *J. Chem. Phys. 5*: 364-367.

Parks, G.S. and Thomas, S.B. 1934. The heat capacities of crystalline, glassy and undercooled liquid glucose. *J. Am. Chem. Soc. 56*: 1423.

Pavlath, A.E. and Gregorski, K.S. 1985. Atmospheric pyrolysis of carbohydrates with thermogravimetric and mass spectrometric analyses. *J. Anal. Appl. Pyrolysis 8*: 41-48.

Precht, D. 1988. Fat crystal structure in cream and butter. Chpt. 8 in *Crystallization and Polymorphism of Fats and Fatty Acids*, ed. N. Garti and K. Sato. Marcel Dekker, New York, pp. 305-361.

Privalov, P.L. 1979. Stability of proteins. Small globular proteins. *Adv. Protein Chem. 33*: 167-241.

Privalov, P.L. and Khechinashvili, N.N. 1974. A thermodynamic approach to the problem of stabilization of globular protein structure: A calorimetric study. *J. Mol. Biol. 86*: 665-684.

Ring, S.G., Gee, J.M., Whittam, M., Orford, P. and Johnson, I.T. 1988. Resistant starch: Its chemical form in foodstuffs and effect on digestibility *in vitro. Food Chem. 28*: 97-109.

Roos, Y. 1993. Melting and glass transitions of low molecular weight carbohydrates. *Carbohydr. Res. 238*: 39-48.

Roos, Y. and Karel, M. 1991a. Amorphous state and delayed ice formation in sucrose solutions. *Int. J. Food Sci. Technol. 26*: 553-566.

Roos, Y. and Karel, M. 1991b. Nonequilibrium ice formation in carbohydrate solutions. *Cryo-Lett. 12*: 367-376.

Roos, Y. and Karel, M. 1991c. Plasticizing effect of water on thermal behavior and crystallization of amorphous food models. *J. Food Sci. 56*: 38-43.

Roos, Y. and Karel, M. 1991d. Water and molecular weight effects on glass transitions in amorphous carbohydrates and carbohydrate solutions. *J. Food Sci. 56*: 1676-1681.

Roos, Y. and Karel, M. 1991e. Phase transitions of mixtures of amorphous polysaccharides and sugars. *Biotechnol. Prog. 7*: 49-53.

Schenz, T.W., Israel, B. and Rosolen, M.A. 1991. Thermal analysis of water-containing systems. In *Water Relationships in Foods*, ed. H. Levine and L. Slade. Plenum Press, New York, pp. 199-214.

Schlichter-Aronhime, J. and Garti, N. 1988. Solidification and polymorphism in cocoa butter and blooming problems. Chpt. 9 in *Solidification and Polymorphism in Cocoa Butter and the Blooming Problems*, ed. N. Garti and K. Sato. Marcel Dekker, New York, pp. 363-393.

Schofield, J.D., Bottomley, R.C., Timms, M.F. and Booth, M.R. 1983. The effect of heat on wheat gluten and the involvement of sulphydryl-disulphide interchange reactions. *J. Cereal Sci. 1*: 241-253.

Shewry, P.R., Tatham, A.S., Forde, J., Kreis, M. and Miflin, B.J. 1986. The classification and nomenclature of wheat gluten proteins: A reassessment. *J. Cereal Sci. 4*: 97-106.

Slade, L. and Levine, H. 1991. Beyond water activity: Recent advances based on an alternative approach to the assessment of food quality and safety. *Crit. Rev. Food Sci. Nutr. 30*: 115-360.

Slade, L. and Levine, H. 1994. Water and the glass transition - Dependence of the glass transition on composition and chemical structure: Special implications for flour functionality in cookie baking. *J. Food Eng. 22*: 143-188.

Slade, L. and Levine, H. 1995. Glass transitions and water-food structure interactions. *Adv. Food Nutr. Res. 38*. In press.

Slade, L., Levine, H. and Finley, J.W. 1988. Protein-water interactions: Water as a plasticizer of gluten and other protein polymers. In *Protein Quality and the Effects of Processing*, ed. D. Phillips and J.W. Finlay. Marcel Dekker, New York, pp. 9-124.

Spies, R.D. and Hoseney, R.C. 1982. Effect of sugars on starch gelatinization. *Cereal Chem. 59*: 128-131.

Templeman, G.J., Scholl, J.J. and Labuza, T.P. 1977. Evaluation of several pulsed NMR techniques for solids-in-fat determination of commercial fats. *J. Food Sci. 42*: 432-435.

van den Berg, C. 1981. Vapour sorption equilibria and other water-starch interactions; a physico-chemical approach. Doctoral thesis, Agricultural University, Wageningen, the Netherlands.

von Eberstein, K., Höpcke, R., Konieczny-Janda, G. and Stute, R. 1980. DSC-Untersuchungen an Stärken. Teil I. Möglichkeiten thermoanalytischer Methoden zur Stärkecharacterisierung. *Starch 32*: 397-404.

von Krüsi, H. and Neukom, H. 1984. Untersuchungen über die Retrogradation der Stärke in konzentriertrn Weizenstärkegelen. Teil 2. Einfluss von Stärkeabbauprodukten auf die Retrogradation der Stärke. *Starch 36*: 300-305.

Walstra, P. 1987. Fat crystallization. Chpt. 5 in *Food Structure and Behaviour*, ed. J.M.V. Blanshard and P. Lillford. Academic Press, Orlando, FL, pp. 67-85.

Weast, R.C. 1986. *CRC Handbook of Chemistry and Physics*. 67th ed. CRC Press, Boca Raton, FL.

White, G.W. and Cakebread, S.H. 1966. The glassy state in certain sugar-containing food products. *J. Food Technol. 1*: 73-82.

Whittam, M.A., Noel, T.R. and Ring, S.G. 1990. Melting behaviour of A- and B-type crystalline starch. *Int. J. Biol. Macromol. 12*: 359-362.

Whittam, M.A., Noel, T.R. and Ring, S.G. 1991. Melting and glass/rubber transitions of starch polysaccharides. Chpt 22 in *Food Poymers, Gels, and Colloids*, ed. E. Dickinson. The Royal Society of Chemistry, Cambridge, pp. 277-288.

Wootton, M. and Bamunuarachchi, A. 1980. Application of differential scanning calorimetry to starch gelatinization. III. Effect of sucrose and sodium chloride. *Starch 32*: 126-129.

Wright, D.J., Leach, I.B. and Wilding, P. 1977. Differential scanning calorimetric studies of muscle and its constituent proteins. *J. Sci. Food Agric. 28*: 557-564.

Yannas, I.V. 1972. Collagen and gelatin in the solid state. *J. Macromol. Sci., Rev. Macromol. Chem. C7(1)*: 49-104.

Young, F.E. and Jones, F.T. 1949. Sucrose hydrates. The sucrose-water phase diagram. *J. Phys. Colloid Chem. 53*: 1334-1350.

Zeleznak, K.J. and Hoseney, R.C. 1987. The glass transition in starch. *Cereal Chem. 64*: 121-124.

Zobel, H.F. 1988a. Starch crystal transformations and their industrial importance. *Starch 40*: 1-7.

Zobel, H.F. 1988b. Molecules to granules: A comprehensive starch review. *Starch 40*: 44-50.

Zobel, H.F. 1992. Starch granule structure. In *Developments in Carbohydrate Chemistry*, ed. R.J. Alexander and H.F. Zobel. The American Association of Cereal Chemists, St. Paul, MN, pp. 1-36.

Zobel, H.F., Young, S.N. and Rocca, L.A. 1988. Starch gelatinization: An X-ray diffraction study. *Cereal Chem. 65*: 443-446.

# CHAPTER 6

# Prediction of the Physical State

## I. Introduction

The physical state of food materials can be related to a number of physico-chemical and textural properties that affect their behavior in processing and storage. Prediction of the physical state may be used to evaluate effects of environmental factors on changes that may occur in processing and storage and the extent of such changes. Predictions of the physical state have been rarely applied in food engineering, although the physical state of synthetic polymers with similar physicochemical properties is of profound importance to their machinability and utility in various applications.

The main difference between food materials and polymers is found in their chemical composition. Foods are complex mixtures of solids and water, while polymers are mostly composed of repeating units of well-characterized molecules. However, there has been a successful introduction of the polymer science principles to characterize food materials as nonequilibrium systems according to their physical state, defined by the thermodynamic quantities of pressure, temperature, and volume (Slade and Levine, 1991). The main requirements in food processing, storage, distribution, and consumer acceptability are related to factors that affect food safety and quality. Food processing conditions should provide desired quality characteristics and improve safety and shelf life. Understanding of food properties in both processing and storage and the various interactions between food components and surroundings requires knowledge of the basic phenomena that may have an effect on observed changes.

Phase transitions affect significantly the physical properties of food materials and they may contribute to the kinetics of various chemical reactions (e.g., White and Cakebread, 1966; Simatos and Karel, 1988; Slade and Levine, 1991; Karel *et al.*, 1993; Nelson and Labuza, 1994). Therefore, attempts have been made to establish methods for the prediction of the physical state of food components and foods. The basic variables that affect food properties are temperature, water content, and time. Sorption isotherms have provided valuable information on the steady state water contents of foods at various temperature and relative humidity environments. Although time-dependent phenomena have been observed from sorption isotherms, e.g., crystallization of amorphous components (Karel, 1973; Roos and Karel, 1990), they have provided no information on the basic phenomena causing the observed time-dependent changes. Introduction of the theories used in the characterization of the phase behavior of polymers has significantly increased understanding of the nonequilibrium behavior of foods and kinetics of various changes observed in food processing and storage (e.g., Levine and Slade, 1986; Slade and Levine, 1988, 1991).

Several relationships for the prediction of the physical state of polymers at various conditions have been introduced (Flory, 1953; Ferry, 1980; Sperling, 1986) and also applied to food components (Soesanto and Williams, 1981; Levine and Slade, 1986; Roos and Karel, 1990; Slade and Levine, 1991; Roos, 1993). State diagrams based on experimental data can also be used to predict food behavior at various conditions (Slade and Levine, 1991; Roos and Karel, 1991a). This chapter provides information on methods that can be applied to predict the physical state of food materials at various conditions and the relationships between the physical state and food properties.

## II.  Prediction of Water Plasticization

Nonfat food solids are often soluble or at least partially soluble in water. This is observed from various effects of water on food stability. The plasticizing effect of water on food solids is often observed from changes in texture or in rates of chemical reactions that may occur above some critical water content or proceed at rates that are dependent on water availability (Labuza *et al.*, 1970; Katz and Labuza, 1981). It is obvious that these changes are at least partially related to the physical changes that occur in foods due to water plasticization (Slade and Levine, 1991).

## A. Plasticization Models

Plasticization of polymers by diluents can be related to free volume or thermodynamic properties such as heat capacity. Several studies have applied equations which are commonly used to predict compositional effects of mixtures of miscible polymers on glass transition temperature of food materials plasticized by water. Theoretical considerations have often failed to predict the effect of water on $T_g$ (e.g., Orford *et al.*, 1989; Kalichevsky *et al.*, 1993a) and predictions of empirical equations have often proved to be more accurate than those based on theory. The prediction of water plasticization, however, is an important tool in evaluating properties of food materials and other biopolymers at various processing and storage conditions. The prediction of the water plasticization can also be applied in evaluating food texture in processing and consumption. Extrusion provides a good example of a food process that applies thermal and water plasticization during processing and allows formation of glassy products with a crispy texture due to cooling and loss of water (Slade and Levine, 1995).

## 1. Gordon and Taylor equation

The Gordon and Taylor equation (Gordon and Taylor, 1952) has been applied in predicting water plasticization of several food components and food materials, including carbohydrates and proteins (Roos and Karel, 1991c,f; Kalichevsky *et al.*, 1993b) as well as pharmaceutical materials (Hancock and Zografi, 1994).

Application of the Gordon and Taylor equation to predict water plasticization requires determination of the empirical constant, $k$. The constant can be derived from experimental data for $T_g$ at various water contents. The data should cover the whole range of water contents and at least water contents from 0 to 50%. The equation written into the form of equation (6.1) can be used to obtain values for the constant $k$ with the experimental data for samples with various weight fractions of solids, $w_1$, and water, $w_2$, and respective glass transition temperatures, $T_{g1}$ and $T_{g2}$. The glass transition temperature for amorphous water is often taken as -135°C based on the data of various studies (see Table 4.2). The Gordon and Taylor equation may also be used to calculate the weight fraction of solids for any $T_g$ value with equation (6.2) when the $k$ and $T_g$ are known.

$$k = \frac{w_1 T_{g1} - w_1 T_g}{w_2 T_g - w_2 T_{g2}}$$
(6.1)

$$w_1 = \frac{k(T_g - T_{g2})}{k(T_g - T_{g2}) + T_{g1} - T_g}$$
(6.2)

The Gordon and Taylor equation is relatively simple and fairly easy to apply in predicting water plasticization. However, use of the Gordon and Taylor equation is restricted to binary mixtures of solids and water. Thus, the equation cannot be applied for predicting effects of three or more components on $T_g$. Addition of the term $qw_1w_2$, where $q$ is another constant, to the Gordon and Taylor equation gives equation (6.3).

$$T_g = \frac{w_1 T_{g1} + kw_2 T_{g2}}{w_1 + kw_2} + qw_1w_2$$
(6.3)

Equation (6.3) is an empirical equation that is known as the Kwei equation (Lin *et al.*, 1989).

## 2. Couchman and Karasz equation

The thermodynamic theory of the glass transition temperature of mixtures relates the heat capacity changes of pure polymer components at their glass transition temperature to the $T_g$ of the mixture (Couchman and Karasz, 1978; Couchman, 1978). The equation derived with thermodynamic considerations is equal to the Gordon and Taylor equation with $k$ obtained with equation (6.4) and it may be written into the form of equation (6.5), where the subscripts 1 and 2 refer to solids and water, respectively.

$$k = \frac{\Delta C_{p2}}{\Delta C_{p1}}$$
(6.4)

$$T_g = \frac{w_1 \Delta C_{p1} T_{g1} + w_2 \Delta C_{p2} T_{g2}}{w_1 \Delta C_{p1} + w_2 \Delta C_{p2}}$$
(6.5)

Prediction of water plasticization by the Couchman and Karasz equation requires that the changes of the heat capacities at $T_g$ of both components are known. The heat capacity change is often obtained with DSC for the pure components. The main shortcomings of applying equation (6.5) in predicting water plasticization in foods are due to difficulties in determination of the $T_g$ for pure water and the divergence between published data (see Table 4.2).

The Couchman and Karasz (1978) equation was used by Orford *et al.* (1989) for the prediction of water plasticization of malto-oligosaccharides. They used -139°C and 1.94 J/g°C for $T_{g2}$ and $\Delta C_{p2}$, respectively, which were those reported by Sugisaki *et al.* (1968). Orford *et al.* (1989) found that the predicted glass transition temperatures were higher than those determined experimentally, but agreed within 25°C at $w_2 = 0.1$ to 0.22. A better agreement was reported for starch using extrapolated data for $T_{g1}$ and $\Delta C_{p1}$ and experimental data of Zeleznak and Hoseney (1987). The $T_g$ values obtained with DSC and DMTA by Kalichevsky *et al.* (1992a) for amylopectin agreed with the prediction of Orford *et al.* (1989), although NMR detected molecular mobility at lower temperatures. Kalichevsky and Blanshard (1992) modified the equation and extended its use to predict water plasticization in tertiary mixtures of water, amylopectin, and casein. The modified equation (6.6) used the weight fractions, $T_g$, and $\Delta C_p$ of all component compounds, referred to with subscripts 1, 2, and 3.

$$T_g = \frac{w_1 \Delta C_{p1} T_{g1} + w_2 \Delta C_{p2} T_{g2} + w_3 \Delta C_{p3} T_{g3}}{w_1 \Delta C_{p1} + w_2 \Delta C_{p2} + w_3 \Delta C_{p3}} \tag{6.6}$$

In another study Kalichevsky and Blanshard (1993) suggested applicability of the modified Couchman and Karasz equation to predict water plasticization of mixtures of amylopectin, fructose, and water. The values used for $T_g$ and $\Delta C_p$ were -139°C and 1.94 J/g°C, 7°C and 0.83 J/g°C, and 229°C and 0.41 J/g°C for water, fructose, and amylopectin, respectively.

Applicability of the Couchman and Karasz equation and its modifications may be difficult for high-$T_g$ biological materials such as proteins and starch. The $\Delta C_p$ values reported for starch (e.g., Orford *et al.*, 1989) are based on extrapolations, since the $T_g$ of the anhydrous polymer cannot be determined due to thermal decomposition. Experimental determination of $\Delta C_p$ values may also suffer from the fact that the $T_g$ temperature range often broadens and the $\Delta C_p$ decreases with decreasing water content. Changes in material properties that occur over a broad temperature range may cause problems in observing the $T_g$ with DSC for many biopolymers. In some cases the $T_g$ can be measured with mechanical methods, but no information on the $\Delta C_p$ has been obtained (Kalichevsky *et al.*, 1993c). In most studies that have applied equation (6.5) or (6.6) to predict water plasticization in food materials the higher value of Sugisaki *et al.* (1968) for $\Delta C_p$ of water has been proved to give the best fit (Kalichevsky *et al.*, 1992b; Kalichevsky and Blanshard, 1993). If the use of the $\Delta C_p$ values in equation (6.6) is not feasible, the equation reduces to the form of the Gordon and Taylor equation, provided that the anhydrous mixture is considered as component 1 and water as component 2, which allows empirical determination of the constant, $k$.

3. Other equations

Several equations other than the Gordon and Taylor and Couchman and Karasz equations are available for predicting compositional effects on $T_g$. Those equations have also been tested for the prediction of water plasticization in foods. One of the first attempts to predict water plasticization-related phenomena in foods was that of Tsourouflis *et al.* (1976) and To and Flink (1978), who found similarities between collapse temperatures, $T_c$, of dehydrated food powders and glass transition temperatures of polymers.

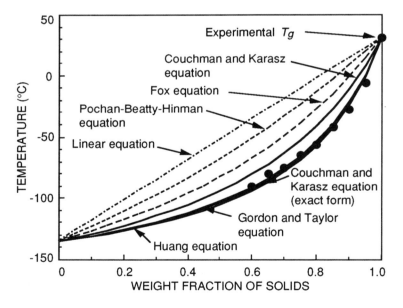

**Figure 6.1** Glass transition temperature, $T_g$, of glucose at various water contents with $T_g$ curves predicted with various models used to predict effects of composition on the $T_g$ of binary polymer blends. $T_g$ values used for glucose and water were 31 and -135°C, respectively. Values for the change in heat capacity at $T_g$, $\Delta C_p$, used in calculations were 0.64 J/g°C for glucose (Roos and Karel, 1991d) and 1.94 J/g°C for water (Sugisaki *et al.*, 1968).

Tsourouflis *et al.* (1976) suggested a linear relationship between log $m$ ($m$ is moisture content in g/100 g of solids) and collapse temperature, $T_c$ or $1/T_c$. They also noticed that the same relationship applied to moisture-content dependence of sticky point temperatures of spray-dried food powders. To and Flink (1978) reported that the relationship was not linear. They proposed that a plot of log $T_g$ against $m$ was composed of two linear segments. However, in most cases the prediction of the effect of composition on $T_g$ is based on fitting one model to experimental data. Predictions of the models may vary significantly as shown in Figure 6.1.

*a. Couchman and Karasz Equation – Exact Form.* The original equation based on thermodynamics that was derived by Couchman and Karasz (1978), in its exact form, assumed the approximation that the $\Delta C_p$ values of the component compounds were temperature-independent. The equation (6.7) reported by Couchman and Karasz (1978), in its more exact form, is equivalent to equation (6.8).

$$\ln T_g = \frac{w_1 \Delta C_{p1} \ln T_{g1} + w_2 \Delta C_{p2} \ln T_{g2}}{w_1 \Delta C_{p1} + w_2 \Delta C_{p2}} \tag{6.7}$$

$$\ln \left( \frac{T_g}{T_{g1}} \right) = \frac{w_2 \Delta C_{p2} \ln \left( T_{g2} / T_{g1} \right)}{w_1 \Delta C_{p1} + w_2 \Delta C_{p2}} \tag{6.8}$$

Equation (6.8) was rearranged into the approximate form given by equation (6.5) with the assumption that $\ln (1 + y) = y$. Thus, the logarithmic values of the glass transition temperatures are not used in the equation that is often referred to as the Couchman and Karasz equation. The use of the approximate Couchman and Karasz equation as shown in Figure 6.1 for glucose tends to underestimate the extent of water plasticization, but the exact form gives about an equal fit to experimental data as the Gordon and Taylor equation with an empirical $k$ value.

Equation (6.7) has been shown to fit $T_g$ data of thermodynamically miscible polymer blends (Sperling, 1986). It should be noticed that $\Delta C_p$ values derived from DSC measurements or extrapolations may cause some error to the predicted $T_g$, but the error caused by the approximation that $\ln (1 + y) = y$ may be more significant, especially in predicting water plasticization due to the large differences between the $T_g$ and $\Delta C_p$ values of solids and water. The fact that the approximation is valid only for small $y$ when $T_{g1}/T_{g2}$ is not greatly different from unity was also pointed out by Sperling (1986). Despite the frequent use of the approximation of the Couchman and Karasz equation in predicting water plasticization of biopolymers and food components, the use of the equation in its exact form should be preferred. However, Kalichevsky *et al.* (1993a) found that equation (6.7) overestimated the $T_g$ depression caused by water plasticization of amylopectin. They used also a modified form of equation (6.8) with the assumption that $\Delta C_p T_g$ was constant, which gave an approximately equal overestimation for the $T_g$ depression.

*b. Fox Equation.* The Fox equation was originally derived for random copolymers (Fox, 1956), but it has not been considered to be applicable to predict water plasticization in food materials. Equation (6.9) may be used in predicting the $T_g$ of binary blends of such materials, which have about constant $\Delta C_{pi} T_{gi}$ (Couchman, 1978).

$$\frac{1}{T_g} = \frac{w_1}{T_{g1}} + \frac{w_2}{T_{g2}} \tag{6.9}$$

*c. Pochan-Beatty-Hinman Equation.* Couchman (1978) reported that the composition dependence of several polymer systems may follow equation (6.10). The equation, known as the Pochan-Beatty-Hinman equation, has not been used in the prediction of the composition dependence of food polymers. As shown in Figure 6.1 the prediction for the $T_g$ depression of glucose by water has a fairly high deviation from experimental data. Equation (6.10) may also be considered to be a modification of equation (6.7) with the assumption that $\Delta C_{p1} = \Delta C_{p2}$ (Couchman, 1978).

$$\ln T_g = w_1 \ln T_{g1} + w_2 \ln T_{g2} \tag{6.10}$$

*d. Linear Equation.* A further simplification of equation (6.10) reduces the equation to the form of equation (6.11), which predicts that the depression of $T_g$ of the higher $T_g$ component decreases linearly with increasing weight fraction of the lower $T_g$ component. It is obvious that the linear equation cannot be used to predict water plasticization of food materials. The linear equation is often used to predict composition dependence of the $T_g$ of polymer blends although the equation usually predicts too high values for $T_g$ (Sperling, 1986).

$$T_g = w_1 T_{g1} + w_2 T_{g2} \tag{6.11}$$

*e. Huang Equation.* The Huang equation is a Couchman and Karasz equation-based empirical equation that has been shown to predict water plasticization of carbohydrates (Huang, 1993). The applicability of equation (6.12) for predicting the $T_g$ depression of glucose is shown in Figure 6.1. The Gordon and Taylor equation, the exact form for the Couchman and Karasz equation, and the Huang equation give about equal fit to the experimental data for glucose. The values predicted by the Couchman and Karasz equation are slightly lower than those predicted by the Gordon and Taylor equation and the Huang equation predicts slightly higher values than the Gordon and Taylor equation at intermediate water contents.

$$T_g = \left[ \frac{w_1 \Delta C_{p1}(T_{g1} + T_{g2}) + 2w_2 \Delta C_{p2} T_{g2}}{w_1 \Delta C_{p1}(T_{g1} + T_{g2}) + 2w_2 \Delta C_{p2} T_{g1}} \right] T_{g1} \qquad (6.12)$$

The advantage of the Huang equation is that $T_g$ and $\Delta C_p$ values are needed only for the anhydrous solids and pure water. The equation is complicated, but can easily be used with computers. Testing the applicability of the equation, however, needs experimental data on glass transition temperatures for the material with various water contents.

The Gordon and Taylor equation can be used without the $\Delta C_p$ values needed for the prediction of the $T_g$ curve with most other equations. Hence, Huang (1993) pointed out that the use of the Gordon and Taylor equation avoids the controversial use of the various available $\Delta C_p$ values for water. Huang (1993) found that the Fox equation and the Pochan-Beatty-Hinman equation were applicable for the prediction of the $T_g$ of maltose, maltotriose, maltohexose, starch, and gluten, but the predicted values were much higher than those determined experimentally. Both the Huang equation and the exact form of the Couchman and Karasz model were found to fit experimental data. It seems that the Gordon and Taylor equation is often a relatively good predictor of water plasticization of food components. The Couchman and Karasz equation and the Huang equation give often an equally good fit, but in some cases, especially if the exact $T_g$ or $\Delta C_p$ values of the components are not known, predictions may differ from experimental data.

## B. Effects of Composition

The physical state of food solids is significantly affected by composition. Fairly few data are available for predicting the effect of food composition on the glass transition temperature and water plasticization of mixtures. Food materials can often be considered as binary mixtures of solids and water, which allows prediction of water plasticization using the Gordon and Taylor equation (Roos and Karel, 1991c,d,f; Roos, 1993). The extended Couchman and Karasz equation may also be used as was suggested and applied for mixtures of amylopectin, casein, gluten, and fructose by Kalichevsky *et al.* (1992b, 1993a,b) and Kalichevsky and Blanshard (1993).

### 1. Water plasticization of mixtures

Several biopolymers can be considered to be mixtures of homopolymers. Such materials include starch and gluten, which show one single glass transition occurring over a temperature range that decreases with increasing water content (Hoseney *et al.*, 1986; Zeleznak and Hoseney, 1987).

The $T_g$ of binary mixtures of amorphous polymers may be assumed to decrease with increasing amount of the lower molecular weight component. Such behavior was shown to apply for glucose polymers by Orford *et al.* (1989) and Roos and Karel (1991e). These studies also suggested that the effect of molecular weight of biopolymers on the $T_g$ can be predicted with the Fox and Flory equation. Maltodextrins composed of starch hydrolysis products contain glucose polymers that may have a large molecular weight distribution. However, the Gordon and Taylor equation has been applicable to predict their water plasticization (Roos and Karel, 1991e) when the material and water have been considered to be a binary mixture of solids and water. The same approach has proved to apply to the prediction of water plasticization of cereal proteins (Kokini *et al.*, 1993, 1994). It is obvious that other equations, especially the Couchman and Karasz and Huang equations, may also be used to predict water plasticization of mixtures of food solids.

Compositional effects on the glass transition temperature are useful in the design of food materials that are subject to well-defined processing and storage conditions. The $T_g$ and water plasticization may affect such properties as stickiness or collapse, e.g., one of the main applications of maltodextrins is their use as drying aids. Maltodextrins are able to decrease stickiness and improve storage stability of food powders, which is due to their ability to increase the $T_g$ of the solids. The glass transition temperature of mixtures of miscible food solids is dependent on the component compounds. Unfortunately, the number of various compounds may be large and the combined effect on the $T_g$ can be unpredictable. However, the effect of composition on the $T_g$ of binary or tertiary mixtures of food solids is often sufficient for the evaluation of compositional effects on physical properties. The glass transition temperature of such mixtures, e.g. sugars and starch, can be successfully predicted (Roos and Karel, 1991b,e; Kalichevsky *et al.*, 1992b). Roos and Karel (1991e) studied glass transition temperatures of sucrose and maltodextrin mixtures. Maltodextrins of various molecular weights were significantly plasticized by sucrose. It was also noticed that addition of maltodextrins to sucrose caused a fairly small and almost molecular weight-independent increase of the $T_g$ up to a concentration of 50% (w/w). Thus, in practical applications such as dehydration of fruit juices fairly high concentrations of maltodextrins are needed. The increase of the $T_g$ also increases the critical water content and water activity for stability during storage (Roos and Karel, 1991e; Roos, 1993).

## 2. Water activity and glass transition

Stability of food materials is significantly affected by water activity, which is defined by the relative vapor pressure ($a_w$ = RVP = $p/p_0$) of water within food solids at steady state conditions. Both the physical state of food solids and stability are extremely sensitive to changes in water availability at low water contents, which often include water contents typical of low-moisture and intermediate-moisture foods.

Water activity has become one of the most significant measures of the state of water in foods. Similarly, glass transition temperature may be considered to be the most significant measure of the state of amorphous food solids. The first relationship between water activity and glass transition temperature was established by Roos (1987). He studied the effect of water on the physical state of freeze-dried strawberries by measuring relationships between glass transition temperature, water content, and water activity. Glass transition temperatures were determined with DSC for the anhydrous material and materials that were rehumidified over saturated salt solutions to various water contents. Strawberry solids showed typical behavior of materials that become plasticized by water, probably due to the high amount of amorphous carbohydrates. Roos (1993) found that the $T_g$ decreased with increasing water content and the water plasticization could be predicted with the Gordon and Taylor equation. Interestingly, a plot that showed $T_g$ against $a_w$ gave a straight line and a linear relationship was established between $T_g$ and water activity at 25°C (Roos, 1987). The linear relationship was later found to apply to several other amorphous food solids and food components. However, the linearity often applies over the $a_w$ range of 0.1 to 0.8 and the relationship over the whole $a_w$ range is sigmoid, as shown in Figure 6.2 (Roos and Karel, 1991e; Roos, 1994).

The linearity between $T_g$ and water activity provides a simple method for prediction of the effect of relative humidity (RH) on the $T_g$ of low- and intermediate-moisture foods. As shown in Figure 6.2 the linear portion of the $T_g$ depression with increasing $a_w$ is dependent on molecular weight or the $T_g$ value of the anhydrous material. The slope of the linear part of the curve

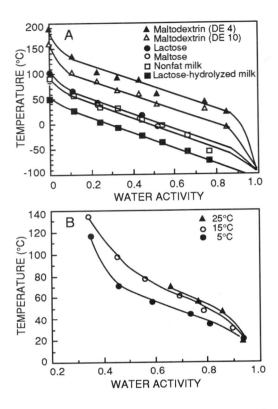

**Figure 6.2** Glass transition temperature, $T_g$, of food materials as a function of water activity, $a_w$. A. $T_g$ against $a_w$ at 25°C gives a sigmoid curve, but shows linearity over a wide $a_w$ range. Data from Roos and Karel (1991e) and Jouppila and Roos (1994b). B. $T_g$ of amylopectin against $a_w$ at 5, 15, and 25°C. Data from Jouppila *et al.* (1995) and Laine and Roos (1994).

seems to depend on temperature. The linear part of the curve for amylopectin is fairly narrow at room temperature. Decreasing temperature extends the linear portion and decreases slope due to the increased water adsorption that occurs at the lower temperature (Figure 6.2). The linear relationship between $T_g$ and $a_w$ is extremely useful in product development applications and in establishing criteria for requirements of product storage conditions and packaging materials. The limitation of the method is that the relationship is temperature-dependent due to changes in water activity and water adsorption with temperature. Results of studies on materials with different water adsorption behavior and glass transition temperatures have suggested that the slope of the line showing depression of the $T_g$ with increasing

$a_w$ is almost material-independent (Figure 6.2). Prediction of the $T_g$ with the linear relationship between $T_g$ and $a_w$ allows a rapid and fairly reliable method for locating the $T_g$ for materials stored at various conditions before experimental verification of the $T_g$ is obtained. Such prediction is often important and needed in the selection of the temperature range to be used in analytical work and in locating the $T_g$ for materials with various water contents and analytical methods.

### 3. Water sorption and plasticization

Water sorption isotherms are important tools in the characterization of relationships between water content and water availability. Various equations can be fit to experimental sorption data (Chirife and Iglesias, 1978), but the most universal equation for the prediction of water content as a function of water activity has proved to be the GAB model (van den Berg *et al.*, 1975; van den Berg and Bruin, 1981).

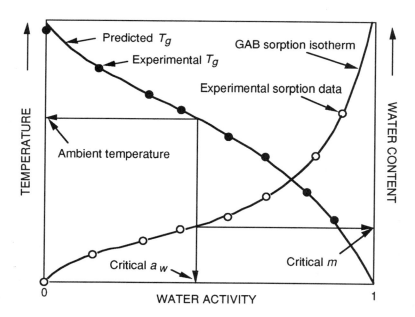

**Figure 6.3** A schematic representation of a modified state diagram showing depression of the glass transition temperature, $T_g$, with increasing water activity, $a_w$. The sorption isotherm shows the water content, $m$, as a function of water activity at ambient temperature. The predicted $T_g$ curve and the GAB sorption isotherm give the critical values for $a_w$ and $m$ that depress the $T_g$ to ambient temperature.

Prediction of the physical state of food materials is often based on modeling water plasticization, i.e., $T_g$ depression with increasing water content or water activity. $T_g$ determination as a function of water content allows establishing state diagrams that describe the concentration and temperature dependence of the physical state. However, sorption data are needed for the description of changes in the physical state that may occur during storage of low- and intermediate-moisture foods. Roos (1993) used the Gordon and Taylor equation and the GAB sorption model to predict water plasticization of various food components and foods. Modeling of water plasticization and water adsorption allowed prediction of food stability at various processing and storage conditions. The depression of the $T_g$ with increasing $a_w$ showed the water activity that was sufficient to depress the $T_g$ to below ambient temperature. Sorption isotherms may be used to predict the corresponding water content. Roos (1993) defined the water content and the water activity that depressed the $T_g$ to ambient temperature to be the critical $m$ and $a_w$, respectively. It may be important to notice that the state diagram can be considered to be a "map" for the selection of food processing, freezing, and frozen storage conditions (Slade and Levine, 1991), but the modified state diagram with water sorption data as shown in Figure 6.3 is more useful as a "map" for the selection of storage conditions for low- and intermediate-moisture foods. The difference between the utility of the diagrams is clearly evident in such applications as locating transition "zones" for cereal proteins (Kokini *et al.*, 1993, 1994) and evaluating critical $a_w$ values for product stability (Roos, 1993). However, the temperature dependence of the water activity must be taken into account in all modeling and interpretation of sorption data.

## III. Mechanical Properties and Flow

Mechanical properties of food solids may be related to phase transitions. It is obvious that first-order transitions such as melting of ice or melting of fat have a tremendous effect on food properties. Melting often occurs over a temperature range, which is an important property of food fats and spreads. The glass transition that affects the physical state of amorphous food materials occurs also over a temperature range, but its effect on the physical state is observed over a much larger temperature range. The main difference between melting and glass transition is that the first-order transition changes the solid material into the free-flowing liquid, but the transformation occurring over the glass transition temperature range is the change of the solid glass to the supercooled, viscous, liquid state. Modeling of mechanical properties and flow as a function of temperature is important to the textural properties of frozen foods and to fats and spreads. Modeling of the mechanical properties in amorphous foods has improved understanding of the basic phenomena, which govern various time-dependent changes that may considerably affect food processability, stability, and shelf life.

### A. Viscosity of Amorphous Foods

Mechanical properties of glassy materials are related to the freezing of the molecules below $T_g$. At temperatures above $T_g$, significant changes are observed with increasing temperature due to the dramatic increase of molecular mobility and decrease of relaxation times of mechanical changes. Although several mechanical properties can be measured above $T_g$, viscosity is probably the most important property that can be predicted and correlated with other mechanical properties above $T_g$.

1. Viscosity of frozen foods

Viscosity of the freeze-concentrated solute phase in frozen foods is an important factor that may affect time-dependent recrystallization phenomena, ice formation, and material properties in such processes as freeze-concentration and freeze-drying.

   Viscosity measurement of freeze-concentrated solute matrices is an extremely difficult and time-consuming task, but the formation of highly viscous amorphous states and vitrification have been recognized early (e.g., Rey, 1960). Bellows and King (1973) introduced the amorphous viscosity theory of collapse in freeze-drying. The theory was based on the fact that freezing of solutions separates water as pure ice and the noncrystalline solutes form a *concentrated amorphous solution*, which Bellows and King (1973) referred to as CAS. They used equation (6.13) to estimate the viscosity, which was sufficiently high to prevent flow, which allowed collapse during freeze-drying.

$$\tau_c = \frac{\mu r}{2\gamma} \tag{6.13}$$

   Reduced viscosity shortens the time allowing collapse, $\tau_c$, which therefore must be longer than the drying time, $\tau_d$, at that viscosity. Surface tension, $\gamma$, was taken as the driving force for the collapse of pores with radius,

$r$, that were left by the ice crystals. Typical values of surface tension and pore size diameter of frozen solutions predicted a CAS viscosity of $10^4$ to $10^7$ Pa s to allow collapse. Bellows and King (1973) pointed out that the viscosity was a strong function of concentration, which increased due to the Arrhenius-type temperature effect and increasing concentration as more water was separated as ice. The reported experimental viscosity data for sugar solutions between 5.5 and -30°C showed that the viscosity was a strong function of temperature and composition. Higher molecular weight sugars had higher viscosities than lower molecular weight sugars at the same temperature. The viscosities for sugars decreased in the order raffinose > sucrose > glucose > fructose. Although Bellows and King (1973) did not discuss the effect of glass transition on the viscosity of the freeze-concentrated solutes, it should be noticed that the $T'_g$ values also decrease in the same order.

The WLF-type temperature dependence of mechanical properties of freeze-concentrated materials has been emphasized by Levine and Slade (1986). The effect of $T_g$ on viscosity is probably the main factor that causes differences in the viscosity values of various compounds in freeze-concentrated materials as was reported by Bellows and King (1973). It has been well documented that the $T'_g$ values are dependent on molecular weight and they increase with increasing molecular weight (Levine and Slade, 1986; Roos and Karel, 1991f). Therefore, solutions of high molecular weight compounds, e.g., starch, become maximally freeze-concentrated at a fairly high temperature and only a few degrees below the melting temperature of pure water. Solutions of simple sugars in turn remain only partially freeze-concentrated well below typical temperatures of frozen food storage.

Levine and Slade (1988) pointed out that the retarding effect of added maltodextrins in ice cream on ice recrystallization is based on the elevation of $T'_g$. Levine and Slade (1988) suggested that mechanical properties of frozen materials follow the WLF-type temperature dependence above $T'_g$. It is obvious that the $T'_g$ has an effect on the viscosity of the unfrozen freeze-concentrated solute phase, but it is believed that the viscosity of the solute phase is controlled by $T_g$ rather than $T'_g$ above $T'_m$, as suggested by Roos and Karel (1991c,f) and Simatos and Blond (1993) and shown in Figure 6.4. Moreover, viscosity measurements of freeze-concentrated sugar solutions that were reported by Kerr and Reid (1994) suggested that the viscosity of freeze-concentrated solutions was governed by $T_g$ rather than by $T'_g$. The WLF-type temperature dependence above $T'_g$ predicts that the viscosity at the onset of ice melting is $10^8$ to $10^9$ Pa s, which agrees fairly well with the viscosity found by Bellows and King (1973) to allow collapse in freeze-drying. The onset of ice melting is also observed with other methods such as DMTA,

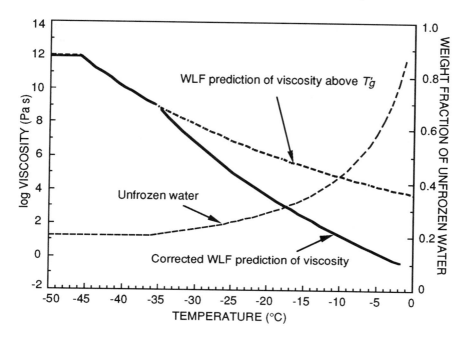

**Figure 6.4** Viscosity of freeze-concentrated sucrose solutions as predicted by the WLF temperature dependence above glass transition temperature, $T_g$. The glass transition temperature of the maximally freeze-concentrated solutes, $T_g'$ and onset temperature of ice melting within the maximally freeze-concentrated matrix, $T_m'$, of -46 and -34°C, respectively, were used in calculations (Roos and Karel, 1991c). The occurrence of ice melting above $T_m'$ increases the amount of unfrozen water, which causes dilution and depresses $T_g$ of the partially freeze-concentrated solute phase. The prediction of viscosity with $T_g'$ as the reference temperature underestimates the decrease of viscosity and the prediction should be based on $T - T_g$ corrected with dilution. The predicted viscosity limit for ice formation in frozen sucrose solutions is $10^8$ to $10^9$ Pa s.

ESR, and TMA, which show changes in mechanical properties, decreasing relaxation times, and increasing mobility above $T_m'$ due to increasing temperature above $T_g$ and ice melting (Maurice *et al.*, 1991; Schenz *et al.*, 1991; Le Meste and Huang, 1992; Hemminga *et al.*, 1993; MacInnes, 1993).

It may be assumed that at a sufficiently low temperature the viscosity of a freeze-concentrated solute matrix becomes high enough to retard diffusion and delay ice formation. Luyet and Rasmussen (1967) postulated that a viscosity of $10^8$ Pa s was sufficiently low to allow ice crystal growth in a relatively short time in poly(vinylpyrrolidone) solutions, which is about the same as the viscosity predicted by the WLF equation to allow ice melting in sucrose solutions, as shown in Figure 6.4. Levine and Slade (1989; Slade and Levine, 1991) have emphasized that ice formation in real time ceases at $T_g'$, since the high viscosity of the freeze-concentrated solute matrix prevents diffusion of a sufficient number of water molecules to the surface of ice crystal lattice and crystal growth. Viscosity seems to be one of the main factors that control ice formation in frozen solutions. Biological materials, especially highly concentrated sugar solutions, may be cooled rapidly to low temperatures without ice formation. In such materials ice formation is kinetically inhibited due to the formation of the glassy state. However, if the quenched solutions are heated to temperatures above $T_g$, ice formation may occur. Ice formation during rewarming is detrimental to frozen biological materials that suffer from freezing injury due to concentration effects and mechanical damage, causing loss of biological activity (e.g., MacFarlane, 1986). It should also be noticed that the Arrhenius equation, the Vogel-Tamman-Fulcher (VTF) equation, and the power-law equation can be fitted to experimental viscosity data (Kerr and Reid, 1994).

## 2. Viscosity of low-moisture foods

Viscosity is an important property of low-moisture foods, since it may be related to flow of amorphous food powders (Downton *et al.*, 1982) and other time-dependent changes in amorphous foods, e.g., crystallization of sugars, that affect storage stability and quality. Most studies that have reported viscosities for low-moisture foods have used amorphous carbohydrates as models.

*a. Effect of Thermal Plasticization.* Parks and Reagh (1937) pointed out that glucose exists in several glassy states due to the metastability and time-dependent nature of its physical properties. They found that annealing of amorphous glucose affected the viscosity of glucose glass. The viscosity in the glassy state was found to range from $10^9$ to $10^{14}$ Pa s, depending on annealing. These values agreed with the typical viscosity of $10^{12}$ Pa s of glassy materials. Williams *et al.* (1955) found that the viscosity of amorphous glucose agreed with viscosities of other glass-forming compounds. The viscosity of amorphous glucose followed similar-type temperature dependence above $T_g$ as amorphous polymers and inorganic materials. The WLF equation derived from the viscosity data of a number of compounds was concluded to be applicable to predict temperature dependence of relaxation times of mechanical changes, including viscosity, with the "universal" constants obtained.

Soesanto and Williams (1981) analyzed experimental data for aqueous mixtures of an amorphous blend of fructose and sucrose (1:7) that was considered to correspond to the typical solids composition of fruit juices. The results showed that the viscosity decreased above $T_g$ and that the WLF model with the universal constants fitted well to the experimental viscosity data. The WLF equation with the universal constants has also been found to predict viscosity of other amorphous sugars, e.g., viscosity of amorphous sucrose and sucrose solutions above $T_g$ (Roos and Karel, 1991c). It is likely that the temperature dependence of viscosity below $T_g$ and at temperatures above about $T_g + 100°C$ follows the Arrhenius model (Levine and Slade, 1986, 1989; Slade and Levine, 1991). However, it should be emphasized that the universal coefficients reported by Williams *et al.* (1955) are not always valid and should be used only when sufficient experimental data are not available to derive specific constants for the material studied (Ferry, 1980; Peleg, 1992).

The viscosity data of amorphous sugars have been found to follow not only the WLF model but also several other models relating viscosity and temperature. Angell *et al.* (1982) found that the VTF equation could be fitted to the viscosity data for sorbitol over the whole temperature range from glass transition temperature to the melting temperature. They pointed out that the

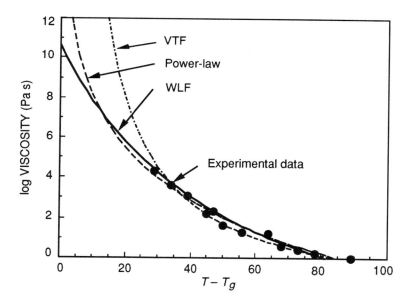

**Figure 6.5** Comparison of viscosity predicted for amorphous sucrose with various models. The experimental data shown are viscosity values for 70 and 75% (w/w) sucrose solutions reported by Bellows and King (1973). The glass transition data and WLF prediction of viscosity are from Roos and Karel (1991c). The VTF and power-law predictions were obtained by fitting the models to the experimental data.

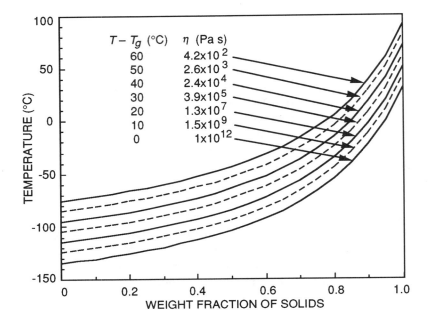

**Figure 6.6** Isoviscosity states of glucose above glass transition temperature, $T_g$, at various water contents. The viscosity, $\eta$, at $T_g$ was assumed to be $10^{12}$ Pa s and to decrease with increasing temperature as predicted by the WLF equation with the "universal" constants.

behavior of fitting of the VTF model over such a wide temperature range of experimental data was atypical of most molecular liquids, but probably a property of polyalcohols such as glycerol and butanediol in addition to sorbitol. Viscosity data of supercooled fructose and glucose were also analyzed by Ollett and Parker (1990). They used the VTF, WLF, and power-law equations and found that the estimates obtained for $T_g$ and $\eta_g$ were relatively insensitive to which equation was used. An Arrhenius plot of the viscosity data showed nonlinearity, which suggested that the temperature dependence of viscosity over the whole temperature range measured was not described by the Arrhenius relationship.

The viscosity data for concentrated sucrose solutions that were reported by Bellows and King (1973) together with measured $T_g$ data have also been found to follow the WLF temperature dependence (Roos and Karel, 1991c). A comparison of the viscosity predictions for amorphous sucrose by the power-law, VTF, and WLF equations is shown in Figure 6.5. The power-law and VTF equations overestimated the viscosity at low $T - T_g$ values. However, the equations fit the data fairly well over the experimental temperature range and indicate the difficulty of extrapolating viscosity data in the vicinity of $T_g$.

Viscosity data are available for only a few sugars above $T_g$ and data for real amorphous food materials are almost nonexistent. The similarities between the physical properties of various materials, however, have revealed that the relaxation times of changes in foods may follow the same type of temperature dependence above $T_g$ as occurs in other amorphous materials, e.g., synthetic polymers (Williams *et al.*, 1955; Slade and Levine, 1991). The use of the WLF and other models in predicting temperature dependence of viscosity allows establishing of state diagrams that show isoviscosity states above $T_g$ as a function of composition. Such diagrams, as shown in Figure 6.6, may be used in the evaluation of compositional effects on relaxation times at a constant temperature or in establishing critical temperatures for the stability of amorphous materials.

*b. Effects of Water Plasticization.* Most amorphous low-moisture foods are stored at a relatively narrow temperature range, although the materials may have been exposed to high temperatures during various manufacturing processes such as dehydration and extrusion. Food materials are plasticized by both temperature and water. Water at a constant temperature may affect the physical properties of amorphous foods similarly as temperature changes material properties at a constant water content.

The glass transition temperature of amorphous food materials is governed by water content as has been clearly demonstrated by the state diagrams. Both the thermal and water plasticization of food materials are based on the increase in free volume (Ferry, 1980). The state diagram with isoviscosity lines (Figure 6.6) can be used to predict the effect of both temperature and water content on the physical state. Assuming that the WLF-type temperature dependence applies, the viscosity at a constant water content decreases with increasing temperature. At a constant temperature an increase in water content depresses the $T_g$, which is observed from decreasing viscosity. In Figure 6.6 the isoviscosity lines can be used to quantify changes in viscosity that may result from either thermal or compositional factors or both. Viscosity data shown in Figure 6.6 emphasize the fact that small changes in temperature or water content in either direction above, but in the vicinity of $T_g$, affect the viscosity by orders of magnitude. Such changes in viscosity or relaxation times of other mechanical properties may have dramatic effects on the physical state of low- and intermediate-moisture foods.

Roos (1994) used the WLF equation to predict viscosity of amorphous sucrose as a function of water content at a constant temperature. In such modeling the viscosity may be assumed to be constant and typical of the glassy state when the water content is not sufficient to depress the $T_g$ to below ambient temperature. Above the critical $m$, the $T_g$ is located below ambient temperature and the viscosity may be obtained with the WLF equation and the proper $T - T_g$ (Figure 6.7). Such modeling can be used to predict the effect of water content on the mechanical properties of amorphous foods at several temperatures without sorption data, which may be useful in predicting viscosity at elevated temperatures during processing, e.g., extrusion of amorphous food materials.

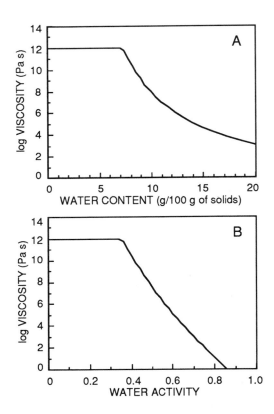

**Figure 6.7** Predicted viscosity for skim milk powder at 24°C. A. Water sorption data was used to predict water content, $m$, with the GAB model. $T_g$ was predicted with the Gordon and Taylor equation. The WLF equation with the "universal" constants was used to obtain viscosity as a function of $m$. Viscosity at temperatures above 24°C was assumed to be $10^{12}$ Pa s. B. The GAB relationship between water activity, $a_w$, and $m$ was applied to show the viscosity prediction as a function of $a_w$. The decrease of viscosity occurs as the $T_g$ is depressed to below ambient temperature at critical $m$ and $a_w$, which. for skim milk, were found to be 7.6 g/100 g of solids and 0.37, respectively (Jouppila and Roos, 1994a,b).

Water activity is often a more important indicator of food stability than water content. This applies especially to amorphous food storage, since small changes in storage relative humidity may have an extensive plasticization effect and cause loss of stability. The effect of water plasticization can be expressed in terms of $a_w$. The effect of water on $T_g$ may be first predicted with the Couchman and Karasz or the Gordon and Taylor equation. Prediction of the effect of $a_w$ on the $T_g$ requires application of the sorption isotherm of the material. The sorption isotherm should be predicted by the GAB model or some other sorption model that has proved to fit experimental data. Both modeling of water plasticization and sorption properties should be done before predicting the effects of $a_w$ on viscosity of a particular low- or intermediate-moisture food.

Figure 6.7 shows the WLF prediction of viscosity for skim milk powder as a function of water content and water activity. At low water contents the viscosity is that typical of the glassy state. An obvious decrease in viscosity occurs as the water content becomes sufficiently high to depress the $T_g$ to below ambient temperature due to plasticization. The decrease of viscosity over a fairly narrow water content range is orders of magnitude, which results from the transformation of the solid material into the free-flowing liquid state. The depression of the viscosity with increasing $a_w$ is even more dramatic, which explains the high sensitivity of the physical state of a number of carbohydrate foods to storage relative humidity. Plasticization above the critical $a_w$ changes relaxation times of mechanical changes rapidly and results in quality changes, e.g., stickiness and loss of crispness.

## B. Viscoelastic Properties

The physical state of food components governs mechanical properties of food materials. In fats and oils the relative amounts of liquid and solid materials can be used to characterize their response to a mechanical stress. In such materials the behavior is related to the first-order transition, which depends on temperature and on the polymorphic form. In amorphous foods, which will be discussed here, the physical state is related to the glass transition and changes in the viscoelastic behavior may be characterized in relation to the $T_g$. Relationships between the physical state and viscoelastic properties have been fairly well established for synthetic polymers (e.g., Ferry, 1980), but only a few studies have used the polymer science approach to predict the viscoelastic behavior of food materials (e.g., Slade and Levine, 1991; Kalichevsky *et al.*, 1993c).

1. Relaxation time and time-temperature superposition principle

Kalichevsky *et al.* (1993c) stated that dynamic mechanical techniques are used to characterize molecular motions and relaxation behavior of viscoelastic materials. They emphasized the utility of such techniques to study molecular motions which, in food materials, give rise to the glass transition and relaxations below $T_g$.

Relaxations in viscoelastic materials occur by molecular motions, which are temperature-dependent. The relaxation time of a polymer is a measure of the time required by molecules to respond to an external stress at a constant temperature. One of the main peculiarities of the WLF equation is that it provides a logarithmic relationship between time and temperature, as was pointed out by Sperling (1986). A simple relationship between relaxation time, $\tau$, viscosity, $\eta$, and modulus, $E$, as given in equation (6.14) can be defined for the Maxwell element, which is composed of a spring (modulus) and dashpot (viscosity) in series.

$$\tau = \frac{\eta}{E} \tag{6.14}$$

The relaxation time definition for the Maxwell element defines that the modulus decreases linearly with time, which is not valid for viscoelastic materials. The Kelvin element is a parallel combination of a spring and a dashpot. Its response to an external stress is defined by retardation time. The stress, $\sigma$, for the Kelvin element is defined by equation (6.15), which states that the stress is due to both the viscous and elastic components.

$$\sigma = \eta \frac{d\varepsilon}{dt} + E\varepsilon \tag{6.15}$$

The stress of the elastic component is a linear function of strain, $\varepsilon$, but the response of the viscous component depends on time. Under constant stress the strain is defined by equation (6.16), which may be written into the form of equation (6.17), where $\tau$ is retardation time.

$$\varepsilon = \frac{\sigma}{E}\left[1 - e^{-(E/\eta)t}\right] \tag{6.16}$$

$$\varepsilon = \frac{\sigma}{E}\left(1 - e^{-t/\tau}\right) \tag{6.17}$$

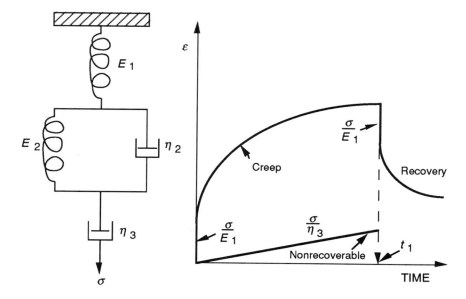

**Figure 6.8**   The combined, four-element Maxwell and Kelvin element model of viscoelastic behavior. The predicted creep behavior of the model shows that after the initial strain due to component 1 the creep results from the Kelvin element and component 3. Relaxation of the stress at time $t_1$, component 1 shows recovery followed by time-dependent recovery of component 2 and partial recovery of component 3. After Sperling (1986).

Combination of the Maxwell and Kelvin elements provides the most simple model for viscoelastic behavior as shown in Figure 6.8. The strain of the combined, four-component model is defined by equations (6.18) and (6.19), where the subscripts 1, 2, and 3 refer to the components 1, 2, and 3 as defined in Figure 6.8.

$$\varepsilon = \varepsilon_1 + \varepsilon_2 + \varepsilon_3 \qquad (6.18)$$

$$\varepsilon = \frac{\sigma}{E_1} + \frac{\sigma}{E_2}\left(1 - e^{-t/\tau_2}\right) + \frac{\sigma}{\eta_3}t \qquad (6.19)$$

The temperature dependence of relaxation times results from increasing molecular mobility with increasing temperature. The time-temperature superposition principle assumes that relaxation data obtained at one temperature can be shifted along the time axis to another temperature. Thus, a typical master curve of an amorphous material can be established at one temperature based on data obtained at other temperatures (Ferry, 1980; Sperling, 1986). The master curve is established using a shift factor that may assume WLF temperature dependence of relaxation times above $T_g$. Ferry (1980) showed that the time-temperature superposition principle applies to storage modulus, loss modulus, and other modulus functions that are also used in the characterization of amorphous food materials (e.g., Cocero, 1993; Kalichevsky *et al.*, 1993c). According to Ferry (1980) the modulus measured at frequency $f$ at temperature $T$ is equivalent to the modulus measured at frequency $fa_T$ at a reference temperature, $T_0$, where $a_T$ refers to the ratio of relaxation times at $T$ and $T_0$. The application of the time-temperature superposition principle based on the WLF temperature dependence of molecular relaxation processes in the characterization of the viscoelastic behavior of amorphous food materials has been emphasized by Levine and Slade (e.g., 1989; Slade and Levine, 1991).

## 2. Master curves of biological materials

Master curves are reported for only a few biopolymers (Nakamura and Tobolsky, 1967; Kalichevsky *et al.*, 1993c) and data for food materials are scarce. However, master curves of polymers are useful in the description of the changes in physicochemical properties that occur in their amorphous state.

A typical master curve of polymers is based on collecting modulus data at several temperatures, e.g., with the stress relaxation method, which measures the modulus as a function of time after deformation. Such measurement of the time dependence of modulus is equivalent to the measurement of modulus as a function of frequency at a constant temperature (Kalichevsky *et al.*, 1993c). The modulus values obtained may then be shifted to another temperature, e.g., 25°C, which predicts the modulus as a function of time or frequency at that temperature. A schematic representation of the change of modulus as a function of time at various temperatures is shown in Figure 6.9. Such curves can be used to predict time- and temperature-dependent behavior of amorphous, viscoelastic materials.

The master curve of modulus of a viscoelastic, amorphous material shows the five regions of viscoelastic behavior. A typical master curve is shown in Figure 6.10. The master curve of modulus is useful in the interpretation of experimental data of various time-dependent changes of the physical properties of biopolymers and foods. It should also be noticed that various mechanical properties of food materials, when measured as a function of frequency, time, temperature, water content, or water activity, should produce a modulus curve that reflects the physical state. The modulus data may also

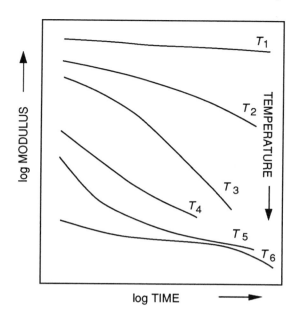

**Figure 6.9** A schematic representation of time and temperature dependence of modulus of amorphous materials. Relaxations in the glassy state at a low temperature, $T_1$, are slow. An increasing temperature increases the rate of relaxation, i.e., the modulus decreases with time. The maximum relaxation rate is observed within the glass transition temperature range at temperature $T_3$. The rate decreases again when the "rubbery plateau" region is approached. At high temperatures or long times as shown for temperature $T_6$ flow of the material may be observed.

be determined at various temperatures and transformed to a master curve using the superpositioning principle. Such curves could be used to show food behavior at one temperature, e.g., at food storage temperature, that would be directly applicable to predict mechanical properties in practical applications. Although sufficient data to establish relationships between moduli at various frequency, time, temperature, water content, and water activity conditions of food materials have not been reported, the mechanical behavior of gluten (Attenburrow *et al.*, 1990), elastin (Lillie and Gosline, 1990, 1993), and cereal-based foods (Attenburrow and Davies, 1993) suggests that the superpositioning principle may apply. The loss and storage modulus data of cereal proteins obtained as a function of temperature were successfully used in the characterization of mechanical properties of cereal proteins by Cocero (1993) and Kokini *et al.* (1993, 1994). A typical value of Young's modulus of

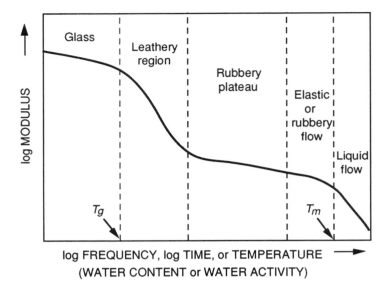

**Figure 6.10** A typical master curve of the modulus of amorphous materials showing the five regions of viscoelastic behavior. The master curve at constant temperature may be obtained by superpositioning of modulus data obtained at various temperatures as a function of frequency or time. For water-plasticized food materials the modulus curve may also be obtained at a constant temperature as a function of water content or water activity. A typical value for Young's modulus in the glassy state is $3 \times 10^9$ Pa. Adapted from Levine and Slade (1990).

glassy polymers is $3 \times 10^9$ Pa, which applies to a number of amorphous polymers (Sperling, 1986) and probably to amorphous foods.

Some amorphous food components exist in a partially crystalline state, which shows different thermomechanical properties from the completely amorphous materials. According to Sperling (1986) the presence of crystalline regions in polymers decreases the depression of the modulus within the leathery region and extends the rubbery plateau region until liquid flow occurs above the melting temperature. Cross-linking of polymers extends the rubbery plateau region as shown in Figure 6.11. Both the behavior of partially crystalline polymers and cross-linking of polymers may affect the thermal behavior of wheat bread dough due to the presence of the partially crystalline starch fraction and the thermosetting protein fraction. However, the thermal behavior is also affected by gelatinization.

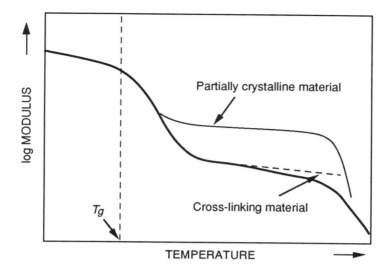

**Figure 6.11** A schematic representation of modulus curves typical of partially crystalline and cross-linking materials as a function of temperature in comparison with the modulus curve of completely amorphous, viscoelastic materials.

## 3. Effect of molecular weight

Molecular weight is one of the main factors that affects the glass transition temperature of anhydrous food components. It may be assumed that food materials, although they are complicated mixtures of solids and water, are characterized by an "effective" molecular weight. Within the glass transition temperature range and the leathery region the effect of molecular weight on the viscoelastic properties of polymers is quite low (Ferry, 1980). The molecular weight and molecular weight distribution, however, have a strong influence on the viscoelastic behavior and the extent of the rubbery plateau region (Ferry, 1980; Sperling, 1986).

The effect of molecular weight on viscoelastic properties of polymers can be observed from the modulus curve. However, it should be remembered that food solids represent a wide range of molecular weights and molecular weight distributions. The viscoelastic behavior of low molecular weight food solids, e.g., those of fruits and vegetables that are composed mostly of sugars, is probably significantly different from food solids that are composed of high molecular weight carbohydrates or proteins. The viscoelastic behavior of low molecular weight amorphous solids such as glucose was discussed by Ferry (1980). A comparison of the modulus curves between low and high

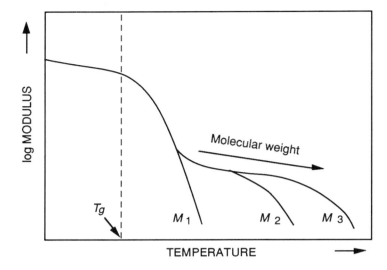

**Figure 6.12** Effect of molecular weight on the rubbery plateau region of the modulus curve. The temperature range of the rubbery plateau region of amorphous polymers increases with increasing molecular weight. For low molecular weight compounds such as amorphous sugars the rubbery plateau region is almost nonexistent.

molecular weight amorphous materials is shown in Figure 6.12. Ferry (1980) pointed out that the viscoelastic properties of glucose can be, in large part, described with a single Maxwell element.

The effect of molecular weight on viscoelastic behavior over the rubbery plateau region is an important factor in the determination of material properties as was discussed by Ferry (1980). Cocero (1993) used storage modulus data of glutenin at various water contents in the analysis of its thermosetting behavior. The analysis was based on the determination of the molecular weight between cross-links, $M_C$, which revealed that $M_C$ decreased during thermosetting that occurred above $T_g$.

Water as a plasticizer of food materials affects the effective molecular weight and the viscoelastic behavior. Kalichevsky *et al.* (1992b) studied the effect of water on the viscoelastic behavior of gluten. The decrease of storage modulus observed at glass transition was found to shift to lower temperatures with increasing water content due to plasticization and depression of the $T_g$. The modulus in the glassy state was almost constant, being about $10^9$ Pa at all water contents, but its value with increasing water content was depressed at the rubbery plateau region. The transition at low water contents was also broader than at high water contents. The effect of water on the storage

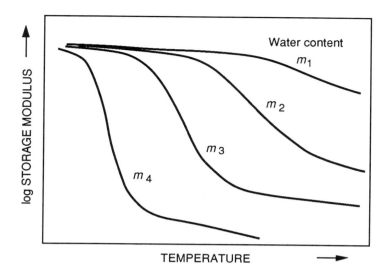

**Figure 6.13** A schematic representation of storage modulus of water-plasticizable biopolymers as a function of temperature. The glass transition temperature, $T_g$, decreases with increasing water content. The depression of the storage modulus above $T_g$ increases with increasing water content, $m$ (water content increases in the order $m_1 > m_2 > m_3 > m_4$), and occurs more steeply with increasing $m$.

modulus is shown in Figure 6.13. Water has similar effects on the viscoelastic behavior of amylopectin-gluten mixtures (Kalichevsky and Blanshard, 1992) and on amylopectin (Kalichevsky *et al.*, 1993a). It should be noticed that the drop in modulus as a function of water activity occurs fairly steeply as the $T_g$ is depressed to below ambient temperature due to the tremendous depression of $T_g$ with increasing $a_w$.

The effective molecular weight of foods can be a complicated result of the molecular weights and properties of component compounds, which may all affect the viscoelastic behavior of amorphous foods above $T_g$. The modulus of amylopectin and gluten has been found to be affected by simple sugars in addition to water. Addition of glucose to gluten or fructose to amylopectin may decrease the modulus at the rubbery plateau (Kalichevsky *et al.*, 1992b, 1993a). The decrease in modulus increases with increasing sugar content. However, fairly few data are available on the effects of various food components on their viscoelastic properties.

# References

Angell, C.A., Stell, R.C. and Sichina, W. 1982. Viscosity-temperature function for sorbitol from combined viscosity and differential scanning calorimetry studies. *J. Phys. Chem. 86*: 1540-1542.

Attenburrow, G. and Davies, A.P. 1993. The mechanical properties of cereal based foods in and around the glassy state. Chpt. 16 in *The Glassy State in Foods*, ed. J.M.V. Blanshard and P.J. Lillford. Nottingham University Press, Loughborough, pp. 317-331.

Attenburrow, G., Barnes, D.J., Davies, A.P. and Ingman, S.J. 1990. Rheological properties of wheat gluten. *J. Cereal Sci. 12*: 1-14.

Bellows, R.J. and King, C.J. 1973. Product collapse during freeze drying of liquid foods. *AIChE Symp. Ser. 69*: 33-41.

Chirife, J. and Iglesias, H.A. 1978. Equations for fitting water sorption isotherms of foods: Part 1 - A review. *J. Food Technol. 13*: 159-174.

Cocero, A.M. 1993. The rheology and phase transitions of wheat glutenin. Ph.D. thesis, Rutgers - The State University of New Jersey, New Brunswick, NJ.

Couchman, P.R. 1978. Compositional variation of glass-transition temperatures. 2. Application of the thermodynamic theory to compatible polymer blends. *Macromolecules 11*: 1156-1161.

Couchman, P.R. and Karasz, F.E. 1978. A classical thermodynamic discussion of the effect of composition on glass-transition temperatures. *Macromolecules 11*: 117-119.

Downton, G.E., Flores-Luna, J.L. and King, C.J. 1982. Mechanism of stickiness in hygroscopic, amorphous powders. *Ind. Eng. Chem. Fundam. 21*: 447-451.

Ferry, J.D. 1980. *Viscoelastic Properties of Polymers*, 3rd ed. John Wiley & Sons, New York.

Flory, P.J. 1953. *Principles of Polymer Chemistry*. Cornell University Press, Ithaca, NY.

Fox, T.G. 1956. Influence of diluent and of copolymer composition on the glass temperature of a polymer system. *Bull. Am. Phys. Soc. 1*: 123.

Gordon, M. and Taylor, J.S. 1952. Ideal copolymers and the second-order transitions of synthetic rubbers. I. Non-crystalline copolymers. *J. Appl. Chem. 2*: 493-500.

Hancock, B.C. and Zografi, G. 1994. The relationship between the glass transition temperature and the water content of amorphous pharmaceutical solids. *Pharm. Res. 11*: 471-477.

Hemminga, M.A., Roozen, M.J.G.W. and Walstra, P. 1993. Molecular motions and the glassy state. Chpt. 7 in *The Glassy State in Foods*, ed. J.M.V. Blanshard and P.J. Lillford. Nottingham University Press, Loughborough, pp. 157-187.

Hoseney, R.C., Zeleznak, K. and Lai, C.S. 1986. Wheat gluten: A glassy polymer. *Cereal Chem. 63*: 285-286.

Huang, V.T. 1993. Mathematical modeling on the glass transition curves of glassy polymers. Paper No. 911. Presented at the Annual Meeting of the Institute of Food Technologists, Chicago, July 10-14.

Jouppila, K. and Roos, Y.H. 1994a. Water sorption and time-dependent phenomena of milk powders. *J. Dairy Sci. 77*: 1798-1808.

Jouppila, K. and Roos, Y.H. 1994b. Glass transitions and crystallization in milk powders. *J. Dairy Sci. 77*: 2907-2915.

Jouppila, K., Ahonen, T. and Roos, Y. 1995. Water adsorption and plasticization of amylopectin glasses. In *Food Macromolecules and Colloids*, ed. E. Dickinson, The Royal Society of Chemistry, London. In press.

Kalichevsky, M.T. and Blanshard, J.M.V. 1992. A study of the effect of water on the glass transition of 1:1 mixtures of amylopectin, casein and gluten using DSC and DMTA. *Carbohydr. Polym. 19*: 271-278.

Kalichevsky, M.T. and Blanshard, J.M.V. 1993. The effect of fructose and water on the glass transition of amylopectin. *Carbohydr. Polym. 20*: 107-113.

Kalichevsky, M.T., Jaroszkiewicz, E.M., Ablett, S., Blanshard, J.M.V. and Lillford, P.J. 1992a. The glass transition of amylopectin measured by DSC, DMTA and NMR. *Carbohydr. Polym. 18*: 77-88.

Kalichevsky, M.T., Jaroszkiewicz, E.M. and Blanshard, J.M.V. 1992b. Glass transition of gluten. 1: Gluten and gluten-sugar mixtures. *Int. J. Biol. Macromol. 14*: 257-266.

Kalichevsky, M.T., Jaroszkiewicz, E.M. and Blanshard, J.M.V. 1993a. A study of the glass transition of amylopectin-sugar mixtures. *Polymer 34*: 346-358.

Kalichevsky, M.T., Blanshard, J.M.V. and Tokarczuk, P.F. 1993b. Effect of water content and sugars on the glass transition of casein and sodium caseinate. *Int. J. Food Sci. Technol. 28*: 139-151.

Kalichevsky, M.T., Blanshard, J.M.V. and Marsh, R.D.L. 1993c. Applications of mechanical spectroscopy to the study of glassy biopolymers and related systems. Chpt. 6 in *The Glassy State in Foods*, ed. J.M.V. Blanshard and P.J. Lillford. Nottingham University Press, Loughborough, pp. 133-156.

Karel, M. 1973. Recent research and development in the field of low-moisture and intermediate-moisture foods. *CRC Crit. Rev. Food Technol. 3*: 329-373.

Karel, M., Buera, M.P. and Roos, Y. 1993. Effects of glass transitions on processing and storage. Chpt. 2 in *The Glassy State in Foods*, ed. J.M.V. Blanshard and P.J. Lillford. Nottingham University Press, Loughborough, pp. 13-34.

Katz, E.E. and Labuza, T.P. 1981. Effect of water activity on the sensory crispness and mechanical deformation of snack food products. *J. Food Sci. 46*: 403-409.

Kerr, W.L. and Reid, D.S. 1994. Temperature dependence of the viscosity of sugar and maltodextrin solutions in coexistence with ice. *Lebensm.-Wiss. u. -Technol. 27*: 225-231.

Kokini, J.L., Cocero, A.M., Madeka, H. and de Graaf, E. 1993. Order-disorder transitions and complexing reactions in cereal proteins and their effect on rheology. In *Proceedings of International Symposium Advances in Structural and Heterogeneous Continua*, Moscow, August 1993. In press.

Kokini, J.L., Cocero, A.M., Madeka, H. and de Graaf, E. 1994. The development of state diagrams for cereal proteins. *Trends Food Sci. Technol. 5*: 281-288.

Labuza, T.P., Tannenbaum, S.R. and Karel, M. 1970. Water content and stability of low-moisture and intermediate-moisture foods. *Food Technol. 24*: 543-544, 546-548, 550.

Laine, M.J.K. and Roos, Y. 1994. Water plasticization and recrystallization of starch in relation to glass transition. In *Proceedings of the Poster Session, International Symposium on the Properties of Water, Practicum II*, ed. A. Argaiz, A. López-Malo, E. Palou and P. Corte. Universidad de las Américas-Puebla, pp. 109-112.

Le Meste, M. and Huang, V. 1992. Thermomechanical properties of frozen sucrose solutions. *J. Food Sci. 57*: 1230-1233.

Levine, H. and Slade, L. 1986. A polymer physico-chemical approach to the study of commercial starch hydrolysis products (SHPs). *Carbohydr. Polym. 6*: 213-244.

Levine, H. and Slade, L. 1988. "Collapse" phenomena - A unifying concept for interpreting the behaviour of low moisture foods. Chpt. 9 in *Food Structure - Its Creation and Evaluation*," ed. J.M.V. Blanshard and J.R. Mitchell. Butterworths, London, pp. 149-180.

Levine, H. and Slade, L. 1989. A food polymer science approach to the practice of cryostabilization technology. *Comments Agric. Food Chem. 1*: 315-396.

Levine, H. and Slade, L. 1990. Influences of the glassy and rubbery states on the thermal, mechanical, and structural properties of doughs and baked products. Chpt. 5 in *Dough Rheology and Baked Product Texture*, ed. H. Faridi and J.M. Faubion. AVI Publishing Co., New York, pp. 157-330.

Lillie, M.A. and Gosline, J.M. 1990. The effects of hydration on the dynamic mechanical properties of elastin. *Biopolymers 29*: 1147-1160.

Lillie, M.A. and Gosline, J.M. 1993. The effects of swelling solvents on the glass transition in elastin and other proteins. Chpt. 14 in *The Glassy State in Foods*, ed. J.M.V. Blanshard and P.J. Lillford. Nottingham University Press, Loughborough, pp. 281-301.

Lin, A.A., Kwei, T.K. and Reiser, A. 1989. On the physical meaning of the Kwei equation for the glass transition temperature of polymer blends. *Macromolecules 22*: 4112-4119.

Luyet, B. and Rasmussen, D. 1967. Study by differential thermal analysis of the temperatures of instability in rapidly cooled solutions of polyvinylpyrrolidone. *Biodynamica 10(209)*: 137-147.

MacFarlane, D.R. 1986. Devitrification in glass-forming aqueous solutions. *Cryobiology 23*: 230-244.

MacInnes, W.M. 1993. Dynamic mechanical thermal analysis of sucrose solutions. Chpt. 11 in *The Glassy State in Foods*, ed. J.M.V. Blanshard and P.J. Lillford. Nottingham University Press, Loughborough, pp. 223-248.

Maurice, T.J., Asher, Y.J. and Thomson, S. 1991. Thermomechanical analysis of frozen aqueous systems. In *Water Relationships in Foods*, ed. H. Levine and L. Slade. Plenum Press, New York, pp. 215-223.

Nakamura, S. and Tobolsky, A.V. 1967. Viscoelastic properties of plasticized amylose films. *J. Appl. Polym. Sci. 11*: 1372-1386.

Nelson, K.A. and Labuza, T.P. 1994. Water activity and food polymer science: Implications of state on Arrhenius and WLF models in predicting shelf life. *J. Food Eng. 22*: 271-289.

Ollett, A.-L. and Parker, R. 1990. The viscosity of supercooled fructose and its glass transition temperature. *J. Texture Stud. 21*: 355-362.

Orford, P.D., Parker, R., Ring, S.G. and Smith, A.C. 1989. Effect of water as a diluent on the glass transition behaviour of malto-oligosaccharides, amylose and amylopectin. *Int. J. Biol. Macromol. 11*: 91-96.

Parks, G.S. and Reagh, J.D. 1937. Studies on glass. XV. The viscosity and rigidity of glucose glass. *J. Chem. Phys. 5*: 364-367.

Peleg, M. 1992. On the use of the WLF model in polymers and foods. *Crit. Rev. Food Sci. Nutr. 32*: 59-66.

Rey, L.R. 1960. Thermal analysis of eutectics in freezing solutions. *Ann. N.Y. Acad. Sci. 85*: 510-534.

Roos, Y.H. 1987. Effect of moisture on the thermal behavior of strawberries studied using differential scanning calorimetry. *J. Food Sci. 52*: 146-149.

Roos, Y.H. 1993. Water activity and physical state effects on amorphous food stability. *J. Food Process. Preserv. 16*: 433-447.

Roos, Y.H. 1994. Water activity and glass transition temperature. How do they complement and how do they differ. In *ISOPOW Practicum II: Food Preservation by Moisture Control*, ed. J. Welti-Chanes. Technomic Publishing Co., Lancaster, PA. In press.

Roos, Y. and Karel, M. 1990. Differential scanning calorimetry study of phase transitions affecting the quality of dehydrated materials. *Biotechnol. Prog. 6*: 159-163.

Roos, Y. and Karel, M. 1991a. Applying state diagrams to food processing and development. *Food Technol. 45*: 66, 68-71, 107.

Roos, Y. and Karel, M. 1991b. Water and molecular weight effects on glass transitions in amorphous carbohydrates and carbohydrate solutions. *J. Food Sci. 56*: 1676-1681.

Roos, Y. and Karel, M. 1991c. Amorphous state and delayed ice formation in sucrose solutions. *Int. J. Food Sci. Technol. 26*: 553-566.

Roos, Y. and Karel, M. 1991d. Nonequilibrium ice formation in carbohydrate solutions. *Cryo-Lett. 12*: 367-376.

Roos, Y. and Karel, M. 1991e. Phase transitions of mixtures of amorphous polysaccharides and sugars. *Biotechnol. Prog. 7*: 49-53.

Roos, Y. and Karel, M. 1991f. Plasticizing effect of water on thermal behavior and crystallization of amorphous food models. *J. Food Sci. 56*: 38-43.

Schenz, T.W., Israel, B. and Rosolen, M.A. 1991. Thermal analysis of water-containing systems. In *Water Relationships in Foods*, ed. H. Levine and L. Slade. Plenum Press, New York, pp. 199-214.

Simatos, D. and Blond, G. 1993. Some aspects of the glass transition in frozen foods systems. Chpt. 19 in *The Glassy State in Foods*, ed. J.M.V. Blanshard and P.J. Lillford. Nottingham University Press, Loughborough, pp. 395-415.

Simatos, D. and Karel, M. 1988. Characterization of the condition of water in foods - physico-chemical aspects. In *Food Preservation by Water Activity Control*, ed. C.C. Seow. Elsevier, Amsterdam, pp. 1-41.

Slade, L. and Levine, H. 1988. Structural stability of intermediate moisture foods - A new understanding? Chpt. 8 in *Food Structure - Its Creation and Evaluation*, ed. J.M.V. Blanshard and J.R. Mitchell. Butterworths, London, pp. 115-147.

Slade, L. and Levine, H. 1991. Beyond water activity: Recent advances based on an alternative approach to the assessment of food quality and safety. *Crit. Rev. Food Sci. Nutr. 30*: 115-360.

Slade, L. and Levine, H. 1995. Glass transitions and water-food structure interactions. *Adv. Food Nutr. Res. 38*. In press.

Soesanto, T. and Williams, M.C. 1981. Volumetric interpretation of viscosity for concentrated and dilute sugar solutions. *J. Phys. Chem. 85*: 3338-3341.

Sperling, L.H. 1986. *Introduction to Physical Polymer Science*. John Wiley & Sons, New York.

Sugisaki, M., Suga, H. and Seki, S. 1968. Calorimetric study of the glassy state. IV. Heat capacities of glassy water and cubic ice. *Bull. Chem. Soc., Jpn. 41*: 2591-2599.

To, W.C. and Flink, J.M. 1978. 'Collapse,' a structural transition in freeze dried carbohydrates. II. Effect of solute composition. *J. Food Technol. 13*: 567-581.

Tsourouflis, S., Flink, J.M. and Karel, M. 1976. Loss of structure in freeze-dried carbohydrates solutions: Effect of temperature, moisture content and composition. *J. Sci. Food Agric. 27*: 509-519.

van den Berg, C. and Bruin, S. 1981. Water activity and its estimation in food systems: Theoretical aspects. In *Water Activity: Influences on Food Quality*, ed. L.B. Rockland and G.F. Stewart. Academic Press, New York, pp. 1-61.

van den Berg, C., Kaper, F.S., Weldring, J.A.G. and Wolters, I. 1975. Water binding by potato starch. *J. Food Technol. 10*: 589-602.

White, G.W. and Cakebread, S.H. 1966. The glassy state in certain sugar-containing food products. *J. Food Technol. 1*: 73-82.

Williams, M.L., Landel, R.F. and Ferry, J.D. 1955. The temperature dependence of relaxation mechanisms in amorphous polymers and other glass-forming liquids. *J. Am. Chem. Soc. 77*: 3701-3707.

Zeleznak, K.J. and Hoseney, R.C. 1987. The glass transition in starch. *Cereal Chem. 64*: 121-124.

# CHAPTER 7

# *Time-Dependent Phenomena*

## I.  Introduction

Food materials exist typically in a metastable, nonequilibrium state, which is subject to various time-dependent changes. The rate of these changes often depends on the physical state and therefore on temperature and water content. In foods having low water contents stability can be attained at normal storage conditions if the water content is kept below a critical value. Water contents above the critical value are known to increase deterioration, which is caused by changes in structure or by chemical changes that may result from an increase in enzymatic activity or nonenzymatic browning. The growth of microorganisms is also time-dependent, but microbiological deterioration is often governed by environmental factors such as temperature and pH rather than the physical state of food solids (Chirife and Buera, 1994).

Relationships between water availability and food stability have been of great interest and the effect of water activity on deteriorative changes in foods has been well documented. Results of studies on the effect of phase transitions which accompany changes promoted by increasing water availability have improved understanding of the physical phenomena causing deterioration. The occurrence and time-dependent character of stickiness and caking have been shown to be related to viscosity (Downton *et al.*, 1982; Wallack and King, 1988) and to be governed by the glass transition temperature (e.g., Levine and Slade, 1986; Roos and Karel, 1991a). Caking and lactose crystallization have also been well-known phenomena that occur in milk powders and are affected by water content and temperature (e.g., Warburton and Pixton, 1978; Saltmarch and Labuza, 1980; Jouppila and Roos, 1994a,b). Crystallization is a first-order phase transition that may occur during storage of a number of amorphous foods containing sugars, polysaccharides, and proteins. Crystallization during storage is often detrimental to food quality and it is the major cause of structural changes that occur in bread and other bakery products during storage.

This chapter describes various aspects of time-dependent changes that are related to phase transitions and which may occur and govern quality changes during processing and storage of foods. Such foods include several low-moisture foods, confectioneries, various cereal products, and frozen foods. In addition structural transitions during dehydration or agglomeration can be described as phase transitions-related phenomena.

## II.  Time-Dependent Properties of the Physical State

The nonequilibrium state in foods is always subject to an existing driving force towards the equilibrium state. The amorphous state below the melting temperature, $T_m$, has a higher energy than the equilibrium, crystalline state. Above $T_g$ the molecular mobility is sufficient to allow reorganization of the molecules. However, the time needed for crystallization may depend on temperature and diffusion, but also on material characteristics including molecular size. At temperatures below $T_g$ molecular mobility is low and due to the high viscosity, diffusion of molecules is not sufficient for molecular rearrangements to crystalline structures. Time-dependent changes do occur also below $T_g$, which are often referred to as *physical aging*.

### A. Glass Formation

Amorphous materials can be obtained by several techniques, which all provide a rapid change of the material from an equilibrium state to a nonequilibrium state without allowing time needed for the material to adjust to changes occurring in its surroundings and to maintain equilibrium. The common methods for producing amorphous materials include rapid cooling from melt to temperatures well below $T_m$ or rapid removal of solvent. Cooling from melt requires that the process occurs with a rate that is faster than the rate of nucleation to avoid crystallization. Thus, molecules may remain in the super-cooled, liquid, amorphous state that has a higher enthalpy and volume than the equilibrium, crystalline material at the same conditions. In solutions molecules exist in a random order. The solute molecules may become "frozen" in the solid, glassy state due to a sufficiently rapid removal of the solvent. Again such removal should occur at a rate that is higher than the rate of nucleation to avoid crystallization. Both cooling from melt and rapid removal of solvent are typical of food manufacturing processes. Food processes that result in amorphous or partially amorphous states in foods include baking, evaporation, extrusion, dehydration, and freezing. A typical feature of all these processes is that they produce a melt of mixtures of food solids and water at a high temperature followed by rapid cooling or removal of water, producing amorphous materials with low water contents.

1. Glass formation from melt

Glass formation from melt is typical of hard sugar candies which are formed by rapid cooling of concentrated syrups. In such materials crystallization during cooling can be avoided by using sugar combinations that retard crystallization and allow formation of clear, transparent glasses. Extrusion is another process that produces a melt of component compounds and water at a high temperature followed by subsequent, rapid cooling to below $T_g$ of the melt, which results in formation of stable, solid amorphous foods (e.g., Slade and Levine, 1991, 1995).

The most important food solids, which form amorphous, solid structures from melt are carbohydrates and proteins. The melting temperature of the higher melting components can be depressed by the addition of lower melting components and water. The adjustment of a proper water content in any melt both in concentration processes and extrusion is essential for the formation of the glassy texture. Too high water contents during processing and, especially, after cooling may result in the formation of rubbery structures, which are difficult to dehydrate. Although lipids may affect the physical properties of amorphous foods they probably do not affect the physical state of the glass-forming solids (Kalichevsky *et al.*, 1992; Jouppila and Roos, 1994a,b). The formation of amorphous structures in foods by various processes is shown in Figure 7.1.

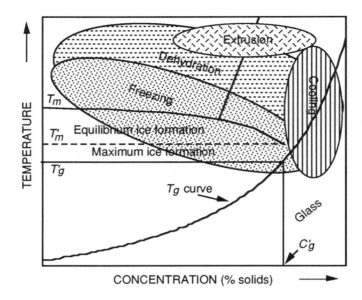

**Figure 7.1** Formation of amorphous structures in food processing. Such processes as evaporation at high temperature and extrusion followed by rapid cooling produce amorphous structures from melt. Formation of amorphous foods by solvent removal occurs in dehydration processes and in freezing.

Various glassy states of the same material may be formed in cooling processes depending on the cooling rate. The physical state of the glass produced may be affected by the time-dependent characteristics of the "freezing" of molecules to the solid, glassy state. The complicated process of the formation of the glassy state of cookies and crackers by solubilization, evaporation, and cooling processes as described by Levine and Slade (1993) is a good example of glass formation in foods. The effect of cooling on the thermodynamic properties of amorphous synthetic polymers has been well established. Cooling rate affects the free volume of molecules within the amorphous matrix, as may be observed from various relaxations when rapidly cooled materials are heated to above $T_g$. The time-dependent nature of the glass formation may also be observed from changes in the physical state that result from annealing in the vicinity of the $T_g$. Food stability is often related to the location of the $T_g$ in relation to storage temperature, but it should also be remembered that the glass transition itself is a time-dependent property of the amorphous state and that the relative rate of cooling from melt may contribute to the textural properties of rapidly cooled amorphous foods.

## 2. Glass formation by solvent removal

Glass formation by the removal of solvent is probably the most common and natural process of food preservation. The solvent in foods is water that can be removed by dehydration or separated from food solids by freezing (Herrington, 1934). It is also well established that completely amorphous food materials can be produced by freeze-drying.

*a. Dehydration.* Dehydration of most foods results in the formation of at least partially amorphous materials. Such materials as fruits and vegetables are often dried in air without extensive crystallization of component sugars. The crystallization of such compounds is probably delayed by interactions of various molecules in the complicated mixtures of sugars and other compounds. DSC curves of dehydrated fruits and vegetables often show a glass transition (Roos, 1987; Sá and Sereno, 1994). The most rapid removal of solvent occurs in drum-drying and spray-drying. These processes probably produce quantitatively the highest amounts of amorphous food solids, e.g., most spray-dried food powders. The proper control of the amorphous state of lactose has been one of the key factors for maintaining stability of dehydrated milk products. As shown in Figure 7.1 dehydration is usually based on supplying sufficient amounts of heat to the material, which increases temperature and allows evaporation. The removal of water concentrates the food solids, which, at low water contents, enter the glassy state.

*b. Freezing.* Ice formation in food materials results in freeze-concentration of solutes. The extent of freeze-concentration is dependent on temperature according to the melting temperature depression of water caused by the solute phase. Eutectic crystallization of solutes in most foods is unlikely during normal freezing processes that cool the material to sufficiently low temperatures. As shown in Figure 7.1 the glass transition temperature of freeze-concentrated solutes is dependent on the extent of freeze-concentration (e.g., Levine and Slade, 1986, 1989; Roos and Karel, 1991c). Maximum freeze-concentration may occur at temperatures slightly below the onset temperature of ice melting, $T_m'$, in the maximally freeze-concentrated material (Roos and Karel, 1991c). Glass transition of the maximally freeze-concentrated material is located at $T_g'$ (Levine and Slade, 1986). Foods that contain high amounts of low molecular weight sugars have low $T_g'$ values. Generally, the $T_g'$ increases with increasing molecular weight of the solute fraction (Levine and Slade, 1986, 1989; Roos and Karel, 1991d,e).

*c. Freeze-Drying.* The formation of amorphous materials in proper freeze-drying processes is based on the separation of water from solution by sufficiently rapid freeze-concentration to the maximally freeze-concentrated state. Thus, frozen materials to be freeze-dried should contain pure ice crystals within a maximally freeze-concentrated solute matrix that has glass transition at $T'_g$ (e.g., Franks, 1990). The solute fraction at temperatures below $T'_g$ remains in the solid, glassy state and supports its own weight, although the ice crystals are removed by sublimation. A successful freeze-drying process produces materials that have $T_g$ close to the $T_g$ of the anhydrous solutes due to the removal of all ice and most of the plasticizing water. It should be noticed that any exposure of a freeze-drying material to temperatures above $T'_m$ allows ice melting and plasticization of the freeze-concentrated solutes. The freeze-concentrated solute fraction at temperatures above $T'_m$ cannot support its own weight, which results in collapse, reduced water removal, and poor product quality.

## B. Relaxation Phenomena in Amorphous Foods

Amorphous materials show various relaxation phenomena, depending on their physical state and the time allowed for the material to adjust to changes in temperature, pressure, or amount of a plasticizer. Relaxation phenomena are often observed within the glass transition temperature range when glassy materials are heated. However, it should be noticed that relaxation phenomena, which reflect the approach of the material towards equilibrium, occur also in the glassy state, but the relaxation times are orders of magnitude larger than above $T_g$.

## 1. Enthalpy relaxations

Enthalpy relaxations of amorphous materials are frequently observed with DSC during heating of glassy materials to above the glass transition temperature. Enthalpy relaxations may be either endothermic or exothermic, depending on the thermal history of the material and the time scale of observation.

Enthalpy, $H$, relaxations of amorphous materials can be described in terms of their heating and cooling rates. Similar relaxations can also be observed from changes in volume, $V$, and entropy, $S$, and they affect changes occurring in heat capacity, $C_p$, the thermal expansion coefficient, $\alpha$, and compressibility, $\beta$. During cooling of glass-forming materials from melt towards $T_g$, enthalpy, entropy, and volume decrease continuously with

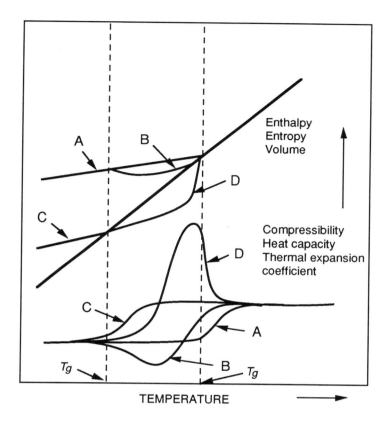

**Figure 7.2** Effect of cooling rate on the physical state of amorphous glasses and relaxation phenomena observed during heating to temperatures above the glass transition temperature, $T_g$. Fast cooling below $T_g$ results in the formation of glasses with relatively high enthalpy, $H$, volume, $V$, and entropy, $S$. Slow cooling allows more time for molecular relaxations before freezing and lower enthalpy, entropy, and volume states are obtained. Heating of the glasses with various rates affects location of the $T_g$ and the thermal properties show relaxations that result from differences in cooling and heating rates. After Weitz and Wunderlich (1974).

decreasing temperature. It is well known that large-scale molecular motions become extremely slow at $T_g$, the molecules "freeze," and the material cannot respond to further changes in temperature within the experimental time scale (e.g., Weitz and Wunderlich, 1974; Tant and Wilkes, 1981; Wunderlich, 1981) or as Perez (1994) pointed out molecular movements become prevented below $T_g$ and the configurational modes are frozen abruptly. Thus, changes in $H$, $V$, and $S$ occur more slowly below $T_g$ than above $T_g$. The cooling rate affects the physical state of amorphous materials in terms of $H$,

$V$, and $S$, which are observed from relaxation phenomena as the glass is heated to temperatures above $T_g$.

The effects of heating and cooling rates on relaxation phenomena are shown in Figure 7.2. Rapid cooling of an amorphous material to below $T_g$ freezes molecules into a physical state with a fairly high enthalpy, entropy, and large volume. Reheating of the material at the same rate changes the slope of the increase of the extensive properties of state, but no relaxations are observed (A). Thus, a step change in compressibility, heat capacity, and the thermal expansion coefficient is observed at $T_g$ according to curve A. Similarly, slow cooling followed by slow heating gives no indication of relaxation phenomena, but the change in thermal properties and $T_g$ are observed at a lower temperature (C). Slow heating of the fast-cooled material results in changes above the lower $T_g$ (or theoretical $T_g$ that would be equivalent to $T_g$ observed at a heating rate of 0°C/min) and decreases in enthalpy, entropy, and volume (B) are evident. These changes result from the longer time allowed for the molecules to relax above $T_g$ than was allowed during the initial cooling over the same temperature range. The decrease of enthalpy indicates loss of heat, which is observed as an exotherm associated with the change in heat capacity occurring over the glass transition temperature range (B). An inverse relaxation phenomenon is observed when a slowly cooled material is heated rapidly over glass transition (D). The rapid heating does not allow sufficient time for the molecules to relax at the glass transition corresponding to the initial cooling rate and the physical state, and extra energy at a higher temperature is needed for the relaxation of molecules to the supercooled liquid-like state. Therefore, DSC scans of amorphous materials that have relaxed to a fairly low-volume glassy state produce an endotherm at $T_g$ (D). The size of the endotherm may be considered to be a quantitative measure of the amount of heat needed for the relaxation process. It should be noticed that the amount of heat required for the relaxation at $T_g$ cannot be higher than the heat of fusion of the same crystalline material would be at $T_g$, since the crystalline state has the lowest enthalpy at equilibrium and requires the exact amount of latent heat to be transferred to the liquid state.

Enthalpy relaxations in amorphous materials may also result from physical aging that occurs in the glassy state (Struik, 1978). It has been observed that annealing below $T_g$, but at a sufficiently high temperature, increases the size of the relaxation endotherm observed by DSC. The size of the endotherm increases with increasing annealing time at a constant temperature. The size of the endotherm decreases if annealing temperature is decreased, suggesting that the relaxation time below $T_g$ increases with decreasing temperature. Enthalpy relaxations of glassy food materials have not been studied systematically and no correlations between physical properties and the physical state of glassy foods are available. However, the relaxations observed during heating of glassy materials may give information on the physical state. Noel *et al.* (1993) pointed out that the enthalpy relaxations are a feature that may give information on time-dependent changes in glass structure, which can be related to embrittlement in the glassy state. Indeed, physical aging in gelatinized starch during sub-$T_g$ annealing has been found to cause embrittlement due to free volume relaxations (Shogren, 1992).

## 2. Structural relaxations

It is obvious that the physical state of glassy materials affects their mechanical behavior. According to Tant and Wilkes (1981) glassy polymers become more brittle during physical aging. They pointed out that the decreasing free volume increases the time needed for the molecular response to stress. Some correlations between changes in mechanical properties and enthalpy relaxations of annealed materials have been reported (Tant and Wilkes, 1981). Structural relaxations may also be related to changes in mobility that can be observed from mechanical and dielectric relaxations (Noel *et al.*, 1992; Kalichevsky *et al.*, 1993).

As was shown in Figure 7.2 the observed $T_g$ depends on the experimental time scale. The transition temperatures as a function of heating rate have been determined for several polymers with DSC. The scanning rate dependence of the observed $T_g$ can be used to obtain information on structural relaxation processes that occur in the vicinity of $T_g$ (Noel *et al.*, 1993). According to Noel *et al.* (1993) the time taken to go through glass transition at a scanning rate of 10°C/min is of the same order as the shear stress relaxation time. Noel *et al.* (1993) suggested the use of equation (7.1) to obtain the heat of activation, $Q_S$, for structural relaxations using DSC.

$$\frac{d \ln |q|}{d(1/T_g)} = -\frac{Q_s}{R} \qquad (7.1)$$

Equation (7.1) defines the relationship between observed glass transition temperature, $T_g$, and heating rate, $q$. Noel *et al.* (1993) pointed out that $Q_S$, which may be obtained from the slope of a plot showing $q$ against $T_g$, is comparable with the heat of activation obtained from viscosity data over the same temperature range. Thus, it may be concluded that the heating rate dependence of the observed $T_g$ is a measure of changes in molecular mobility that occur within the $T_g$ temperature range.

The relationships between enthalpy relaxations and structural relaxations have not been applied in the characterization of physical properties of amorphous foods. One of the problems in applying such information to foods is perhaps their porosity and its effect on physical properties in the macroscopic scale. However, information on structural relaxations that reflect the physical state of amorphous foods would probably be useful in evaluating the brittleness or cracking behavior of glassy foods and food components.

## III.   Collapse Phenomena

Levine and Slade (1988a) considered collapse phenomena to include various time-dependent structural transformations, which may occur in amorphous foods and other biological materials at temperatures above the glass transition temperature. Such phenomena include or have an effect on (1) stickiness and caking of food powders; (2) plating of particles on amorphous granules; (3) crystallization in powders; (4) structural collapse of dehydrated structures; (5) loss and oxidation of encapsulated lipids and flavors; (6) enzymatic activity; (7) nonenzymatic browning; (8) graining of boiled sweets; (9) sugar bloom in chocolate; (10) ice recrystallization; and (11) solute crystallization during frozen storage (Levine and Slade, 1988a).

### A. Stickiness and Caking

Stickiness and caking are phenomena that may occur when amorphous food products are heated or exposed to high humidities. It has been realized that stickiness and caking are typical of amorphous food powders. Stickiness and caking may occur both during production of dried foods and during food storage. The difference between stickiness and caking is not always clearly defined, but it may be assumed that *stickiness* refers to the tendency of a material to adhere on a surface of similar or different type. Such adhesion may occur temporarily and it does not necessarily involve caking. Caking may be considered as a collapse phenomenon that occurs due to stickiness of particles that form permanent aggregates and harden into a mass, which results in loss of free-flowing properties of powders.

### 1. Stickiness

Stickiness of food materials may occur and affect material behavior and quality during both production and storage. Stickiness is also an important factor in the production of granular materials with instant properties from fine powders with agglomeration techniques. The main cause of stickiness is plasticization of particle surfaces, which allows a sufficient decrease of surface viscosity for adhesion.

Peleg (1977) divided powders into cohesive powders and noncohesive powders. Cohesive powders were considered to be those in which interparticle forces are active and may reduce flowability and cause stickiness and caking. Noncohesive powders, in accordance, have negligible interparticle forces and they are free flowing, although interparticle forces may result from an increase in water content or temperature. Downton *et al.* (1982) referred to the considerations of White and Cakebread (1966) and defined sugar-containing powders such as dehydrated fruit juices to be amorphous materials, which have an extremely high viscosity in the glassy state. They suggested that an increase in temperature or water content is the cause of the formation of an incipient liquid state of a lower viscosity at the particle surface, which results in stickiness. Downton *et al.* (1982) proposed that particles may stick together if sufficient liquid can flow to build strong enough bridges between the particles and that the driving force for the flow is surface tension. The relationship between viscosity and surface tension was considered to follow equation (7.2), where $\eta$ is viscosity, $k$ is a constant, $\gamma$ is surface tension, $t$ is contact time, $K$ is a constant, and $d$ is the distance over which flow must occur.

$$\eta = \frac{k\gamma t}{Kd} \tag{7.2}$$

Equation (7.2) suggests that stickiness is a time-dependent property of amorphous powders. Since the viscosity in the glassy state is extremely high, the contact time must be very long for the occurrence of stickiness. The dramatic decrease of viscosity above $T_g$ obviously reduces the contact time and causes stickiness that can be related to the time scale of observation. The Lazar *et al.* (1956) method for measuring stickiness was used by Downton *et al.* (1982). They estimated that the method detected stickiness which occurred after a contact time of 1 to 10 s. The estimated contact time predicted that a surface viscosity lower than $10^6$ to $10^8$ Pa s was sufficient to cause stickiness. The hypothesis was tested using an amorphous powder, which was composed of a mixture of fructose (12.5% of total solids) and sucrose (87.5% of total solids). Downton *et al.* (1982) determined viscosities and sticky points for the material at various water contents. The sticky point was found to decrease with increasing water content. The critical viscosity that caused stickiness was almost independent on water content, ranging from 0.3 x $10^7$ to 4.0 x $10^7$ Pa s, which agreed well with the predicted viscosity

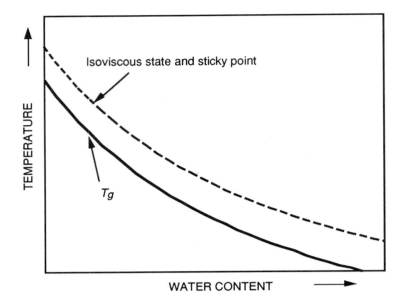

**Figure 7.3**  Relationships between glass transition temperature, $T_g$, water content, viscosity, and occurrence of stickiness in food powders. The $T_g$ decreases with increasing water content due to plasticization. Stickiness occurs at an isoviscous state, which is located above $T_g$ with a constant temperature difference, $T - T_g$. The free-flowing properties of amorphous powders are maintained at temperatures below the $T_g$ curve. Stickiness may occur above $T_g$ and instantly at the sticky point temperature.

range. Wallack and King (1988) reported that the critical viscosity range applied also to other amorphous powders, including coffee extract and a mixture of maltodextrin, sucrose, and fructose.

Soesanto and Williams (1981) studied the viscosity of a fructose-sucrose mixture of composition equal to that used by Downton *et al.* (1982). Soesanto and Williams (1981) found that the viscosity followed the WLF temperature dependence above $T_g$. It may be assumed that stickiness, when determined using the method of Lazar *et al.* (1956), is observed to occur at an isoviscosity state that is a consequence of the $T_g$ and therefore defined by a constant $T - T_g$ (Slade and Levine, 1991). Roos and Karel (1991a) used DSC for the determination of the $T_g$ of the fructose-sucrose mixture as a function of water content. They found that stickiness occurred at about 20°C above the $T_g$ and that the sticky point temperature corresponded with the endset of the glass transition temperature range. The relationship between temperature, stickiness, viscosity, and water content is shown in Figure 7.3. The WLF equation with the constants $C_1 = -17.44$ and $C_2 = 51.6$ predicts an isoviscosity state of $10^7$ Pa s at about 20°C above $T_g$, which agrees with the experimental and predicted critical viscosity values of Downton *et al.* (1982). The main importance of the relationship between the sticky point and $T_g$ is that the $T_g$ of amorphous food powders can be used as a stability indicator. Thus, knowledge of the $T_g$ and its dependence on water content can be used to evaluate causes of stickiness problems, especially in the production of spray-dried powders.

## 2. Caking

Caking of sticky powders occurs when sufficient time is allowed for surface contact of sticky particles. According to Peleg (1977) liquid bridging is one of the main interparticle phenomena that cause caking in food powders. Factors that may cause liquid bridging are (1) water adsorption; (2) melting of component compounds (e.g., lipids); (3) chemical reactions that produce liquids (e.g., nonenzymatic browning); (4) excessive liquid ingredient; (5) water released due to crystallization of amorphous sugars; and (6) wetting

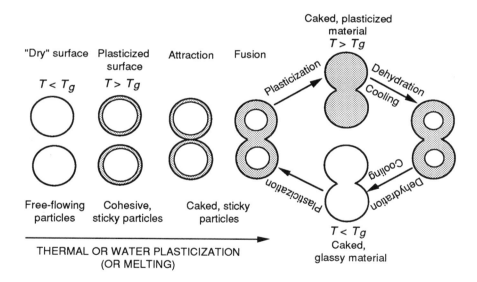

**Figure 7.4** Mechanism of caking in amorphous food powders or by melting of solid food components. Plasticization, either thermal or by water, that depresses $T_g$ at the particle surface to below ambient temperature decreases viscosity and allows liquid bridging between particles. Transitions between the solid, glassy state and the liquid-like plasticized state may occur after initial caking. In some cases melting phenomena at the particle surface, e.g., melting of lipids, may cause caking. After Peleg (1983).

of the powder or equipment. Food powders were considered to have a high complexity of flow behavior due to hygroscopicity and time-dependent physicochemical changes. The most common caking mechanism in food powders is plasticization due to water adsorption and subsequent interparticle fusion (Peleg and Mannheim, 1977; Peleg, 1983).

Caking of amorphous powders results from the change of the material from the glassy to the less viscous liquid-like state, which allows liquid flow and formation of interparticle liquid bridges. Peleg (1983) pointed out that "humidity caking" is the most common mechanism of caking in food powders. Humidity caking is a consequence of an increasing water content, plasticization, and depression of $T_g$ to below ambient temperature (e.g., Slade and Levine, 1991). A modification of the schematic representation of the caking mechanism suggested by Peleg (1983) is shown in Figure 7.4. The caking mechanism was schematically described as a surface wetting process. However, the effect of temperature was also considered.

## B. Collapse

*Collapse* is used here to refer to liquid flow and subsequent loss of quality of food materials, which may occur during dehydration processes or due to a change in the physical state of dehydrated foods (e.g., Tsourouflis *et al.*, 1976; Flink, 1983). *Collapse* as a physicochemical phenomenon can be defined as viscous flow that results from decreasing viscosity above glass transition. The loss of structure occurs since the material is unable to support its own weight. Such flow is time-dependent and it can be observed from reducing of the macroscopic volume towards that typical of the liquid state and concomitant loss of porosity.

### 1. Collapse and glass transition

Collapse, similarly to stickiness and caking, can be related to viscosity and glass transition of amorphous foods. Collapse may be considered also as an extended caking phenomenon, i.e., loss of structure that results in liquefaction, which occurs within the experimental time scale.

Tsourouflis *et al.* (1976) suggested that the loss of structure is related to a decrease in viscosity. They found that the decrease of viscosity allowing sufficient flow for collapse could be achieved by various combinations of water content and temperature. Tsourouflis *et al.* (1976) observed that collapse of freeze-dried materials occurred above a specific temperature within 45 min and more rapidly above the collapse temperature, $T_c$. The definition of the collapse temperature allowed its measurement for amorphous foods with various collapse properties. Tsourouflis *et al.* (1976) reported that the collapse temperatures were high at low water contents and they decreased with increasing water content. They concluded that sufficient flow for collapse was possible above a critical viscosity, which was a function of temperature and water content. Another factor that was important in defining the collapse behavior was molecular weight of the solids. Tsourouflis *et al.* (1976) noticed that the $T_c$ of maltodextrins increased with increasing molecular weight. They also reported that the addition of high molecular weight polysaccharides to such products as orange juice increased the $T_c$ of the dehydrated material. A comparison between the $T_c$ values and sticky point temperatures reported by Brennan *et al.* (1971) showed that the $T_c$ and sticky point were related to water content in a similar manner. The relationships between $T_c$ and water content were confirmed in a comprehensive study of collapse in dehydrated materials by To and Flink (1978a,b,c). To and Flink (1978b) also found significant similarities between the collapse temperature and glass transition

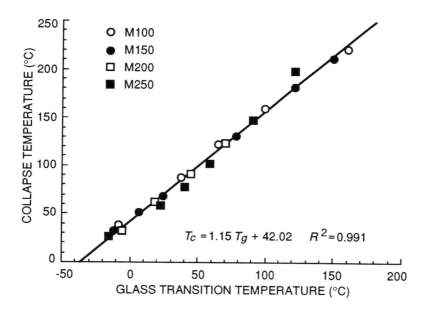

**Figure 7.5** Correlation between collapse temperature, $T_c$, and glass transition temperature, $T_g$. The relationship is linear, showing that the $T_c$ as determined by Tsourouflis *et al.* (1976) for maltodextrins (Maltrin) with dextrose equivalent values of 10 (M100), 15 (M150), 20 (M200), and 25 (M250) occurs about 40 to 70°C above $T_g$. The $T_g$ was predicted with the Gordon and Taylor equation with data from Roos and Karel (1991b) and Roos (1993a).

temperature of polymers. The similarities included the molecular weight dependence according to the Fox and Flory equation of both phenomena and the effect of composition on $T_c$ and $T_g$ in mixtures.

Levine and Slade (1986) considered collapse of dehydrated materials to be a $T_g$-governed phenomenon. Roos and Karel (1991b) determined glass transition temperatures for maltodextrins. The $T_c$ values for corresponding materials were reported by Tsourouflis *et al.* (1976). Roos and Karel (1991b) observed that the $T_g$ values determined with DSC were lower than the corresponding $T_c$ values, but showed similar decrease with increasing water content. The relationship between $T_g$ and $T_c$ is shown in Figure 7.5. The high correlation between the experimental $T_c$ and $T_g$ values suggests that the determination of $T_c$ is a method applicable for $T_g$ determination of amorphous foods. The temperature difference between $T_c$ and $T_g$ is a measure of the difference between the experimental time scale of the $T_g$ observation with DSC and the $T_c$ determination methods, including determination of the sticky point temperature, as was also pointed out by Slade and Levine (1991) and Chuy and Labuza (1994).

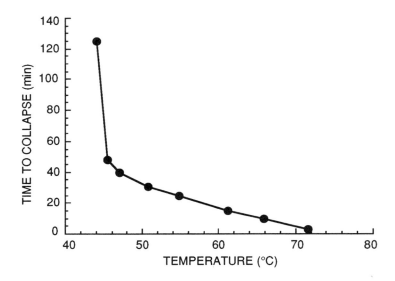

**Figure 7.6** Collapse of freeze-dried sucrose as a function of temperature. Time to collapse indicates time required for the loss of structure at each temperature. Data from Tsourouflis *et al.* (1976).

## 2. Collapse time

The close relationships between collapse temperature and glass transition suggest that collapse occurs above $T_g$ with a rate that is defined by the temperature difference, $T - T_g$ (Levine and Slade, 1986). To and Flink (1978a,b,c) assumed that the degree of collapse was not a time-dependent phenomenon. However, the occurrence of collapse at about a constant $T - T_g$ indicates that the time to collapse increases significantly with decreasing $T - T_g$. Thus, collapse may occur at temperatures close to $T_g$, but the time of its observation should be orders of magnitude longer than is normally used for the determination of collapse temperatures.

Tsourouflis *et al.* (1976) determined the time required for collapse of amorphous sucrose at various temperatures. The results suggested that time to collapse increased with decreasing temperature, as shown in Figure 7.6. They concluded that the temperature required for collapse became relatively constant as the time to collapse increased exponentially at low temperatures. Thus, $T_c$ was taken as the temperature at which a significant decrease of the rate of the phenomenon was obvious. However, it should be noticed that

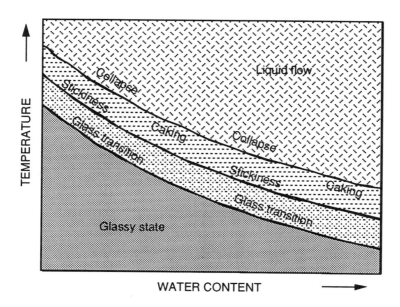

**Figure 7.7** Glass transition temperature, $T_g$, related collapse phenomena and their dependence on water content. The transitions observed are time-dependent and occur at isoviscosity states above $T_g$ according to the experimental time scale.

amorphous sucrose may crystallize during collapse, which probably alters the collapse behavior in comparison with collapse behavior of noncrystallizing materials such as maltodextrins.

The results of Tsourouflis *et al.* (1976) and To and Flink (1978a) showed that an equal extent of collapse may be observed when amorphous materials are kept a long time at a low temperature or a short time at a high temperature. The time-temperature relationships of the extent of collapse were studied by To and Flink (1978a,b). The extent of collapse increased with increasing temperature, which most likely reflected the time-dependent nature of the phenomenon and longer relaxation times at the lower temperature. It may be concluded from the studies of Tsourouflis *et al.* (1976) and To and Flink (1978a,b) that, as shown in Figure 7.7, there was a critical viscosity above which the collapse time substantially increased. In other words there was a critical temperature difference above $T_g$, which must be exceeded for initiation of collapse within a time of practical importance. The time dependence of the collapse phenomenon was also discussed by Flink (1983). He pointed out the fact that the glass transition is reversible, but a noncollapsed structure cannot be reproduced by cooling.

## IV.    Crystallization and Recrystallization

Crystallization and recrystallization in amorphous foods are time-dependent phase transitions, which are governed by the physical state and glass transition. These processes include (1) crystallization of amorphous components such as sugars; (2) retrogradation phenomena in gelatinized starch; (3) solute crystallization in frozen foods; (4) time-dependent ice formation at low temperatures; and (5) recrystallization of ice during frozen storage. Both crystallization and recrystallization processes belong to phase transitions, which may dramatically decrease food quality during storage.

### A. Crystallization of Amorphous Sugars

Amorphous materials, including amorphous sugars, exist in a thermodynamically metastable, nonequilibrium state below the equilibrium melting temperature. Thus, there is always a driving force towards the equilibrium, crystalline state. Molecules of amorphous solids in the glassy state are not able to change their spatial arrangement to the highly ordered crystalline equilibrium state. However, as the temperature is increased to above glass transition the molecular mobility increases, which often allows crystallization. Crystallization of amorphous sugars is probably the most typical crystallization phenomenon of low-moisture foods.

## 1. Crystallization of pure sugars

One of the most significant differences between glassy and crystalline sugars is their water sorption behavior. Amorphous sugars are hygroscopic and they may adsorb high amounts of water at low relative humidities. Crystalline sugars show little adsorption of water at low humidities and their water adsorption becomes significant only at high relative humidities due to solubilization.

*a. Effect of Water.* Makower and Dye (1956) published the first systematic study on the crystallization behavior of amorphous glucose and sucrose. Amorphous sucrose was produced by spray-drying of a 20% solution. The amorphous state of glucose was obtained by rapid cooling from melt, since spray-drying was not possible due to problems caused by stickiness and crystallization. The glassy melt was powdered by crushing and grinding at -25°C. Samples of the powdered sugars were exposed to various relative humidities at 25°C and their water uptake was determined periodically. Makower and Dye (1956) found that an equilibrium water content was reached with sucrose at up to 24% RH. The corresponding relative humidity for glucose was only 4.6%. At higher relative humidities the water content increased to a maximum and then decreased. The rate of the loss of adsorbed water increased with increasing storage RH.

Makower and Dye (1956) suggested that the increase in the rate of crystallization observed from the weight loss with increasing RH was due to a decrease in viscosity, which affected the rate of orientation of the molecules. Stability of amorphous glucose and sucrose could be maintained at 25°C for more than 2 years below 4.6 and 11.8% RH, respectively. At higher humidities water adsorption caused crystallization and the adsorbed water was released. The results revealed that crystallization of amorphous sugars was a time-dependent phenomenon and that the rate of crystallization increased with increasing storage relative humidity. Makower and Dye (1956) reported also that the experiments were carried out at constant humidity chambers with humectants, which could maintain a constant humidity during storage, although additional water was released from samples due to crystallization. They pointed out that crystallization in closed containers would have resulted in a more rapid crystallization process due to a concomitant increase in humidity caused by the water released and subsequent higher adsorption of water by the remaining amorphous material.

Karel (1973) established a sorption isotherm for amorphous sucrose that was based on the results of Makower and Dye (1956). The sorption isotherm, similar to that shown schematically in Figure 7.8, showed that water adsorption of amorphous sugars at a constant temperature increased with increasing storage relative humidity. The "equilibrium" water content can be maintained at low water contents for long periods of time. According to the sorption isotherm time-dependent crystallization of amorphous sugars may occur above a critical storage RH. Crystallization of amorphous sucrose was predicted to occur within 1000 days at about 20% RH. The time required for crystallization decreased rapidly with increasing relative humidity. The sorption isotherm suggested that amorphous sucrose was totally unstable at intermediate relative humidities due to instant crystallization. At high relative humidities the crystallized material gained water due to solubilization. Karel (1973) pointed out that water sorption in amorphous food components is complicated and may require consideration of kinetics.

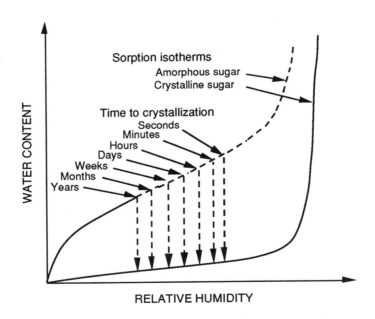

**Figure 7.8** Effect of relative humidity on water adsorption by amorphous sugars. The adsorption is significantly higher than that of the crystalline material at low relative humidities. Above a critical relative humidity the amorphous material releases water due to crystallization and the water content decreases to that of the crystalline material. Time to crystallization decreases with increasing relative humidity.

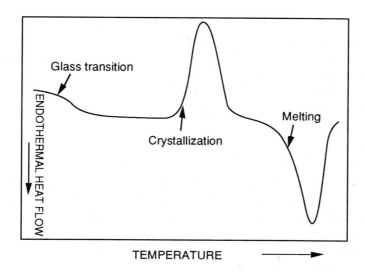

**Figure 7.9** Schematic representation of a DSC thermogram that is typical of freeze-dried amorphous sugars. The amorphous materials show glass transition followed by a crystallization exotherm. Instant crystallization occurs and releases heat as the molecular mobility becomes sufficiently high to allow molecular arrangements to the crystalline form within the experimental time scale. The crystals formed during heating melt and produce the melting endotherm at the typical melting temperature of the material.

Simatos and Blond (1975) used X-ray and electron diffraction techniques to study the physical state of sucrose that was 1) freeze-dried, 2) melted + supercooled, or 3) freeze-dried + crystallized by exposures to high relative humidities. The X-ray diffraction results showed that freeze-dried sucrose and sucrose supercooled from melt had diffraction patterns similar to and typical of amorphous materials. Sucrose that had crystallized due to water adsorption showed the characteristic X-ray diffraction patterns of crystalline sucrose. The electron diffraction study revealed that the freeze-dried materials probably contained some nuclei, which probably decreased its stability.

*b. Effect of Temperature.* An increase in temperature increases the rate of collapse phenomena of amorphous foods. Simatos and Blond (1975), in their study on the physical state of freeze-dried sucrose, found that electron diffraction patterns after heating the material to 120°C suggested that the material contained numerous small crystals. Another indication that amorphous sugars may crystallize due to heating can be revealed from the results of Tsourouflis *et al.* (1976). They actually measured the rate of sucrose crystallization instead of pure collapse, as was already shown in Figure 7.6. The collapse of sucrose probably involved the loss of structure due to the volume change that occurred as the material crystallized. Therefore, the results suggested that the time to crystallization for a material that had a constant water content decreased dramatically as the temperature was increased.

Simatos and Blond (1975) used also DTA to observe the thermal behavior of freeze-dried sucrose. The thermogram obtained for the freeze-dried material showed a crystallization exotherm, which did not appear in crystallized materials. Similar behavior of freeze-dried sucrose was reported by To and Flink (1978a). They found that freeze-dried sucrose had a glass transition, a recrystallization exotherm, and a melting endotherm. These transitions are observed in DSC thermograms of crystallizing amorphous materials and they are typical of amorphous sugars (Roos and Karel, 1990; Roos, 1993b). A schematic DSC thermogram typical of amorphous sugars is shown in Figure 7.9. Roos and Karel (1990) reported that instant crystallization of amorphous lactose and sucrose occurred at about 40 to 60°C above $T_g$, when a heating rate of 5°C/min was used. The instant crystallization temperature decreased with increasing water content in a manner similar to when the $T_g$ was decreased due to water plasticization.

Roos and Karel (1990) observed that heating of amorphous sugars above $T_g$ to a predetermined temperature with DSC and isothermal holding could be used to study the time required for crystallization. They observed that the isothermal holding time before an exotherm occurred increased exponentially with decreasing temperature. The thermogram for the isothermal process also showed that the crystallization exotherm became broader with decreasing temperature, probably because of the longer time required for crystallization. Roos and Karel (1990) concluded that crystallization of amorphous sugars occurs above $T_g$. The time required for crystallization depends on temperature and water content. The relationships between the physical state, temperature, and water content of amorphous sugars are schematically described in Figure 7.10. The results of Roos and Karel (1992) on rates of crystallization of amorphous lactose suggested that $T_g$ was the main factor that controlled crystallization of the amorphous sugar at various temperatures and water contents.

*c. Crystallization Kinetics.* The effects of glass transition and the WLF behavior on the rates of relaxation processes in amorphous food components, including sugar crystallization, have been emphasized by Levine and Slade (1988a). The effect of the glassy state on the crystallization behavior of sugars was also recognized by White and Cakebread (1966), but only a few studies have reported crystallization rates for amorphous sugars that can be used to analyze the temperature and water content dependence of the phenomenon.

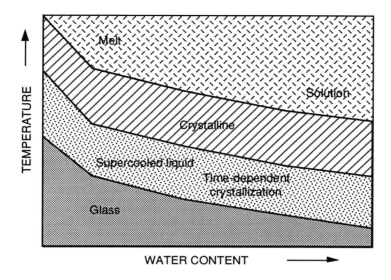

**Figure 7.10** Effect of temperature and water content on the physical state of amorphous sugars. The materials exist as glasses below the water content-dependent glass transition temperature, $T_g$. Heating or an increase in water content may transform the glassy materials into the supercooled liquid state that allows time-dependent crystallization. The crystals formed melt above the melting temperature, $T_m$, that is depressed with increasing water content.

Roos and Karel (1990) fitted the WLF equation to crystallization data obtained with the isothermal DSC method. The equation predicted time to crystallization of amorphous lactose over the experimental temperature range and the extrapolated estimate for time to crystallization at the $T_g$ was several hundred years. Although time to crystallization was a function of $T - T_g$ water-plasticized lactose crystallized slightly faster at equal $T - T_g$ than anhydrous lactose. However, the crystallization data were obtained over a fairly narrow temperature range and the Arrhenius equation could also have fitted to the data (Roos and Karel, 1990, 1991a).

Crystallization of amorphous lactose at various water contents and $T - T_g$ conditions were studied also by Roos and Karel (1992). They stored the materials in sealed containers and at constant humidity conditions over saturated salt solutions. The suggestion of Makower and Dye (1956) that crystallization in a sealed package would result in a rapid crystallization due to the transfer of water from the crystallizing material to the remaining amorphous regions was experimentally confirmed. Crystallization at constant relative humidity conditions allowed crystallization to occur at a constant $T - T_g$. Roos and

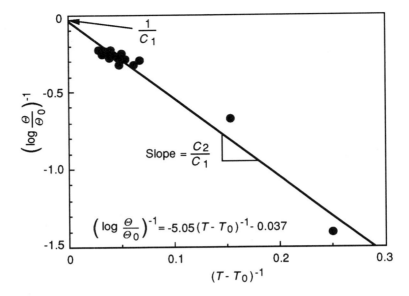

**Figure 7.11** Time to crystallization, $\Theta$, of amorphous lactose plotted against $(T - T_0)^{-1}$, where $T$ is temperature and $T_0$ is a reference temperature. The time to crystallization at $T_0$ is $\Theta_0$. The constants $C_1$ and $C_2$ for the WLF equation were obtained from the slope and intercept of the regression line according to Peleg (1992). Data from Roos and Karel (1992).

Karel (1992) confirmed that the WLF equation fitted to the crystallization data, although the universal constants were used in predicting time to crystallization, $\Theta$. The experimental time to crystallization at low $T - T_g$ values was longer than would have been predicted by the Arrhenius-type temperature dependence. Although Roos and Karel (1992) used the universal WLF constants in predicting the temperature dependence of time to crystallization, they could have derived other constants from the experimental data, as was suggested by Ferry (1980) and Peleg (1992). Figure 7.11 shows the crystallization data reported by Roos and Karel (1992), transformed and linearized according to Peleg (1992). Figure 7.11 can be used to obtain the constants $C_1$ and $C_2$ for the WLF equation when $T_g + 10°C$ is used as the reference temperature, $T_0$, with the experimental time to crystallization at $T_0$ being $\Theta_0$. The WLF constants obtained were -28.8 and 124.8 for $C_1$ and $C_2$, respectively, when $T_g$ was used as the reference temperature. Figure 7.12 shows the WLF prediction of time to crystallization of amorphous lactose and the difference between the predictions obtained with the universal and calculated constants.

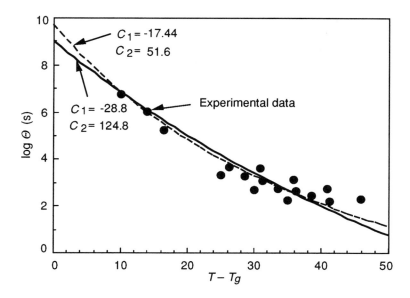

**Figure 7.12** WLF prediction of time to crystallization, $\Theta$, for amorphous lactose at temperatures above the glass transition temperature, $T_g$.

The above results suggested that the time required for total crystallization of amorphous sugars was defined by $T - T_g$. However, it may be assumed that crystallization occurs at all temperatures above $T_g$ but below $T_m$ (Levine and Slade, 1990; Roos and Karel, 1991a, 1992; Hartel, 1993). The rate of crystallization of polymers is often assumed to follow the Avrami equation, which has a sigmoid shape, showing that the rate of crystallization is initially slow. The rate increases as crystallization proceeds and decreases again as the crystallinity approaches unity. Roos and Karel (1992) reported that the rate of lactose crystallization at various $T - T_g$ conditions did not follow the Avrami equation. As shown in Figure 7.13 for lactose crystallization at $T - T_g = 10°C$, the initial rate of crystallization followed the equation, but if all data were used, the rate was overestimated at the beginning of the crystallization process and underestimated when the crystallinity exceeded about 0.5. Interestingly, the model based on Fermi's distribution, which was suggested by Peleg (1993) to be used in modeling of stiffness and modulus data instead of the WLF equation (see Chapter 8), gives a better prediction for the rate of crystallization over the whole experimental range than the Avrami equation. It should be noticed that the fit of the WLF model to time to crystallization data suggested that the rate of crystallization was related to viscosity and to the

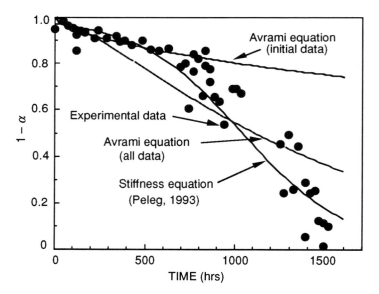

**Figure 7.13** Crystallinity, $\alpha$, of amorphous lactose as a function of time at 10°C above the glass transition temperature, $T_g$. The Avrami equation may be fitted to the crystallization data at the beginning of the crystallization process, but not over the total time of crystallization. The "stiffness equation" suggested by Peleg (1993) to be used in modeling of modulus data fitted to the experimental data for crystallinity better than the Avrami equation. Data from Roos and Karel (1992).

relaxation times of mechanical properties above $T_g$. Perhaps, the increase in crystallinity at a constant temperature has the shape of a modulus curve as a function of time. If the time-temperature superposition principle applies to the increase in crystallinity, it is also possible that the increase in crystallinity has the shape of a modulus curve as a function of temperature. Unfortunately, there are not enough experimental data available to test this hypothesis.

Crystallization kinetics of amorphous sugars seem to be controlled by glass transition and melting temperatures. The time to crystallization at constant $T - T_g$ conditions may be related to viscosity and diffusion. Thus, the temperature dependence of crystallization kinetics of amorphous sugars may follow the WLF relationship rather than the Arrhenius equation (Levine and Slade, 1986). It should be noticed, however, that the kinetics may also be affected by the macroscopic structure of the amorphous material, e.g., the kinetics may be different in sugars that have been freeze-dried from the crystallization kinetics found in sugar melts. Different kinetic behavior in such materials may also be due to differences in diffusion and moisture transfer between the material and its surroundings. It is obvious that water released by crystallization is easier to remove from a porous structure than from a continuous matrix of a supercooled liquid. The kinetics of crystallization at a constant temperature above $T_g$ can be related to water content and water activity, which define $T - T_g$ due to plasticization. Therefore, crystallization may occur above a critical $m$ or $a_w$ at a constant temperature with a rate defined by $T - T_g$. It should also be noticed that sugars may crystallize into various anomeric forms and that they may exist in various hydrate forms depending on the availability of water and crystallization temperature.

## 2. Crystallization of sugars in amorphous foods

Sugars exist in the amorphous state in a number of low-moisture and frozen foods. The solubility in water as well as the crystallization behavior is different for each sugar. Crystallization can be controlled during food processing, which is done in the production of milk powders containing precrystallized lactose. Crystallization during storage is often uncontrolled. Such crystallization of amorphous sugars in low-moisture foods is detrimental to quality and it may significantly decrease shelf life. Amorphous sugars may also crystallize from freeze-concentrated solute matrices, which is known to be the primary cause of "sandiness" in ice cream.

*a. Low-Moisture Foods.* Low-moisture foods which contain amorphous sugars include a wide range of food products. These products include several baked goods, confectionery, dairy powders, various dehydrated foods and food powders, and dry food mixes. The first reports on crystallization of amorphous sugars in foods were those suggesting that crystallization of amorphous lactose in milk powders and in ice cream during storage caused loss of quality (Supplee, 1926; Troy and Sharp, 1930).

Supplee (1926) among others found that dehydrated milk powders adsorbed fairly high amounts of water at low relative humidities. The adsorbed water was lost at higher humidities, which caused a break in the sorption isotherm. The adsorption and desorption isotherms were also significantly different from each other. Troy and Sharp (1930) pointed out that dehydration of skim milk and whey by spray-drying and by roller-drying produces a glass that is composed of a noncrystalline mixture of $\alpha$- and $\beta$-lactose. They suggested that the water adsorption by hygroscopic milk and whey powders caused stickiness and caking of the powders. Adsorption of a sufficiently high amount of water resulted in lactose crystallization and loss of the adsorbed water. Water adsorption in whey powders allowed the material to become plastic and as crystallization occurred the material became hard.

Herrington (1934) concluded that lactose glasses are stable at room temperature if they are protected from moisture. The existence of lactose in the glassy state in dehydrated milk products and lactose crystallization due to increased molecular mobility in the presence of sufficient water have been confirmed by studies which have determined the physical state using polarized light microscopy and X-ray techniques (King, 1965). The water adsorption of most dehydrated milk products which contain lactose show a break in the sorption isotherm as lactose crystallizes (Berlin *et al.*, 1968a). Berlin *et al.* (1968b) showed that the break in the sorption isotherm was due to amorphous lactose crystallization and it did not appear in sorption isotherms of lactose-free powders.

Water sorption properties and stability of food powders are known to be affected by temperature. It is obvious that the crystallization behavior of amorphous sugars in food powders is temperature-dependent. Berlin *et al.* (1970) observed that the relative humidity at which the break in sorption isotherms of milk powders appeared was dependent on temperature, which was confirmed by Warburton and Pixton (1978). An increase in storage temperature shifted the break in the isotherm to a lower relative humidity. The temperature dependence of water adsorption properties of crystallizing amorphous sugars can be explained by changes in their physical state. DSC heating scans of dehydrated milk powders show a glass transition and a crystallization exotherm that are due to the phase transitions occurring in the amorphous lactose fraction (Jouppila and Roos, 1994b). Water plasticization of dehydrated milk products decreases the $T_g$ of lactose. Thus, a lower water content at a higher temperature may allow lactose crystallization. Water plasticization and depression of the $T_g$ explain also the observed temperature dependence of the break in the sorption isotherms. However, lactose crystallization during water adsorption may occur either into the anhydrous $\beta$-form or into $\alpha$-lactose monohydrate. The crystalline form produced depends on the relative humidity and temperature. According to Vuataz (1988) lactose crystallization occurs into the anhydrous $\beta$-form at relatively low water activities and into $\alpha$-lactose monohydrate above 0.57 $a_w$ at room temperature. At higher temperatures crystallization behavior may change according to the stability of the crystalline form at the crystallization temperature.

The rate of lactose crystallization in dairy powders increases with increasing storage relative humidity (e.g., Saltmarch and Labuza, 1980; Vuataz, 1988). The increasing relative humidity increases the water adsorption of amorphous lactose, which in turn increases water plasticization and the temperature difference, $T - T_g$. It may be assumed that the $T - T_g$ of lactose defines the rate of crystallization. Jouppila and Roos (1994b) determined glass transition temperatures for freeze-dried milk powders, which contained

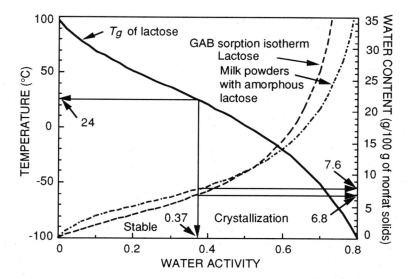

**Figure 7.14** Effect of water adsorption on the glass transition temperature, $T_g$, and stability of lactose and milk powders that contain amorphous lactose at 24°C. The amorphous lactose is in the glassy state below the critical water activity, $a_w$, of 0.37. Time-dependent crystallization occurs above the critical $a_w$. The rate of crystallization increases with increasing $a_w$ due to plasticization and increasing temperature difference, $T - T_g$. The critical water contents can be obtained from the sorption isotherms. Data from Jouppila and Roos (1994b).

various amounts of fat. They found that the $T_g$ of the materials was almost equal to that of lactose. Thus, the $T_g$ of amorphous lactose and milk powders, which had the same water activity, had also about the same $T_g$. The water sorption properties of the nonfat solids were not affected by the fat component. Jouppila and Roos (1994b) established state diagrams for milk powders, which defined critical values for water content and water activity for stability as shown in Figure 7.14. Combined $T_g$ and water sorption data suggested that a water content of 7.6 g/100 g of nonfat solids depressed $T_g$ to 24°C. The corresponding water content for pure lactose was 6.8 g/100 g of solids. The critical $a_w$ was 0.37. These values were in good agreement with several studies that had reported critical water contents and storage relative humidities for milk powders based on water adsorption properties (e.g., Warburton and Pixton, 1978). Jouppila and Roos (1994b) found that the crystallization process was slow and independent of the fat content at low water activities above the critical $a_w$. However, the rate increased with increasing $a_w$ and the crystallization process was delayed by fat at water activities higher than 0.67. It may be concluded that crystallization of lactose in dehydrated milk products occurs as the $T_g$ is depressed to below ambient temperature. The rate of crystallization is probably defined by $T - T_g$ as was shown for pure sugars, but the crystallization process may be affected by other components. The delayed crystallization is also observed from the instant crystallization temperature, which can be determined by DSC. Instant crystallization of amorphous sugars in the presence of other compounds occurs often at a higher temperature and produces a broader exotherm (Roos and Karel, 1991a; Jouppila and Roos, 1994b).

The decreased rate of crystallization is probably the main difference between the crystallization behavior of pure amorphous sugars and sugars in amorphous foods. Crystallization of lactose in milk products is fairly easy to observe. Its crystallization behavior is close to that of pure lactose, probably because of almost nonexistent interactions between lactose and fat or lactose and proteins. Foods that contain mixtures of sugars show more complicated crystallization behavior. Iglesias and Chirife (1978) studied crystallization kinetics of sucrose in amorphous food models that were composed of sucrose and various polysaccharides. The crystallization kinetics were derived from the time-dependent water sorption properties at 35°C and 54% RH. Amorphous sucrose adsorbed a high amount of water, which was lost rapidly. Materials which contained added polysaccharides showed a fairly rapid water uptake. However, the loss of adsorbed water occurred more slowly and the rate decreased with an increase in the polysaccharide content. Iglesias and Chirife (1978) concluded that the delayed crystallization of sucrose was due to interactions of sucrose with the added compounds, increased viscosity, and decreased mobility of the sucrose molecules. Milk powders, which contain lactose hydrolyzed into galactose and glucose, show no break in their sorption isotherms (San Jose *et al.*, 1977; Jouppila and Roos, 1994a). Therefore, crystallization of sugars in the glucose-galactose mixture is probably delayed in comparison with lactose. The delayed crystallization is also observed from the thermal behavior of such powders with DSC. Milk powders with hydrolyzed lactose show a $T_g$ well below that of amorphous lactose and a broad exotherm above 100°C. The $T_g$ decreases with increasing water content, but the exotherm occurs at the same temperature even at higher water activities (Jouppila and Roos, 1994b). It is obvious that sugar crystallization in many dried foods such as dehydrated fruits and vegetables occurs very slowly at temperatures above $T_g$. However, such materials show stickiness and collapse above $T_g$, which may significantly decrease their quality.

Delayed crystallization of amorphous sucrose is observed in the presence of other sugars (Saleki-Gerhardt and Zografi, 1994). The retardation of sucrose crystallization by other sugars is technologically applied in foods by the addition of dextrose or syrups. Such materials may show a glass transition and a significant decrease in modulus above $T_g$ (McNulty and Flynn, 1977; Herrington and Branfield, 1984b), but the crystallization of sucrose becomes significantly delayed by the addition of fructose or glucose (Herrington and Branfield, 1984a). Hartel (1993) pointed out that proper control of crystallization kinetics to promote sugar crystallization or to prevent crystallization requires establishing of a state diagram for the material and defining of regions for stability and instability for the process.

*b. Frozen Foods.* Sugar crystallization in frozen foods may occur due to freeze-concentration of the solutes and supersaturation of the freeze-concentrated solute matrix. At sufficiently low temperatures the solutes vitrify into the glassy state and crystallization ceases. Thus, sugar crystallization may occur above the glass transition temperature of the maximally freeze-concentrated solute matrix, $T'_g$ (Levine and Slade, 1986).

Young *et al.* (1951) studied hydrate formation in sucrose solutions. They found that hydrate formation was greatly reduced at temperatures below -34°C. It is known that the maximally freeze-concentrated state of sucrose forms below -34°C, which is the onset temperature of ice melting, $T'_m$, in maximally freeze-concentrated sucrose solutions (Roos and Karel, 1991c). At temperatures above $T'_m$ ice melting dilutes the freeze-concentrated solute matrix and its viscosity decreases significantly. Young *et al.* (1951) found that the rate of the formation of sucrose hydrates had a maximum at about -23°C. They reported that the formation of hydrates could be reduced by the addition of other sugars, corn syrup, or invert sugar. The formation of hydrates may occur at temperatures typical of frozen storage of foods, particularly when they contain high amounts of sucrose. However, hydrate crystallization is probably unlikely in common food materials.

Lactose is one of the least soluble sugars that is common in milk-containing frozen desserts. It was recognized fairly early that the crystallization of amorphous lactose in ice cream was the main cause of sandiness (Troy and Sharp, 1930; White and Cakebread, 1966). The solubility of lactose at 0°C is only about 12 g/100 g of water and it decreases further below the freezing temperature of water (Nickerson, 1974). The solubility of lactose decreases also in the presence of other sugars, e.g., sucrose (Nickerson and Moore, 1972), which, in such products as ice cream, may significantly facilitate lactose crystallization. However, crystallization of freeze-concentrated solutes can be retarded and greatly reduced by the addition of sugar blends and syrups and by the addition of polysaccharides.

Although sugar crystallization in frozen foods may occur during frozen storage it is probably not a significant problem in foods which are stored at sufficiently low temperatures. Ice recrystallization is another phenomenon that frequently reduces the quality of frozen desserts and other foods. Solute crystallization from the freeze-concentrated matrices is most detrimental in products that are consumed in the frozen state and also in solutions that are dehydrated by freeze-drying. Solute crystallization prior to freeze-drying may reduce the quality of products, which contain sugars as cryoprotectants or as encapsulating agents.

### B. Ice Formation and Recrystallization

Ice formation and recrystallization are important phenomena that may significantly affect viability of organisms in cryopreservation or stability of food materials during frozen storage. Ice formation in various biological materials and foods is a time- and temperature-dependent process. Most of the work on relating ice formation to the physical state of the freeze-concentrated material has been done with simple binary solutions such as glycerol and water or sugars and water. Recrystallization of ice is often observed in frozen foods. The occurrence of recrystallization phenomena is typical of frozen foods which have been stored at fairly high temperatures and those which have been exposed to significant temperature fluctuations. Recrystallization of ice may produce large ice crystals and a nonhomogeneous structure within the frozen product.

### 1. Ice formation

Ice formation in simple solutions and in foods occurs as the temperature of the material is depressed to below the initial concentration-dependent equilibrium freezing temperature. The nucleation process of ice requires that the extent of supercooling is sufficient to allow formation of stable nuclei and subsequent crystal growth. The amount of ice formed at the temperature of observation is a strong function of composition. In simple solutions the extent of ice formation at typical freezing temperatures of foods is mainly defined by the molecular weight of the solute, since the $T'_g$ and $T'_m$ of solutes increase with increasing molecular weight. Thus, the kinetics of ice formation in solutions and in biological materials depend on temperature and on the phase behavior of the freeze-concentrated solutes.

It is well known that the Plank equation, which is given by equation (7.3), can be used in calculations of food freezing times, $t$.

$$t = \frac{\rho \Delta H_m}{T_m - T} \left[ \frac{Pa}{h} + \frac{Ra^2}{k} \right] \tag{7.3}$$

Equation (7.3), where $T_m$ is the equilibrium freezing temperature, $\Delta H_m$ is the latent heat of melting, $\rho$ is density, $T$ is the freezing temperature, $h$ is the convective heat transfer coefficient, $k$ is thermal conductivity, $a$ is diameter or thickness, and $P$ and $R$ are geometry-dependent constants, may be used in engineering applications to estimate freezing times for materials with various geometries (Heldman and Singh, 1981).

Plank equation assumes that the freezing time of a food material is a function of the extent of supercooling and heat transfer within the material. However, the time-dependent phenomena that define the physical state of simple solutions and frozen foods are more complicated. It may be assumed that Plank equation applies at fairly high freezing temperatures, where the melting temperature depression and the kinetic restrictions caused by freeze-concentration are not significant. Several foods that contain high molecular weight carbohydrates or proteins probably form ice according to the relationship between the glass curve and freezing temperature. Such materials contain ice and unfrozen water within a glassy, but partially freeze-concentrated, solute matrix. On the other hand food materials with low molecular weight solutes, especially sugars, have low $T_g'$ and $T_m'$ values. Therefore, the solutes form a glass at fairly low temperatures and such materials may be considered to be composed of ice within a liquid-like but supercooled, partially freeze-concentrated solute matrix.

The relationships between temperature and the physical state of freeze-concentrated food materials can be obtained from state diagrams. Blanshard and Franks (1987) pointed out that the diffusion of water molecules a distance of 1 cm would take about 300 years within a glassy matrix. It is obvious that ice formation in freeze-concentrated solute matrices below $T_g$ is kinetically ceased. Heating to above $T_g$ decreases viscosity most probably according to the WLF-type temperature dependence (Levine and Slade, 1988b) and ice formation in partially freeze-concentrated solute matrices becomes likely. However, heating of a frozen food to above $T_m'$ causes ice melting and formation of the partially freeze-concentrated state.

The time-dependent ice formation is observed in most supercooled binary sugar solutions. Ice formation at low solute concentrations is not significantly delayed, but increasing the solute concentration to more than 50% is often sufficient to delay and reduce ice formation during rapid cooling. Heating of such solutions in DSC produces a low-temperature glass transition, which is followed by an exothermal event that is often referred to as devitrification (e.g., Luyet and Rasmussen, 1968; Simatos and Turc, 1975). The devitrification exotherm indicates ice formation during rewarming of rapidly cooled solutions. Ice formation during rewarming can be confirmed with a simple

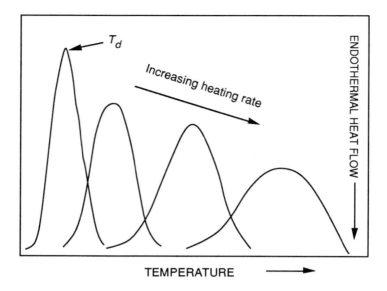

**Figure 7.15** Effect of heating rate, $q$ on the devitrification temperature, $T_d$ and the shape of the devitrification exotherm of rapidly cooled solutions as determined with DSC.

annealing procedure consisting of heating the solution to the observed devitrification temperature, which is followed by recooling. Reheating thermograms of the annealed material show no devitrification or an exotherm with a significantly reduced size (Simatos and Turc, 1975; Roos and Karel, 1991c, d). The devitrification temperature is often located slightly above the observed $T_g$. Thus, the occurrence of ice formation during a dynamic change of temperature is observed as the viscosity above $T_g$ decreases and allows molecular mobility, which, at the devitrification temperature, becomes sufficiently high for ice formation within the experimental time scale.

The devitrification exotherm is fairly sharp at low heating rates. An increase in the heating rate increases devitrification temperature and the exotherm occurs over a broader temperature range. MacFarlane (1986) used the Arrhenius equation in the analysis of the kinetics of the devitrification process. He showed that a plot of $\ln q$ against $1/T_d$ for devitrification of propylene glycol solutions gave a straight line, as was suggested by equation (7.1), where $T_g$ was replaced by $T_d$, which is the observed temperature at the maximum of the devitrification exotherm as shown in Figure 7.15.

The time dependence of ice formation, which is associated with heating of rapidly cooled solutions, can also be observed from annealing treatments.

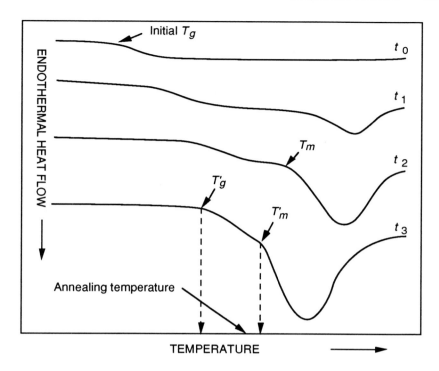

**Figure 7.16** Schematic representation of ice formation during annealing of concentrated solutions as observed with DSC. A rapidly cooled, nonfrozen material shows a glass transition at initial $T_g$. Annealing at a temperature above the onset of glass transition of the maximally freeze-concentrated solute matrix, $T_g'$, but below onset of ice melting in the maximally freeze-concentrated material, $T_m'$, results in time-dependent ice formation. The amount of ice formed increases with annealing time $t_0 < t_1 < t_2 < t_3$ until the maximally freeze-concentrated state is formed after annealing time of $t_3$. Ice formation is observed from increasing $T_g$, from increasing size of the melting endotherm, and from the decreasing onset temperature for ice melting, $T_m$.

Such annealing may be accomplished by heating of a rapidly cooled solution to a predetermined temperature above $T_g$, which is followed by isothermal holding at that temperature and recooling. Roos and Karel (1991c,d) studied time-dependent ice formation in fructose, glucose, and sucrose solutions. Annealing at a temperature slightly below $T_m'$ showed that the observed $T_g$ of the annealed material increased with increasing annealing time. The ascending $T_g$ suggested that plasticizing water was removed from the partially freeze-concentrated solute matrix during annealing. The increase in ice formation was also observed from the size of the ice melting endotherm, which enlarged

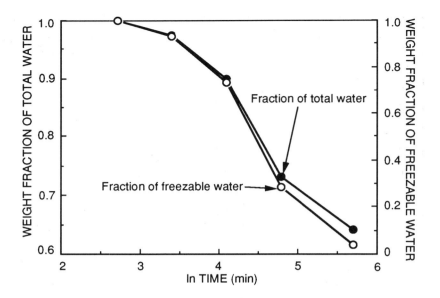

**Figure 7.17** Time-dependent ice formation in a 68% (w/w) sucrose solution as a function of annealing time at -35°C. The maximally freeze-concentrated state was achieved after about 5 hrs of annealing. Data from Roos and Karel (1991c).

with increasing annealing time. A schematic representation of the effect of annealing on ice formation is given in Figure 7.16. It is obvious that the time needed for ice formation in frozen materials is defined by relaxation times above $T_g$. The rate of ice formation in a 68% (w/w) sucrose solution is shown in Figure 7.17. The rate is fairly slow at the beginning of annealing, but it increases with annealing time until the rate decreases as the solution approaches the maximally freeze-concentrated state. The decrease in the unfrozen water content follows fairly well the Avrami equation as well as the stiffness equation with the shape of a modulus curve. However, it should be remembered that although annealing occurs isothermally the $T_g$ increases and the $T - T_g$ decreases with time due to removal of the plasticizing water as ice.

Roos and Karel (1991d) showed that the rate of ice formation above $T_g$ of 65% (w/w) fructose and glucose solutions was defined by $T - T_g$ also at temperatures above $T'_m$. However, the maximally freeze-concentrated state cannot be formed above $T'_m$ and the temperature must be decreased with increasing ice formation. The results of Roos and Karel (1991d) suggested that holding of concentrated solutions close to the equilibrium melting temperature, $T_m$, followed by gradual decrease of temperature to below $T'_m$ can be used to enhance the formation of the maximally freeze-concentrated solute matrix. It should be noticed that the formation of the maximally freeze-concentrated state is preferred in frozen food storage (Levine and Slade, 1988b), but maximum freeze-concentration must be avoided in cryopreservation (Chang and Baust, 1991a,b). Ice formation that occurs during reheating after cryopreservation of viable organisms is one of the main causes of loss of viability and mechanical damage.

Annealing of concentrated solutions below $T_g$ may cause relaxation phenomena associated with $T_g$. Chang and Baust (1990, 1991a,b) found that an endotherm, which was associated with the glass transition temperature of supercooled aqueous glycerol and sorbitol solutions, increased with annealing time below $T_g$ and that the effect was most pronounced when annealing was accomplished at temperatures immediately below the $T_g$. Annealing was also found to decrease the devitrification temperature. Chang and Baust (1991a,b) concluded that in cryopreservation temperatures well below $T_g$ are needed during storage to minimize the effects of annealing on material structure and to avoid the detrimental effects of ice formation during rewarming.

The time-dependent ice formation in sugar solutions has revealed valuable knowledge which can be applied in evaluating and predicting the complicated kinetics of ice formation in food materials. The above kinetic phenomena have been reported for several sugars by various authors. Simatos and Blond (1991) reported glass transitions for amorphous galactose solutions which were obtained for samples with a concentration higher than 55% (w/w). Ablett et al. (1992) reported that sucrose solutions with a concentration of 65% (w/w) or higher vitrified when they were rapidly cooled in DSC. Ablett et al. (1993) reported that fructose glasses were formed when solutions with the initial fructose concentration higher than 60% (w/w) were rapidly cooled in a DSC. Although these studies suggest that fairly high initial solute concentrations are needed to avoid ice formation during freezing, it should be pointed out that freeze-concentration of solutes during the freezing process results in much higher solute concentrations and delayed ice formation at final stages of the freezing process.

The time-dependent nature of ice formation in carbohydrate solutions has been well described, but the kinetic analysis of ice formation at low temperatures requires more experimental data. Blanshard and Franks (1987) pointed out that the overall effect of glass transition and crystallization temperature on crystallization time can be described with time-temperature-transformation (TTT) curves. Such diagrams show the crystallization time at various temperatures and solute concentrations with contour lines. However, kinetic data for ice formation in binary solutions are scarce and establishing such diagrams for more complicated systems, including foods and other biological materials, would require rigorous studies with several techniques. One of the promising techniques for detecting ice formation is NMR imaging (McCarthy and McCarthy, 1994). The technique may provide significant new data on time-dependent characteristics of phase transitions in foods and in particular of factors that affect ice formation at low temperatures.

## 2. Recrystallization of ice

Frozen food materials contain an unfrozen freeze-concentrated solute matrix, which may be in the glassy state or in the liquid-like viscous state within ice crystals that have formed during the freezing process. The ice crystal size that is formed during freezing depends on the freezing rate. Rapid freezing at a low temperature produces a large number of small ice crystals while slow freezing at a high temperature results in the formation of relatively few large crystals. The ice crystals which form during freezing are not stable and various recrystallization mechanisms have been observed in frozen foods depending on temperature (Fennema and Powrie, 1964). Recrystallization phenomena are one of the main reasons for loss of quality during frozen storage. Recrystallization is a temperature-dependent process, which is enhanced by temperature fluctuations. As defined by Fennema (1973a) the term *recrystallization* refers to a change in number, size, shape, orientation, or perfection of ice crystals, which follows completion of the initial freezing.

*a. Recrystallization Mechanisms.* The mechanisms of recrystallization include isomass, migratory, accretive, pressure-induced, and irruptive recrystallization (Fennema, 1973a). However, *irruptive recrystallization* refers to ice formation from highly supercooled freeze-concentrated solute matrices rather than to recrystallization within an existing crystalline phase.

*Isomass recrystallization* refers to a change in the crystal structure, which occurs as a crystal enters a lower energy state without mass transfer with surroundings. Isomass crystallization causes a change in the crystal shape or volume, which occurs without any change in mass. Rounding of the crystal surface is often considered to indicate isomass recrystallization. Donhowe and Hartel (1994a) found significant rounding of ice crystal surfaces in ice cream-type frozen desserts which were stored at -5°C.

*Migratory recrystallization* involves the growth of large ice crystals at the expense of small crystals. Migratory recrystallization results in a decrease of the number of crystals and in an increase of the average crystal size. Migratory recrystallization is often referred to as "Ostwald ripening," which results from differences between surface and free energies. It should also be remembered that small crystals have lower melting temperatures than large crystals.

*Accretive recrystallization*, aggregation of crystals, or sintering of ice results from fusion of contacting crystals, which causes a decrease in the number of crystals and increases crystal size. Rerystallization by accretion is an important recrystallization mechanism in ice cream during storage at constant or fluctuating temperature conditions and particularly at high storage temperatures (Donhowe and Hartel, 1994a).

*Pressure-induced recrystallization* may occur when an external force is applied to crystals. The applied force causes growth of crystals, which have their basal planes aligned with the direction of the force. The growth of the crystals occurs at the expense of crystals that have other orientations.

The main recrystallization mechanisms which affect frozen food quality are migratory recrystallization and accretive recrystallization (Fennema and Powrie, 1964; Fennema, 1973a). However, Donhowe and Hartel (1994a) reported that migratory recrystallization was rarely observed during frozen storage of desserts at several temperatures between -20 and -5°C. Melt-re-freeze crystallization, which involves melting of ice and refreezing of un-frozen water may occur under fluctuating temperatures, especially at relatively high temperatures. Donhowe and Hartel (1994a) found that melt-refreeze crystallization occurred in frozen desserts during exposures to oscillating temperatures between -6 and -8°C.

*b. Recrystallization in Frozen Foods.* Fennema (1973b) pointed out that recrystallization during frozen food storage causes the disappearance of the quality advantages obtained by rapid freezing. Recrystallization of ice in ice cream and other products which are consumed in the frozen state produces a coarse, icy, undesirable texture.

The time-dependent characteristics of recrystallization phenomena in frozen foods have been well established. Differences between the extent of recrystallization at typical temperatures of food storage have suggested that materials with a high natural or added sugar content, e.g., berries and fruits, show considerable recrystallization at -18°C, while recrystallization in frozen fish or in solutions that contain polysaccharides or starch requires higher temperatures (Fennema and Powrie, 1964). The effect of storage temperature and time on the ice crystal size is shown in Figure 7.18.

Ice recrystallization in ice cream has been the subject of a number of studies, probably due to the tremendous effect of the crystal size on mouth-feel. Blanshard and Franks (1987) stated that ice cream with an average ice crystal size of 40 $\mu$m with a distance of 6 to 8 $\mu$m between the crystals is acceptable. The critical size, which produces a noticeable grainy texture, is 40 to 55 $\mu$m. Stabilizers have been found to reduce the growth rate of ice in sugar solutions, but they do not affect the amount of ice formed (Muhr and

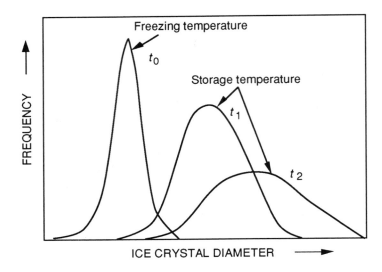

**Figure 7.18** Effect of storage temperature and time on the size of ice crystals in frozen foods. Freezing occurs at a low temperature at time $t_0$ and produces small ice crystals. Freezing is followed by storage at a higher temperature, which allows recrystallization. The size of the ice crystals increases and the range of the crystal sizes increases with increasing storage time from $t_1$ to $t_2$.

Blanshard, 1986; Blanshard and Franks, 1987; Buyong and Fennema, 1988). Levine and Slade (1986) suggested that recrystallization of ice in frozen foods, which often occurs above a critical storage temperature for recrystallization, is governed by glass transition. The cryostabilization technology emphasizes that the glass transition of the maximally freeze-concentrated solute, $T'_g$, governs ice recrystallization rates and that stability is maintained during storage below $T'_g$ (Levine and Slade, 1988a,b, 1989). The effect of $T_g$ on the rate of ice recrystallization is supported by the fact that sugar-containing foods suffer from recrystallization at lower temperatures than those containing polysaccharides. Levine and Slade (1989) suggested that the temperature dependence of ice recrystallization follows the WLF relationship above $T'_g$, which was supported by data on a WLF-type increase of sensory iciness in ice cream products and frozen novelties with increasing $T - T'_g$. The improved cryostabilization and ice recrystallization-reducing effects of polysaccharides in frozen foods accounted for their $T'_g$-elevating properties.

Evaluation of kinetic data on ice recrystallization should consider the effect of ice melting above $T'_m$ on the depression of $T_g$, increase of $T - T_g$, and the additional decrease in relaxation times in comparison to the decrease caused by the lower temperature difference of $T - T'_g$. The results of Blanshard *et al.* (1991) on diffusion of water in sucrose solutions showed that the diffusion coefficient decreased significantly with increasing sucrose concentration, but was not affected by the addition of an alginate stabilizer. This finding suggests that the rate of ice recrystallization in freeze-concentrated solutions may be related to the solute concentration and viscosity of the unfrozen matrix. The viscosity of the unfrozen matrix is probably increased by stabilizers, which decrease the rate of recrystallization. Goff *et al.* (1993) reported that polysaccharide stabilizers did not significantly affect the $T'_g$ of ice cream mixes. Indeed, it would be expected that fairly high amounts of polysaccharides would be needed to have a significant effect on the $T'_g$ or onset of ice melting of frozen sugar matrices, since both water and low molecular weight sugars are plasticizers for the added polysaccharide. Thus, the control of ice crystal growth in stabilized ice cream above $T'_g$ is a function of kinetic properties of the unfrozen solute matrix and the mobility of water within the unfrozen matrix (Goff, 1992; Goff *et al.*, 1993).

Donhowe and Hartel (1994a) studied the rate of ice formation in frozen desserts at constant and oscillating temperatures between -20 and -5°C. They found that the rate of the increase in crystal size was temperature-dependent and followed the modified power-law equation (7.4), where $L_{1,0}$ is the number-based mean crystal size, $L_{1,0}^0$ is the number-based mean crystal size at time $t = 0$, $n$ is the power-law exponent, $\Psi^*$ is the recrystallization rate, and $t$ is time.

$$L_{1,0} = L_{1,0}^0 + \Psi^* t^{1/n} \tag{7.4}$$

Donhowe and Hartel (1994b) fitted the Arrhenius and WLF equations to the temperature dependence of the recrystallization rate of ice cream. The temperature dependence of the rate constant of equation (7.4) followed the Arrhenius equation and gave apparent activation energies of 112 to 126 kJ/mol, which were reported to be high but to agree with those found in previous studies of ice recrystallization kinetics. The data followed also the WLF kinetics, although the differences in the unfrozen water contents between various storage temperatures were not considered. Hagiwara and Hartel (1994) found a linear relationship between ice recrystallization rates and freezing temperatures of ice cream that was made with various sweeteners. The rate of recrystallization at -15.2, -9.4, and -6.3°C increased with increasing temperature and increased at each temperature with decreasing freezing temperature. An increase in the amount of unfrozen water increased the recrystallization rate, but stabilizers decreased the recrystallization rate. Hagiwara and Hartel (1994) reported that the rate of ice recrystallization increased with increasing $T - T'_g$, but the decreased rate due to the use of stabilizers could not be explained by the glass transition theory. It may be concluded that rates of ice recrystallization in frozen foods are affected by composition. The $T'_g$ and $T'_m$ values indicate temperatures for changes in the physical state which control kinetic phenomena. The amount of unfrozen water above $T'_m$ increases significantly with increasing temperature, which decreases viscosity and increases the rate of ice recrystallization. Therefore, the effective $T_g$ that depends on the extent of freeze-concentration should be considered in kinetic studies of ice recrystallization instead of $T'_g$.

*c. Control of Recrystallization.* Ice crystallization during freezing and recrystallization during frozen storage are important factors that affect frozen food quality. Knowledge of the crystallization mechanisms and effects of temperature and time on the physical state can be used to control ice formation and recrystallization in frozen foods. Blanshard and Franks (1987) proposed four itemized strategies for the control of ice crystals in foods. These included (1) the inhibition of nucleation; (2) the control of nucleation; (3) the control of ice crystal growth; and (4) exploitation of the glassy state.

According to Blanshard and Franks (1987) the nucleation of ice in food materials can be inhibited by addition of high amounts of osmotically active materials such as sugars. Then, the unfrozen state together with a low temperature reduces rates of chemical and physical changes, but the detrimental effects of freezing and freeze-concentration can be avoided. The control of nucleation is based on the fact that ice nucleation and crystal growth are both temperature-dependent phenomena. However, their optimum temperatures are different. Extensive nucleation at low freezing temperatures can be used to produce a large number of small ice crystals or less supercooling and nucleation at higher freezing temperatures can be used to produce large ice crystals. The control of ice crystal growth involves addition of compounds, which limit extensive ice crystal growth or ice recrystallization. The added micro- or macromolecular additives such as polysaccharides and proteins may modify diffusion at the ice crystal water interface and thereby control crystal growth. The control of ice nucleation, crystal growth, and ice recrystallization by exploitation of the glassy state is based on the rate-controlling effect of the $T_g$ without added compounds.

The above suggestions for the control of ice formation and ice recrystallization in food materials are used frequently to control the size of ice crystals and the extent of ice recrystallization in frozen foods. Glass transition of the freeze-concentrated solutes and unfrozen water content at temperatures of frozen storage are obviously the most important factors which can be used to control ice formation and ice recrystallization. The $T_g$ and thereby unfrozen water content and kinetics in freeze-concentrated matrices can be modified by proper changes in food composition.

## C. Retrogradation of Starch

*Retrogradation* refers to changes which occur in starch paste, gel, or starch-containing foods on aging. Retrogradation of starch has been of great interest due to its possible role in staling of bread and other textural changes in cereal and other starch-based foods. Starch retrogradation is a temperature- and time-dependent phenomenon, which involves at least partial crystallization of starch components. It can be observed from (1) an increased resistance to hydrolysis by amylolytic enzymes; (2) decreased light transmission; (3) loss of the ability to form the blue complex with iodine; (4) decreased solubility in water; (5) decreased gel compliance or increased modulus of elasticity; and (6) increasing X-ray diffraction (Collison, 1968). It may also be theoretically related to crystallization phenomena, which occur in synthetic polymers (Flory, 1953). According to the definition of Atwell *et al.* (1988) starch retrogradation occurs when molecules comprising gelatinized starch begin to reassociate in an ordered structure, which, under favorable conditions, results in a crystalline order.

### 1. Starch and starch components

Retrogradation of starch has been considered to be basically a crystallization process of gelatinized starch, which can be detected from a change in the appearance of a starch gel or paste, from changes in X-ray diffraction patterns, or from an increasing size of a melting endotherm in DSC scans with increasing storage time. Generally cereal starches retrograde more rapidly than tuber starches. Waxy starches, which contain high amounts of amylopectin, retrograde slowly in comparison with the rapid retrogradation of starches which contain high amounts of linear amylose molecules (Collison, 1968). The rate of retrogradation is also dependent on the ratio of amylopectin and amylose, and on the molecular weight of the starch components. It is generally agreed that crystallization of amylose occurs rapidly after cooling of gelatinized starch (Miles *et al.*, 1984), while amylopectin crystallization is a considerably slower process (Miles *et al.*, 1985).

Based on the data on bread-staling kinetics reported by Cornford *et al.* (1964), which suggested similarities between the temperature dependence of polymer crystallization and changes in the elastic modulus of bread crumb, McIver *et al.* (1968) considered starch retrogradation to be a polymer crystallization process. Therefore, they used the Avrami equation in fitting kinetic data on starch retrogradation. The increase in crystallinity in wheat starch gels was studied with DTA. McIver *et al.* (1968) used the area of the endothermic peak, which appeared during heating over the gelatinization temperature range as a measure of crystallinity. The Avrami equation, which was written into the form of equation (7.5), where $\Delta H_0$ is the initial heat of gelatinization, $\Delta H_f$ is the enthalpy change after crystallization to maximum extent, and $\Delta H_t$ is the enthalpy change after time $t$, was fitted to the data on crystallinity as a function of time.

$$\frac{\Delta H_f - \Delta H_t}{\Delta H_f - \Delta H_0} = e^{-kt^n} \qquad (7.5)$$

However, McIver *et al.* (1968) used the DTA peak area in equation (7.5) instead of the $\Delta H$ values, which have been more applicable for results obtained with DSC in later studies (Fearn and Russell, 1982). McIver *et al.* (1968) found that the DTA peak area of gelatinized wheat starch gels increased with increasing storage time. Data on the increase in crystallinity as determined from the change in the peak area with increasing storage time followed equation (7.5). The constant $n$ at 21°C was found to be about 1, which suggested that the crystallization process was instantaneous and that the growth of the crystallites occurred in one dimension only. However, the use of the value obtained for $n$ as an indicator of the crystal morphology of starch has been criticized (Fearn and Russell, 1982). McIver *et al.* (1968) pointed out that gelatinized starch may contain ordered regions which form nuclei for crystal growth at the crystallization temperature. A subsequent study by Colwell *et al.* (1969) showed that the constant $n$ of equation (7.5) was about unity at several crystallization temperatures between -1 and 43°C. Colwell *et al.* (1969) found also that the temperature at the DTA peak maximum increased with increasing storage temperature. Based on the results of the extent of crystallization at several temperatures Colwell *et al.* (1969) concluded that the extent of retrogradation in starch gels above 21°C was less than was suggested by the results of Cornford *et al.* (1964) for firming of bread crumb.

Russell (1987) found that the Avrami exponent, $n$, was less than unity for retrogradation of starch gels of various starches. The water content of the gels was 57% (w/w) and they were stored at 21°C. The limiting enthalpies at maximum crystallization increased in the order amylomaize < wheat ≈ potato < waxy maize. The order was related to the amylopectin content of the starches and suggested that the increase of $\Delta H_t$ with time was due to the amylopectin fraction. Russell (1987) found also that the rate constant, $k$, of equation (7.5) was sensitive to the time of observation and especially to the number of data points at the beginning and at the end of retrogradation. However, rate constants of amylomaize, waxy maize, and potato starch gels were significantly higher than the rate constant of wheat starch gel, which suggested that retrogradation was faster in gels, which had a lower amylose content or contained no lipids. Russell (1987) concluded that DSC may be used to get information on the crystallization of the amylopectin fraction in starch gels, but the total crystallinity should be determined by X-ray diffraction measurements.

The rate of starch retrogradation is dependent on water content. Zeleznak and Hoseney (1986) reported that retrogradation of gelatinized wheat starch with water contents between 20 and 80% (w/w) was slow at low and high water contents. An increase of melting enthalpy during storage at 25°C showed a maximum at water contents between 50 and 60% (w/w). Levine and Slade (e.g., 1990; Slade and Levine, 1989) have emphasized the use of the fringed-micelle model in the description of the partially crystalline structure of starch. The model is particularly useful in the description of the partially crystalline structure of starch gels, which contain amorphous regions and microcrystalline regions as junction zones between molecules (Slade and Levine, 1989). Slade and Levine (1989) considered retrogradation to be a nonequilibrium polymer crystallization process in completely amorphous starch-water melts, which proceeds with a rate that is determined by temperature and water content. The slow crystallization of amylopectin was referred to as a nucleation-limited growth process, which occurs above glass transition in a mobile, viscoelastic, fringed-micelle network. The crystallization mechanism is thermally reversible above melting temperature and applies to both amylose and amylopectin. The effect of $T_g$ on the rate of crystallization in starch has not been experimentally verified, but the fact that retrogradation is faster at temperatures close to 0°C than at room temperature suggests that the rate approaches a maximum at a temperature between $T_g$ and $T_m$ according to polymer crystallization theories (Flory, 1953; Slade and Levine, 1989).

The glass transition temperature which controls crystallization in 50% (w/w) starch gels is approximately at -75°C and the melting temperature may be assumed to be at 50 to 60°C. It is probable that the rate of crystallization increases below $T_m$ and reaches a maximum well above $T_g$. The above suggestion for the $T_g$ and $T_m$ values agrees fairly well with the observed higher crystallization rates at temperatures close to 0°C than at 25°C. Obviously freeze-concentration below the freezing temperature reduces the rate of crystallization. Laine and Roos (1994) studied the effect of water content and $T - T_g$ on the extent of crystallization of gelatinized corn starch at fairly low water contents and high storage temperatures. Crystallization was found to occur above $T_g$, but the extent of crystallization as determined from the limiting enthalpy value after 10 days with DSC increased as the $T - T_g$ increased from 20 to 75°C. It is interesting to notice that the melting temperature of retrograded starch gels has also been observed to depend on storage temperature. These differences have been attributed to the perfection of the crystals formed (Slade and Levine, 1989), which may also explain the differences in melting enthalpies observed by Laine and Roos (1994). It may be concluded that starch retrogradation is a typical polymer crystallization process that occurs most likely to various extents depending on the amount of plasticizer and the temperature in relation to $T_g$ and $T_m$.

## 2. Staling of bread

Studies on bread staling have often concluded that staling occurs due to changes in the starch fraction, which shows increasing crystallinity, which can be measured by X-ray diffraction (e.g., Cornford *et al.*, 1964; Zobel, 1973). Cornford *et al.* (1964) pointed out that the rate of the increase in starch crystallinity increases with decreasing temperature. The crystallization process in high-polymer systems was stated to be slow just below the melting temperature, to become faster with increasing supercooling, and then to decrease to zero at lower temperatures, where the molecular mobility is insufficient to allow crystallization. Basically, the principles found to affect time-dependent crystallization in starch gels apply also to starch retrogradation in bread. However, staling may be affected by other food components.

The similarities between polymer crystallization and bread staling according to Cornford *et al.* (1964) suggested that the Avrami equation could be used in modeling the rate of bread staling. They considered the crumb modulus, $E$, to be a linear measure of the degree of crystallization and that the fraction of the uncrystallized material, $1 - \alpha$, was obtained from equation (7.6), where $E_0$ and $E_f$ refer to the initial and final elastic modulus, respectively, and $E_t$ is the modulus at time $t$. Cornford *et al.* (1964) found that the elastic modulus had a limiting value that was used as $E_f$.

$$1 - \alpha = \frac{E_f - E_t}{E_f - E_0} \qquad (7.6)$$

$$\frac{E_f - E_t}{E_f - E_0} = e^{-kt^n} \qquad (7.7)$$

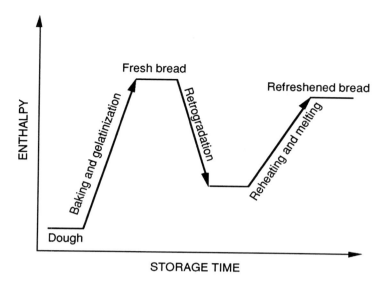

**Figure 7.19** Effect of baking, storage, and reheating on the enthalpy changes observed in bread. The initial enthalpy change occurs during baking due to crystallite melting in the starch fraction. After cooling of bread heat is released due to crystallization in starch. Crystallites formed during retrogradation melt during reheating, which causes the appearance of an endotherm in DSC thermograms.

Cornford *et al.* (1964) gave the Avrami equation in the form of equation (7.7) and stated that equation (7.7) fitted to modulus data with $n = 1$. Therefore, a plot of log $(E_f - E_t)$ against $t$ gave a straight line and a time constant equal to $1/k$ was used in the characterization of the temperature dependence of the staling process. The time constant increased with increasing storage temperature, which showed that the rate of the change in modulus decreased with increasing storage temperature over the temperature range of -1 to 32°C.

Although the Avrami equation was found to fit to the data both on change in modulus and on the DTA peak area (Cornford *et al.*, 1964; McIver *et al.*, 1968) the change in modulus was not necessarily a direct measure of crystallinity. Fearn and Russell (1982) stated that the modulus is a measure of a macroscopic property, while DSC measures a microscopic property of bread. The modulus contains contributions from various bread crumb components and the cell structure, which obviously are independent of crystallinity. It should also be pointed out that several studies have suggested that the value $n$ < 1 in the Avrami equation applies for retrogradation of both starch gels and bread (Russell, 1987). In DSC studies of bread staling equation (7.5) may be fitted to kinetic data. The relationships between melting enthalpy change and storage time are shown schematically in Figure 7.19.

Bread staling as primarily a consequence of time-dependent amylopectin crystallization is governed by glass transition, although it is also affected by composition (Kim and D'Appolonia, 1977; Kulp and Ponte, 1981). Therefore, accelerated staling can be achieved by allowing nucleation to occur at a low temperature followed by storage close to melting temperature, which increases the rate of crystal growth (Slade and Levine, 1989). Laine and Roos (1994) showed that crystallization in white bread occurred above $T_g$ at low water contents. However, experimental data on the relationships between storage $T - T_g$, water content, and storage time for bread staling are scarce.

# References

Ablett, S., Izzard, M.J. and Lillford, P.J. 1992. Differential scanning calorimetric study of frozen sucrose and glycerol solutions. *J. Chem. Soc., Faraday Trans. 88*: 789-794.

Ablett, S., Izzard, M.J., Lillford, P.J., Arvanitoyannis, I. and Blanshard, J.M.V. 1993. Calorimetric study of the glass transition occurring in fructose solutions. *Carbohydr. Res. 246*: 13-22.

Atwell, W.A., Hood, L.F., Lineback, D.R., Varriano-Marston, E. and Zobel, H.F. 1988. The terminology and methodology associated with basic starch phenomena. *Cereal Foods World 33*: 306-311.

Berlin, E., Anderson, A.B. and Pallansch, M.J. 1968a. Water vapor sorption properties of various dried milks and wheys. *J. Dairy Sci. 51*: 1339-1344.

Berlin, E., Anderson, B.A. and Pallansch, M.J. 1968b. Comparison of water vapor sorption by milk powder components. *J. Dairy Sci. 51*: 1912-1915.

Berlin, E., Anderson, B.A. and Pallansch, M.J. 1970. Effect of temperature on water vapor sorption by dried milk powders. *J. Dairy Sci. 53*: 146-149.

Blanshard, J.M.V. and Franks, F. 1987. Ice crystallization and its control in frozen-food systems. Chpt. 4 in *Food Structure and Behaviour*, ed. J.M.V. Blanshard and P. Lillford. Academic Press, Orlando, FL, pp. 51-65.

Blanshard, J.M.V., Muhr, A.H. and Gough, A. 1991. Crystallization from concentrated sucrose solutions. In *Water Relationships in Foods*, ed. H. Levine and L. Slade. Plenum Press, New York, pp. 639-655.

Brennan, J.G., Herrera, J. and Jowitt, R. 1971. A study of the factors affecting the spray drying of concentrated orange juice, on a laboratory scale. *J. Food Technol. 6*: 295-307.

Buyong, N. and Fennema, O. 1988. Amount and size of ice crystals in frozen samples as influenced by hydrocolloids. *J. Dairy Sci. 71*: 2630-2639.

Chang, Z.H. and Baust, J.G. 1990. Effects of sub-$T_g$ isothermal annealing on the glass transition of D-sorbitol. *Cryo-Lett. 11*: 251-256.

Chang, Z.H. and Baust, J.G. 1991a. Physical aging of glassy state: DSC study of vitrified glycerol systems. *Cryobiology 28*: 87-95.

Chang, Z.H. and Baust, J.G. 1991b. Physical aging of the glassy state: Sub-$T_g$ ice nucleation in aqueous sorbitol systems. *J. Non-Cryst. Solids 130*: 198-203.

Chirife, J. and Buera, M.P. 1994. Water activity, glass transition and microbial stability in concentrated/semimoist food systems. *J. Food Sci. 59*: 921-927.

Chuy, L.E. and Labuza, T.P. 1994. Caking and stickiness of dairy-based food powders as related to glass transition. *J. Food Sci. 59*: 43-46.

Collison, R. 1968. Starch retrogradation. Chpt. 6 in *Starch and Its Derivatives*, ed. J.A. Radley, 4th ed. Chapman and Hall, London, pp. 194-202.

Colwell, K.H., Axford, D.W.E., Chamberlain, N. and Elton, G.A.H. 1969. Effect of storage temperature on the ageing of concentrated wheat starch gels. *J. Sci. Food Agric. 20*: 550-555.

Cornford, S.J., Axford, D.W.E. and Elton, G.A.H. 1964. The elastic modulus of bread crumb in linear compression in relation to staling. *Cereal Chem. 41*: 216-229.

Donhowe, D.P. and Hartel, R.W. 1994a. Influence of temperature on ice recrystallization in frozen desserts. I. Accelerated storage studies. *Food Struct.* In press.

Donhowe, D.P. and Hartel, R.W. 1994b. Influence of temperature on ice recrystallization in frozen desserts. II. Bulk storage studies. *Food Struct.* In press.

Downton, G.E., Flores-Luna, J.L. and King, C.J. 1982. Mechanism of stickiness in hygroscopic, amorphous powders. *Ind. Eng. Chem. Fundam. 21*: 447-451.

Fearn, T. and Russell, P.L. 1982. A kinetic study of bread staling by differential scanning calorimetry. The effect of loaf specific volume. *J. Sci. Food Agric. 33*: 537-548.

Fennema, O.R. 1973a. Nature of the freezing process. Chpt. 4 in *Low-Temperature Preservation of Foods and Living Matter*, ed. O.R. Fennema, W.D. Powrie and E.H. Marth. Marcel Dekker, New York, pp. 150-239.

Fennema, O.R. 1973b. Freeze-preservation of foods - technological aspects. Chpt. 11 in *Low-Temperature Preservation of Foods and Living Matter*, ed. O.R. Fennema, W.D. Powrie and E.H. Marth. Marcel Dekker, New York, pp. 504-550.

Fennema, O.R. and Powrie, W.D. 1964. Fundamental of low-temperature food preservation. *Adv. Food Res. 13*: 219-347.

Ferry, J.D. 1980. *Viscoelastic Properties of Polymers*, 3rd ed. John Wiley & Sons, New York.

Flink, J.M. 1983. Structure and structure transitions in dried carbohydrate materials. Chpt. 17 in *Physical Properties of Foods*, ed. M. Peleg and E.B. Bagley. AVI Publishing Co. Westport, CT, pp. 473-521.

Flory, P.J. 1953. *Principles of Polymer Chemistry*. Cornell University Press, Ithaca, NY.

Franks, F. 1990. Freeze drying: From empiricism to predictability. *Cryo-Lett. 11*: 93-110.

Goff, H.D. 1992. Low-temperature stability and the glassy state in frozen foods. *Food Res. Intl 25*: 317-325.

Goff, H.D., Caldwell, K.B., Stanley, D.W. and Maurice, T.J. 1993. The influence of polysaccharides on the glass transition in frozen sucrose solutions and ice cream. *J. Dairy Sci. 76*: 1268-1277.

Hagiwara, T. and Hartel, R.W. 1994. Influence of storage temperature on ice recrystallization in ice cream. Paper No. 71C-3. Presented at the Annual Meeting of the Institute of Food Technologists, Atlanta, GA, June 25-29.

Hartel, R.W. 1993. Controlling sugar crystallization in food products. *Food Technol. 47(11)*: 99-104, 106-107.

Heldman, D.R. and Singh, R.P. 1981. *Food Process Engineering*, 2nd ed. AVI Publishing Co., Westport, CT.

Herrington, B.L. 1934. Some physico-chemical properties of lactose. I. The spontaneous crystallization of supersaturated solutions of lactose. *J. Dairy Sci. 17*: 501-518.

Herrington, T.M. and Branfield, A.C. 1984a. Physico-chemical studies on sugar glasses. I. Rates of crystallization. *J. Food Technol. 19*: 409-425.

Herrington, T.M. and Branfield, A.C. 1984b. Physico-chemical studies on sugar glasses. II. Glass transition temperature. *J. Food Technol. 19*: 427-435.

Iglesias, H.A. and Chirife, J. 1978. Delayed crystallization of amorphous sucrose in humidified freeze dried model systems. *J. Food Technol. 13*: 137-144.

Jouppila, K. and Roos, Y.H. 1994a. Water sorption and time-dependent phenomena of milk powders. *J. Dairy Sci. 77*: 1798-1808.

Jouppila, K. and Roos, Y.H. 1994b. Glass transitions and crystallization in milk powders. *J. Dairy Sci. 77*: 2907-2915.

Kalichevsky, M.T., Jaroszkiewicz, E.M. and Blanshard, J.M.V. 1992. Glass transition of gluten. 2: The effect of lipids and emulsifiers. *Int. J. Biol. Macromol. 14*: 267-273.

Kalichevsky, M.T., Blanshard, J.M.V. and Marsh, R.D.L. 1993. Applications of mechanical spectroscopy to the study of glassy biopolymers and related systems. Chpt. 6 in *The Glassy State in Foods*, ed. J.M.V. Blanshard and P.J. Lillford. Nottingham University Press, Loughborough, pp. 133-156.

Karel, M. 1973. Recent research and development in the field of low-moisture and intermediate-moisture foods. *CRC Crit. Rev. Food Technol. 3*: 329-373.

Kim, S.K. and D'Appolonia, B.L. 1977. The role of wheat flour constituents in bread staling. *Baker's Dig. 51(1)*: 38-44, 57.

King, N. 1965. The physical structure of dried milk. *Dairy Sci. Abstr. 27*: 91-104.

Kulp, K. and Ponte, J.G. 1981. Staling of white pan bread: Fundamental causes. *CRC Crit. Rev. Food Sci. Nutr. 15*: 1-48.

Laine, M.J.K. and Roos, Y. 1994. Water plasticization and recrystallization of starch in relation to glass transition. In *Proceedings of the Poster Session, International Symposium on the Properties of Water, Practicum II*, ed. A. Argaiz, A. López-Malo, E. Palou and P. Corte. Universidad de las Américas-Puebla, pp. 109-112.

Lazar, M.E., Brown, A.H., Smith, G.S., Wong, F.F. and Linquist, F.E. 1956. Experimental production of tomato powder by spray drying. *Food Technol. 10*: 129-134.

Levine, H. and Slade, L. 1986. A polymer physico-chemical approach to the study of commercial starch hydrolysis products (SHPs). *Carbohydr. Polym. 6*: 213-244.

Levine, H. and Slade, L. 1988a. 'Collapse' phenomena - A unifying concept for interpreting the behaviour of low moisture foods. Chpt. 9 in *Food Structure - Its Creation and Evaluation*, ed. J.M.V. Blanshard and J.R. Mitchell. Butterworths, London, pp. 149-180.

Levine, H. and Slade, L. 1988b. Principles of "cryostabilization" technology from structure/property relationships of carbohydrate/water systems - A review. *Cryo-Lett. 9*: 21-63.

Levine, H. and Slade, L. 1989. A food polymer science approach to the practice of cryostabilization technology. *Comments Agric. Food Chem. 1*: 315-396.

Levine, H. and Slade, L. 1990. Influences of the glassy and rubbery states on the thermal, mechanical, and structural properties of doughs and baked products. Chpt. 5 in *Dough Rheology and Baked Product Texture*, ed. H. Faridi and J.M. Faubion. AVI Publishing Co., New York, pp. 157-330.

Levine, H. and Slade, L. 1993. The glassy state in applications for the food industry, with an emphasis on cookie and cracker production. Chpt. 17 in *The Glassy State in Foods*, ed. J.M.V. Blanshard and P.J. Lillford. Nottingham University Press, Loughborough, pp. 333-373.

Luyet, B. and Rasmussen, D. 1968. Study by differential thermal analysis of the temperatures of instability of rapidly cooled solutions of glycerol, ethylene glycol, sucrose and glucose. *Biodynamica 10(211)*: 167-191.

MacFarlane, D.R. 1986. Devitrification in glass-forming aqueous solutions. *Cryobiology 23*: 230-244.

Makower, B. and Dye, W.B. 1956. Equilibrium moisture content and crystallization of amorphous sucrose and glucose. *J. Agric. Food Chem. 4*: 72-77.

McCarthy, M.J. and McCarthy, K.L. 1994. Quantifying transport phenomena in food processing with nuclear magnetic resonance imaging. *J. Sci. Food Agric. 65*: 257-270.

McIver, R.G., Axford, D.W.E., Colwell, K.H. and Elton, G.A.H. 1968. Kinetic study of the retrogradation of gelatinized starch. *J. Sci. Food Agric. 19*: 560-564.

McNulty, P.B. and Flynn, D.G. 1977. Force-deformation and texture profile behavior of aqueous sugar glasses. *J. Texture Stud. 8*: 417-431.

Miles, M.J., Morris, V.J. and Ring, S.G. 1984. Some recent observations on the retrogradation of amylose. *Carbohydr. Polym. 4*: 73-77.

Miles, M.J., Morris, V.J., Orford, P.D. and Ring, S.G. 1985. The roles of amylose and amylopectin in the gelation and retrogradation of starch. *Carbohyr. Res. 135*: 271-281.

Muhr, A.H. and Blanshard, J.M.V. 1986. Effect of polysaccharide stabilizers on the rate of growth of ice. *J. Food Technol. 21*: 683-710.

Nickerson, T.A. 1974. Lactose. Chpt. 6 in *Fundamental of Dairy Chemistry*, ed. B.H. Webb, A.H. Johnson and J.A. Alford, 2nd ed. AVI Publishing Co., Westport, CT, pp. 273-324.

Nickerson, T.A. and Moore, E.E. 1972. Solubility interrelations of lactose and sucrose. *J. Food Sci. 37*: 60-61.

Noel, T.R., Ring, S.G. and Whittam, M.A. 1992. Dielectric relaxations of small carbohydrate molecules in the liquid and glassy states. *J. Phys. Chem. 96*: 5662-5667.

Noel, T.R., Ring, S.G. and Whittam, M.A. 1993. Relaxations in supercooled carbohydrate liquids. Chpt. 8, in *The Glassy State in Foods*, ed. J.M.V. Blanshard and P.J. Lillford. Nottingham University Press, Loughborough, pp. 173-187.

Peleg, M. 1977. Flowability of food powders and methods for its evaluation. *J. Food Process Eng. 1*: 303-328.

Peleg, M. 1983. Physical characteristics of food powders. Chpt. 10 in *Physical Properties of Foods*, ed. M. Peleg and E.B. Bagley. AVI Publishing Co., Westport, CT, pp. 293-323.

Peleg, M. 1992. On the use of the WLF model in polymers and foods. *Crit. Rev. Food Sci. Nutr. 32*: 59-66.

Peleg, M. 1993. Mapping the stiffness-temperature-moisture relationship of solid biomaterials at and around their glass transition. *Rheol. Acta 32*: 575-580.

Peleg, M. and Mannheim, C.H. 1977. The mechanism of caking of powdered onion. *J. Food Process. Preserv. 1*: 3-11.

Perez, J. 1994. Theories of liquid-glass transition. *J. Food Eng. 22*: 89-114.

Roos, Y.H. 1987. Effect of moisture on the thermal behavior of strawberries studied using differential scanning calorimetry. *J. Food Sci. 52*: 146-149.

Roos, Y.H. 1993a. Water activity and physical state effects on amorphous food stability. *J. Food Process. Preserv. 16*: 433-447.

Roos, Y. 1993b. Melting and glass transitions of low molecular weight carbohydrates. *Carbohydr. Res. 238*: 39-48.

Roos, Y. and Karel, M. 1990. Differential scanning calorimetry study of phase transitions affecting the quality of dehydrated materials. *Biotechnol. Prog. 6*: 159-163.

Roos, Y. and Karel, M. 1991a. Plasticizing effect of water on thermal behavior and crystallization of amorphous food models. *J. Food Sci. 56*: 38-43.

Roos, Y. and Karel, M. 1991b. Phase transitions of mixtures of amorphous polysaccharides and sugars. *Biotechnol. Prog. 7*: 49-53.

Roos, Y. and Karel, M. 1991c. Amorphous state and delayed ice formation in sucrose solutions. *Int. J. Food Sci. Technol. 26*: 553-566.

Roos, Y. and Karel, M. 1991d. Nonequilibrium ice formation in carbohydrate solutions. *Cryo-Lett. 12*: 367-376.

Roos, Y. and Karel, M. 1991e. Water and molecular weight effects on glass transitions in amorphous carbohydrates and carbohydrate solutions. *J. Food Sci. 56*: 1676-1681.

Roos, Y. and Karel, M. 1992. Crystallization of amorphous lactose. *J. Food Sci. 57*: 775-777.

Russell, P.L. 1987. The ageing of gels from starches of different amylose/amylopectin content studied by differential scanning calorimetry. *J. Cereal Sci. 6*: 147-158.

Sá, M.M. and Sereno, A.M. 1994. Glass transitions and state diagrams for typical natural fruits and vegetables. *Thermochim. Acta 246*: 285-297.

Saleki-Gerhardt, A. and Zografi, G. 1994. Nonisothermal and isothermal crystallization of sucrose from the amorphous state. *Pharm. Res. 11*: 1166-1173.

Saltmarch, M. and Labuza, T.P. 1980, Influence of relative humidity on the physicochemical state of lactose in spray-dried sweet whey powders. *J. Food Sci. 45*: 1231-1236, 1242.

San Jose, C., Asp, N.-G., Burvall, A. and Dahlquist, A. 1977. Water sorption in hydrolyzed dry milk. *J. Dairy Sci. 60*: 1539-1543.

Shogren, R.L. 1992. Effect of moisture content on the melting and subsequent physical aging of cornstarch. *Carbohydr. Polym. 19*: 83-90.

Simatos, D. and Blond, G. 1975. The porous structure of freeze dried products. Chpt. 24 in *Freeze Drying and Advanced Food Technology*, ed. S.A. Goldblith, L. Rey and W.W. Rothmayr. Academic Press, New York, pp. 401-412.

Simatos, D. and Blond, G. 1991. DSC studies and stability of frozen foods. In *Water Relationships in Foods*, ed. H. Levine and L. Slade. Plenum Press, New York, pp. 139-155.

Simatos, D. and Turc, J.M. 1975. Fundamentals of freezing in biological systems. Chpt. 2 in *Freeze Drying and Advanced Food Technology*, ed. S.A. Goldblith, L. Rey and W.W. Rothmayr. Academic Press, New York, pp. 17-28.

Slade, L. and Levine, H. 1989. A food polymer science approach to selected aspects of starch gelatinization and retrogradation. In *Frontiers in Carbohydrate Research - 1: Food Applications*, ed. R.P. Millane, J.N. BeMiller and R. Chandrasekaran. Elsevier, London, pp. 215-270.

Slade, L. and Levine, H. 1991. Beyond water activity: Recent advances based on an alternative approach to the assessment of food quality and safety. *Crit. Rev. Food Sci. Nutr. 30*: 115-360.

Slade, L. and Levine, H. 1995. Glass transitions and water-food structure interactions. *Adv. Food Nutr. Res. 38*. In press.

Soesanto, T. and Williams, M.C. 1981. Volumetric interpretation of viscosity for concentrated and dilute sugar solutions. *J. Phys. Chem. 85*: 3338-3341.

Struik, L.C.E. 1978. *Physical Aging in Amorphous Polymers and Other Materials.* Elsevier, Amsterdam.

Supplee, G.C. 1926. Humidity equilibria of milk powders. *J. Dairy Sci. 9:* 50-61.

Tant, M.R. and Wilkes, G.L. 1981. An overview of the nonequilibrium behavior of polymer glasses. *Polym. Eng. Sci. 21:* 874-895.

To, E.T. and Flink, J.M. 1978a. 'Collapse,' a structural transition in freeze dried carbohydrates. I. Evaluation of analytical methods. *J. Food Technol. 13:* 551-565.

To, E.T. and Flink, J.M. 1978b. 'Collapse,' a structural transition in freeze dried carbohydrates. II. Effect of solute composition. *J. Food Technol. 13:* 567-581.

To, E.T. and Flink, J.M. 1978c. 'Collapse,' a structural transition in freeze dried carbohydrates. III. Prerequisite of recrystallization. *J. Food Technol. 13:* 583-594.

Troy, H.C. and Sharp, P.F. 1930. $\alpha$ and $\beta$ lactose in some milk products. *J. Dairy Sci. 13:* 140-157.

Tsourouflis, S., Flink, M. and Karel, M. 1976. Loss of structure in freeze-dried carbohydrates solutions: Effect of temperature, moisture content and composition. *J. Sci. Food Agric. 27:* 509-519.

Vuataz, G. 1988. Preservation of skim-milk powders: Role of water activity and temperature in lactose crystallization and lysine loss. In *Food Preservation by Water Activity Control*, ed. C.C. Seow. Elsevier, Amsterdam, pp. 73-101.

Wallack, D.A. and King, C.J. 1988. Sticking and agglomeration of hygroscopic, amorphous carbohydrate and food powders. *Biotechnol. Prog. 4:* 31-35.

Warburton, S. and Pixton, S.W. 1978. The moisture relations of spray dried skimmed milk. *J. Stored Prod. Res. 14:* 143-158.

Weitz, A. and Wunderlich, B. 1974. Thermal analysis and dilatometry of glasses formed under elevated pressure. *J. Polym., Sci. Polym. Phys. Ed. 12:* 2473-2491.

White, G.W. and Cakebread, S.H. 1966. The glassy state in certain sugar-containing food products. *J. Food Technol. 1:* 73-82.

Wunderlich, B. 1981. The basis of thermal analysis. Chpt. 2, in *Thermal Characterization of Polymeric Materials*, ed. E.A. Turi. Academic Press, New York, pp. 91-234.

Young, F.E., Jones, F.T. and Lewis, H.J. 1951. Prevention of the growth of sucrose hydrates in sucrose sirups. *Food Res. 16:* 20-29.

Zeleznak, K.J. and Hoseney, R.C. 1986. The role of water in the retrogradation of wheat starch gels and bread crumb. *Cereal Chem. 63:* 407-411.

Zobel, H.F. 1973. A review of bread staling. *Baker's Dig. 47(10):* 52-61.

# CHAPTER 8

# *Mechanical Properties*

## I. Introduction

Mechanical properties of foods are important to their behavior in processing, storage, distribution, and consumption. Information on various compositional factors on mechanical properties and their dependence on temperature and water content is needed for choosing proper equipment that is required, e.g., for transportation, mixing, and size reduction, as well as for the evaluation of the tolerance of mechanical stress during manufacturing. Food packaging is often used to protect foods from mechanical stress or transfer of water between the product and its environment. Mechanical properties affect also the perceived texture of foods during consumption.

Several compositional and processing parameters affect mechanical properties of foods. The overall mechanical properties may be defined by cell structure which is typical of various vegetables and fruits or they may result from the physical state, flow properties, or porosity. However, the physical state of food solids is one of the most important factors that affect mechanical properties of low-moisture and frozen foods in which small changes in temperature or water content may significantly affect the physical state due to phase transitions. Therefore, the physical state may affect the overall quality of products starting from changes that may occur during processing, e.g., cookies (Levine and Slade, 1993; Slade *et al.*, 1993), to the final stage of consumption, since the mechanical properties define product texture and sensory properties, e.g., perceived crispness (Katz and Labuza, 1981; Sauvageot and Blond, 1991). It should also be remembered that the mechanical properties of fatty foods are almost exclusively defined by the physical state and crystallinity of the lipid fraction.

This chapter provides information on factors that affect mechanical properties of food materials. The relationships between mechanical properties and the physical state and phase transitions during food storage will be discussed with an emphasis on the effects of mechanical properties on food quality. The main emphasis is on those mechanical properties which are related to phase transitions occurring in the nonequilibrium state of nonfat food solids and their time-dependent characteristics.

## II.  Stiffness

*Stiffness* is used here as a general term that refers to the response of food materials to an external stress. The various effects of melting and glass transitions on relaxation times of mechanical properties suggest that changes in stiffness are obvious and occur over such transition temperature ranges. Changes in stiffness may be observed from modulus curves which may describe changes in mechanical properties as a function of frequency, time, temperature, water activity, or water content. Water plasticization is often equal to thermal plasticization and therefore water at a constant temperature has an effect on mechanical properties similar to that caused by an increase in temperature at a constant water content. It is also well known that food materials give a modulus curve when a mechanical property is plotted against water content or water activity (Slade and Levine, 1993; Peleg, 1993, 1994a,b).

### A. Modulus Curves of Food Materials

A number of studies of mechanical properties of foods have not related observed changes to phase transitions, although changes in mechanical properties have been observed to occur as a function of temperature, water activity, and water content. The first studies which reported temperature, water content, and time dependence of mechanical properties of biological, food-related materials were those by Nakamura and Tobolsky (1967) on starch, the study by Hammerle and Mohsenin (1970) on corn endosperm, and the results on mechanical properties of collagen and gelatin as summarized by Yannas (1972). These studies suggested that temperature, diluent content, and time have an equal effect on the modulus, which can be superimposed with any of the variables. Hammerle and Mohsenin (1970) used the WLF relationship for the calculation of time-temperature and time-water content shift factors. In a later study the time-temperature and time-moisture content relationships of rice kernels were found to follow the general behavior of viscoelastic materials (Chattopadhyay et al., 1979).

1. Effect of water on mechanical properties

Water in foods is the main plasticizer which may significantly affect mechanical properties during processing and storage. Changes in the mechanical properties as a function of water activity or water content can be described with modulus curves, which reflect the change in viscoelastic properties above glass transition.

The viscoelastic behavior of cereals has been studied as a function of temperature and water content due to its importance to the rheological behavior of grains in milling. Hammerle and Mohsenin (1970) determined modulus data for corn endosperm as a function of time at several temperatures. The WLF method was used to determine shift factors for time-temperature superpositioning of the modulus data. They collected modulus data also as a function of water content and used the same method for time-moisture superpositioning. The data were superimposed using a time-temperature and time-moisture superpositioning principle to a single master curve at 25°C and 13% (w/w) water. Hammerle and Mohsenin (1970) did not report the $T_g$ for corn endosperm, but the shape of the master curve as well as the effect of temperature and water content on the modulus suggested that the material was in the glassy state at low water contents. An increase in temperature or water content resulted in plasticization and changes in modulus, which were typical of viscoelastic amorphous polymers. Similar viscoelastic behavior has also been observed to occur in other cereal materials. Multon et al. (1981) reported data on elastic modulus for wheat grains as a function of water activity and water content. The modulus curves showed that water activity and water content had a significant effect on modulus, which decreased with increasing water activity or water content. No glass transitions were reported for the material, but Multon et al. (1981) concluded that a homogeneous and vitreous structure with high modulus was more difficult to grind and resulted in a larger particle size and a higher protein content of the flour.

Both water and temperature plasticization decrease the modulus of biological materials in a similar manner and the modulus curves may be superimposed as was suggested by Hammerle and Moshenin (1970). The effect of water plasticization on the modulus of elastin was reported by Lillie and Gosline (1990). They concluded that modulus data obtained at various water contents and temperatures could be superimposed and that the WLF equation was applicable in predicting isoshift curves. Modulus data for wheat gluten as a function of water content were reported by Attenburrow *et al.* (1990). The modulus values indicated that the material was glassy at water contents below about 10% (w/w) and a rapid decrease of the modulus occurred at higher water contents as the material was transformed to the rubbery state due to plasticization. Attenburrow and Davies (1993) used ice cream wafers as a model system to examine the influence of water content on the mechanical properties of low-moisture foods. The effect of water on Young's modulus had the typical shape of a modulus curve. A schematic representation of the effect of water activity or water content on modulus is given in Figure 8.1. Additional evidence of the effects of water on the modulus of

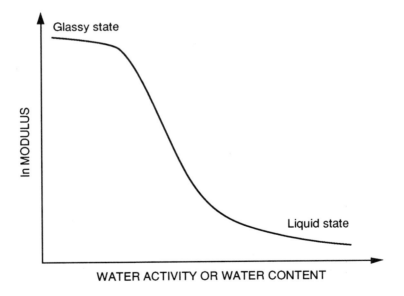

**Figure 8.1** Effect of water plasticization on the modulus of amorphous biopolymers and food materials. Water plasticization decreases the modulus as the glass transition is depressed to below ambient temperature.

food materials has been reported by Kalichevsky *et al.* (1992, 1993) and it is obvious that the decrease of modulus of amorphous food materials occurs due to plasticization. At a sufficiently high water content the material loses the glassy structure and becomes a viscoelastic liquid. The effect of water on the modulus of food solids is similar to the effect of plasticizers on mechanical properties of synthetic polymers at a constant temperature.

## 2. Mathematical analysis of stiffness

The effect of water on the modulus of food materials suggests that stiffness as a function of modulus is related to the physical state and plasticization of food solids. Foods with a glassy structure have a high modulus and viscosity and obviously their stiffnesses are high. A dramatic change in stiffness is observed when plasticization of a glassy material becomes sufficient to cause the transformation into the supercooled liquid state. A master curve of modulus may be obtained with superimposed data and used in the characterization of effects of temperature and water on the mechanical properties. However, characterization of the mechanical properties directly in terms of temperature or water is often more important in practical applications.

Peleg (1993, 1994a,b) pointed out that mechanical changes in amorphous foods are characterized by a modulus curve showing downward concavity. The curve is obtained when the change in a mechanical property is plotted against temperature, water activity, or water content. Peleg (1993, 1994a,b) emphasized that the change could be observed in any mechanical property such as the extent of stickiness, caking, collapse, or modulus that could be generally referred to as stiffness. He analyzed parameters of mechanical properties and modulus data from various studies and found that the modulus curves of a glucose glass, proteins, wheat grains, other food materials, and synthetic polymers were of a shape that was similar as a function of temperature or water content. He suggested that a single equation could be used to model the stiffness data.

$$Y/Y_s = 1 \Big/ \left\{ 1 + \exp\left[ \frac{T - T_c(W)}{a'(W)} \right] \right\} \tag{8.1}$$

Equation (8.1), where $Y$ is the stiffness parameter (e.g., modulus), $Y_s$ is the value of the stiffness parameter in a reference state (e.g., in the glassy state), $T_c(W)$ is temperature that characterizes the transition region, and $a'(W)$ is a constant that indicates the steepness of the stiffness curve [$(W)$ refers to the constant water content], assumes that the change in stiffness occurs at a constant water content and it can be used to model data as a function of temperature. Equation (8.1) could be expressed also in the form of equation (8.2) with values for water content, $W$, water content that characterizes the transition region, $W_c(T)$, and constant $a''(T)$ [($T$)refers to the constant temperature] which is applicable in modeling stiffness as a function of water content.

$$Y/Y_s = 1 \bigg/ \left\{ 1 + \exp\left[ \frac{W - W_c(T)}{a''(T)} \right] \right\} \tag{8.2}$$

Stiffness as a function of water activity, $a_w$, is given by equation (8.3), where $a_{wc}(T)$ is the water activity that characterizes the transition region and $a'''(T)$ is a constant [($T$)refers to the constant temperature].

$$Y/Y_s = 1 \bigg/ \left\{ 1 + \exp\left[ \frac{a_w - a_{wc}(T)}{a'''(T)} \right] \right\} \tag{8.3}$$

Peleg (1994a) stressed that equations (8.1.), (8.2), or (8.3) are probably more applicable for modeling stiffness than the WLF model, since information on the changes in mechanical properties of foods is needed over the $T_g$ range. The WLF model assumes a steady state below $T_g$ and a dramatic decrease of viscosity above $T_g$. Peleg (1994a) pointed out that the proposed stiffness model allowed a moderate drop in the magnitudes of mechanical properties at the onset of the transition, which is typical of observed changes in stiffness instead of the sudden dramatic decrease in relaxation times suggested by the WLF model. However, it should be noticed that the WLF model is extremely useful if modulus data are superimposed in the form of the master curve. The model proposed by Peleg (1993) is also limited to the quantification of observed data with no reference to the viscoelastic properties or time-dependent behavior of stiffness (Wollny and Peleg, 1994).

Equations (8.1), (8.2), and (8.3) can be combined into the form of equation (8.4), where $X$ refers to $a_w$, $T$, or $W$ and $X_c$ to $a_{wc}(T)$, $T_c(W)$, or $W_c(T)$, respectively.

$$Y \bigg/ Y_s = 1 \bigg/ \left\{ 1 + \exp\left[ \frac{X - X_c}{a(X)} \right] \right\} \tag{8.4}$$

Equation (8.4) can be rewritten into the form of equation (8.5). Equation (8.5) may also be written into the form of equation (8.6), which suggests that a plot showing $\ln[(Y_s/Y) - 1]$ against $X$ gives a straight line. Thus, the constant $a(X)$ can be obtained from the slope of the line and $X_c$ gives the value for $X$, which corresponds to the decrease of stiffness by 50%. Stiffness data can also be analyzed in terms of relative stiffness, $Y/Y_S$, which normalizes experimental stiffness data to values between 0 and 1.

$$\ln\left(\frac{Y_s}{Y} - 1\right) = \frac{X - X_c}{a(X)} \tag{8.5}$$

$$\ln\left(\frac{Y_s}{Y} - 1\right) = \frac{-X_c}{a(X)} + \frac{1}{a(X)} X \tag{8.6}$$

The reference value, $X_c$, is useful in establishing criteria for storage conditions of amorphous foods, since it is a manifestation of a change in mechanical properties caused by the glass transition. Equation (8.6) provides a convenient means for modeling stiffness data for practical applications that require knowledge and predictability of changes in mechanical properties in terms of temperature, water content, or water activity. In addition to the two-dimensional analysis of temperature and water content dependence of stiffness it can be represented in three-dimensional surfaces (Peleg, 1993, 1994b), which describe the combined effects of water and temperature on stiffness in a single plot.

It may be concluded that mathematical characterization of stiffness is important in evaluating the effect of temperature and water content on the mechanical properties of food materials. Modeling experimental data on stiffness at a constant water content or at a constant temperature may be successful without numerical values for the reference temperature or glass transition temperature, but combined modeling of effects of both temperature and water on stiffness requires the use of the reference temperature, which may be the $T_g$. However, knowledge of the relationships between temperature and water content is often sufficient for locating critical values for temperature and water content above which mechanical properties are dramatically changed due to glass transition and the change in the physical state into the rubbery state.

## B. Stiffness and Food Properties

Mechanical properties of food powders, cereal foods, snack foods, candies, etc., are important in defining various quality parameters, including free-flowing properties of powders, stickiness, and perceived texture. Modeling of mechanical properties or stiffness as a function of temperature and water content provides a valuable tool for estimating changes, which may occur during a probable shelf life or exposures to abusing environmental conditions.

## 1. Mechanical properties of food powders

Food powders are often produced by rapid dehydration and they contain amorphous carbohydrates and proteins, which become plasticized by water. Changes in mechanical properties may be caused by depression of the glass transition to below storage temperature, which results in dramatic changes in flow properties and loss of acceptable quality. Such changes may be related to the viscous flow of amorphous compounds in the rubbery state and it is often assumed that product quality can be maintained in the glassy state.

The effects of water content on the mechanical properties are well known. Hygroscopic food powders become sticky, decrease in volume, and lose their free-flowing properties above a critical water content or water activity. The mechanical properties of food powders have been related to glass transition of the powder in a few studies. Aguilera *et al.* (1993) studied the effect of water activity on the collapse of powders of fish protein hydrolysates. They defined *collapse* as the deformation of a fixed powder volume, which occurred due to gravitational forces. The deformation of the material with various water activities was determined from the change in volume after a fixed storage time. The curve that showed the extent of collapse as a function of water activity had the typical shape of a modulus curve, which suggested that no collapse occurred at low water activities. Above a critical water activity the change in the extent of collapse was dramatic over a fairly narrow $a_w$ range between about 0.3 and 0.5 $a_w$. At higher water activities collapse became almost instant and independent of $a_w$. Aguilera *et al.* (1993) assumed that the WLF equation was applicable to predict viscosity for the occurrence of the observed collapse at $T - T_g = 35.8°C$. The estimated viscosity for collapse was $10^5$ to $10^7$ Pa s.

Glass transition of amorphous food powders is the main cause of changes in mechanical properties during storage. The results of Aguilera *et al.* (1993) suggested that the free-flowing properties of hygroscopic food powders were lost even by short exposures of such powders to high humidities. Such changes in mechanical properties during storage can be avoided or reduced by establishing and applying critical values for temperature and water content.

## 2. Diffusivity

Diffusivity in low-moisture foods is significantly affected by water content. The diffusivity in food solids has been found to be low at low water contents and to decrease above some critical water content. Diffusivity can be related to the physical state of food solids, but it is also dependent on temperature and a number of other factors, especially on the physical structure and particularly on the porosity of the food material (Karel and Saguy, 1991).

Fish (1958) reported diffusivity values for water in starch gels as a function of temperature and water content. The diffusion coefficient increased with increasing temperature and equation (8.7), where $D$ is the diffusion coefficient, $D_0$ is the frequency factor for diffusion, $E_D$ is the activation energy of diffusion, and $R$ is the gas constant, could be fitted to its temperature dependence.

$$D = D_0 e^{-\frac{E_D}{RT}} \tag{8.7}$$

Although the diffusivity data reported by Fish (1958) followed the Arrhenius-type temperature dependence they found that the activation energy for diffusion increased significantly with decreasing water content. Fish (1958) reported that the diffusion characteristics of water were normal when the water content was higher than 15 g/100 g of solids. At lower water contents the diffusion coefficient decreased more than expected with decreasing water content. Fish (1958) pointed out that similar behavior had been reported for a wide variety of natural and synthetic macromolecules. Karathanos *et al.* (1990) and Marousis *et al.* (1991) reported diffusion coefficients for water in various starch matrices as a function of water content at relatively high temperatures. The results for gelatinized starch indicated that diffusivity was significantly decreased at water contents below 15 g/100 g of solids, in agreement with the results of Fish (1958). Diffusion in other starch matrices, including pressed, pressed granular, granular, and expanded extrudate, was higher due to the differences in the macroscopic structure. However, diffusion in the expanded extrudate seemed to decrease at low water contents. Marousis *et al.* (1991) concluded that the diffusion behavior of water was related to physical or physicochemical changes in stored foods. The results of these studies suggest that glass transition, as shown in Figure 8.2, may cause the observed changes in diffusivity, which is obvious from the shape of the curve, showing diffusion as a function of water content.

Diffusion is affected by temperature as shown in Figure 8.2. Equation (8.7) defines the typical temperature dependence of the diffusion coefficient. However, glass transitions seem to have additional effects on the temperature dependence of diffusion. Glass transition in polymers has been found to cause a change in the activation energy of diffusion, which is often higher

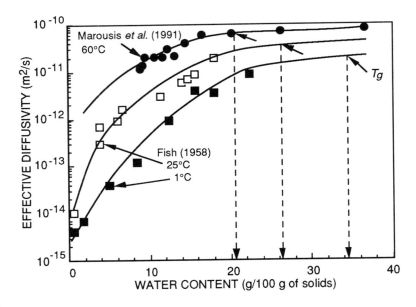

**Figure 8.2**  Diffusivity of water in gelatinized starch at 1, 25, and 60°C as a function of water content. The water content, which depresses the glass transition temperature, $T_g$, to below the experimental temperature is shown with an arrow for each set of data. The $T_g$ was predicted with the Gordon and Taylor equation according to Roos and Karel (1991). The data for diffusivity are from the studies indicated.

above $T_g$ than below $T_g$ (Hopfenberg and Stannett, 1973). Such behavior has also been found to occur in starch (Arvanitoyannis *et al.*, 1994). Therefore, Arrhenius plots that show ln $D$ against $1/T$ may have two linear regions, which are separated by a step change indicating that $E_D$ reaches a maximum value in the vicinity of the glass transition temperature (Vrentas and Duda, 1978). The effect of glass transition on the temperature dependence of diffusion coefficients is shown schematically in Figure 8.3. In polymers the differences between diffusion below and above $T_g$ have been proposed to result from the ability of the penetrant molecules to create "holes" during sorption and diffusion of the molecules between such holes (Hopfenberg and Stannett, 1973). The effect of water plasticization on diffusivity, as was shown in Figure 8.2 for starch, has a typical behavior of polymer-solvent systems (Vrentas *et al.*, 1982). Vrentas *et al.* (1982) reported that the amount of free volume available for molecular transport near the $T_g$ is relatively small. The addition of solvent or an increase in temperature causes a substantial increase in free volume, which allows an enhanced molecular transfer within the polymer matrix.

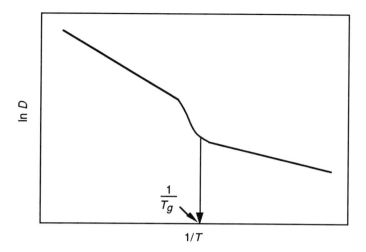

**Figure 8.3** A schematic representation of the effect of the glass transition temperature, $T_g$, on diffusivity, $D$. The diffusivity below $T_g$ and above $T_g$ may follow the Arrhenius-type temperature dependence. However, a step change in the vicinity of $T_g$ may be observed and the activation energy for diffusion is higher above $T_g$ than below $T_g$.

The change in diffusion, which occurs due to the change in polymer structure and results in anomalous diffusion effects, has been characterized with a diffusion Deborah number, *De*, (Vrentas and Duda, 1978; Peppas and Brannon-Peppas, 1994). The diffusion Deborah number is given by equation (8.8), where $\tau$ is a characteristic relaxation time for the material and $t_D$ is a characteristic diffusion time of the system.

$$De = \frac{\tau}{t_D} \tag{8.8}$$

A small Deborah number suggests that the relaxation time is much smaller than the diffusion time. Therefore, structural changes occur during the diffusion process, but the rates of the structural changes are much faster than diffusion and the diffusion process follows Fickian diffusion. A large Deborah number describes a diffusion process in which diffusion is considerably faster than structural changes and the process also follows Fickian diffusion. However, Deborah numbers on the order of 1 indicate that molecular relaxations and diffusion occur in comparable time scales and the changes in molecular structure may cause deviation from Fickian diffusion. Diffusion with a large Deborah number occurs below $T_g$ and diffusion with a small

Deborah number is likely to be observed well above $T_g$. Food materials which are plasticized by water may have a large Deborah number at low water contents. Water adsorption or an increase in temperature results in plasticization, which increases mobility, and a change in diffusivity may be observed as the ratio of relaxation time and diffusion time approaches unity due to glass transition (Peppas and Brannon-Peppas, 1994).

Diffusivity is probably one of the most important properties of several compounds in food materials. Diffusivity is affected by changes in stiffness and also in crystallinity. As shown above, it is obvious that diffusion of water as well as other compounds is retarded in glassy foods. Diffusivity is also affected by changes in stiffness above glass transition. The effective diffusion coefficient may be assumed to increase above $T_g$, but it should also be noticed that the changes in the macroscopic structure, e.g., due to stickiness, caking, collapse, and crystallization of amorphous components, may significantly affect diffusion. It is well known that diffusion in collapsed matrices is retarded in comparison with porous noncollapsed materials. The effects of phase transitions on diffusion are of considerable importance in the evaluation of the effects of the physical state on rates of deteriorative reactions, retention of volatiles in dehydration, and kinetics of quality changes in low-moisture and frozen foods.

3. Encapsulation

Amorphous food solids may encapsulate compounds which are not miscible with other food solids. Compounds that may become encapsulated in solids which are miscible with water include various nonpolar compounds such as lipids, e.g., oils and fats, volatile compounds, and flavors. Encapsulation is common in dehydration processes and extrusion due to the rapid removal of solvent water and formation of the amorphous matrix.

Flink and Karel (1972) studied loss of encapsulated tert-butanol and 2-propanol from freeze-dried maltose after rehumidification at various relative humidities. As shown in Figure 8.4 the loss occurred above a critical relative humidity and it was accompanied with structural transitions. The curve for loss of the volatiles has the typical shape of a modulus curve and it also followed equation (8.6). Based on the results of Flink and Karel (1972) and several other studies on flavor retention in dehydrated foods it is obvious that the loss of the encapsulated compounds occurs as the food matrix is transformed from the glassy state into the rubbery state. Chirife *et al.* (1973) studied the effect of exposure time and relative humidity on the retention of n-propanol in freeze-dried PVP. The results showed that the loss increased

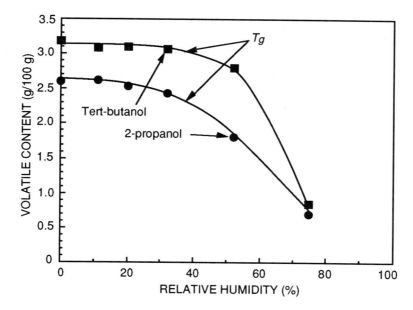

**Figure 8.4** Loss of volatiles from freeze-dried maltose as a function of storage relative humidity. The volatiles are retained encapsulated in the amorphous, glassy sugar matrix at relative humidities which maintain the glass transition temperature, $T_g$, above storage temperature. Loss of the encapsulated volatiles occurs as the relative humidity is sufficient to depress the $T_g$ to below ambient temperature due to collapse and possible crystallization.

with increasing water activity in accordance with the results of Flink and Karel (1972). Chirife and Karel (1974) studied also the effect of temperature on 1-propanol retention in freeze-dried maltose. They found that 1-propanol was retained at 82°C, but loss occurred at 100°C during an exposure of 24 hrs. Roos and Karel (1991) determined the $T_g$ of dry maltose with DSC to be 87°C. This value supports the conclusion that volatiles are likely to be retained in amorphous matrices below $T_g$ and structural changes including the time-dependent crystallization above $T_g$ may cause the loss of encapsulated compounds and flavors in dehydrated foods.

Encapsulated compounds may become lost from amorphous carbohydrate matrices due to the crystallization of the encapsulating compound, e.g., lactose (Flink and Karel, 1972). To and Flink (1978) studied oxidation of encapsulated linoleic acid. The lipid was encapsulated by freeze-drying oil-in-water emulsions containing various encapsulating carbohydrates. After freeze-drying the dried materials were washed with hexane to remove the nonencapsulated fat. Powders with various compositions and collapse temperatures were exposed to different temperatures and oxidation of the fat was followed as a function of time. Release and oxidation of the lipid occurred as carbohydrates in the encapsulating matrix crystallized. Similar results were reported by Shimada *et al.* (1991), who studied oxidation of methyl linoleate which was encapsulated in an amorphous lactose matrix. Exceeding the matrix $T_g$ due to an increase in either temperature or water content resulted in time-dependent crystallization and release of the oil. Thus, oxidation of the oil proceeded with a rate that was obviously correlated with the rate of lactose crystallization. It should also be noticed that several food matrices which contain encapsulated lipids and flavors do not exhibit rapid crystallization of component compounds. However, collapse above $T_g$ cannot be avoided due to the viscous flow above $T_g$. The results of Labrousse *et al.* (1992) on the effect of collapse on oxidation of an encapsulated lipid showed that encapsulated compounds may be lost due to crystallization, but in noncrystallizing matrices such compounds may be partially lost and also become reencapsulated during collapse.

### 4. Crispness

A number of instrumental (Andersson *et al.*, 1973; Bruns and Bourne, 1975), acoustic (Vickers and Bourne, 1976; Vickers, 1981), and sensory methods have been used to evaluate crispness of food materials. Bruns and Bourne (1975) found that crisp foods have a characteristic force-deformation curve which can be identified by a linear slope of the initial portion of the curve. Such curves can be obtained with a three-point snap test. The best correlations of sensory and instrumental data are often obtained by using both acoustic and force-deformation measurements (Vickers, 1988). It should also be noticed that Young's modulus has a high correlation with perceived crispness (Vickers and Christensen, 1980).

In various studies crispness of low-moisture foods has been related to water content or relative humidity and water plasticization of structure (Quast and Karel, 1972; Zabik *et al.*, 1979; Katz and Labuza, 1981; Sauvageot and Blond, 1991). Katz and Labuza (1981) suggested that crispy snack foods have a critical water content, where the addition of water enables mobility of macromolecules and the consumer perceives the resultant loss of crispness. They found that the initial slope of the force-deformation curve was almost constant at low water contents. Katz and Labuza (1981) assumed that at low water activities the water content was insufficient to affect mechanical properties. Above a critical water content a significant decrease of the initial slope occurred, and the products lost their crispness and became texturally unacceptable. The steady state water activity at which crispness is lost is

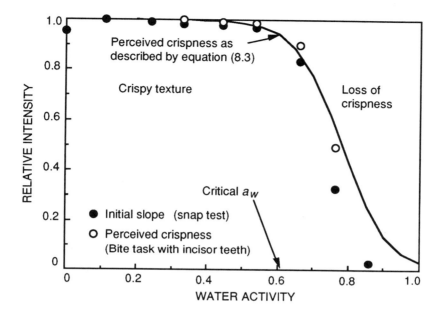

**Figure 8.5** Effect of water plasticization on crispness of amorphous, crispy foods. Crispness, which was determined instrumentally using a snap test and with sensory evaluation by biting with incisor teeth (bite task), was maintained at low water activities, $a_w$, due to the glassy state of the material. Both instrumentally measured and perceived crispness suggested that crispness was lost at a critical $a_w$ above which the $T_g$ was depressed to below ambient temperature. Data from Kärki *et al.* (1994).

specific for each product, but several studies have suggested that a change in crispness occurs often between water activity values of 0.35 and 0.50 (Katz and Labuza, 1981; Hsieh *et al.*, 1990).

Sauvageot and Blond (1991) studied the crispness of breakfast cereals as a function of water activity using sensory techniques. The perceived crispness intensity was found to depend on the product water activity. When perceived crispness was plotted against water activity a typical modulus curve was obtained. Sauvageot and Blond (1991) divided the curve into three parts. At low water activities the crispness intensity was high and varied only slightly with $a_w$. The second portion of the curve indicated a dramatic decrease of crispness within a fairly narrow water activity range. At high water activities the crispnesses of the products were totally lost and they were unacceptable. Sauvageot and Blond (1991) found a good correlation between the results of the sensory methods and an instrumental penetration test. It should also be noticed that equation (8.3) fitted to the sensory data, as was shown by Peleg (1994c).

Kärki *et al.* (1994) studied the effect of water activity and water content on the crispness of an extruded snack food model. Crispness was determined instrumentally with the snap test and the initial slope of the force-deformation curve was used as an indicator of the product texture. They also used three sensory panels, which evaluated crispness using fingers, by biting with incisor teeth, and by biting and chewing. The initial slope and perceived crispness reported by the sensory panels that used the biting techniques had a high correlation. It was found that the loss of crispness occurred as the product water activity was sufficient to depress $T_g$ to below ambient temperature as shown in Figure 8.5. It is obvious that crispy foods such as breakfast cereals, extruded snacks, and other crispy cereal foods are often amorphous and lose the crispy texture due to thermal or water plasticization. However, it should be noticed that the sensory scores obtained for such mechanical properties of extruded foods as crunchiness, crispness, and hardness are affected by the macroscopic structure of the material and in particular cellularity and bulk density of the solids (Barrett *et al.*, 1994).

## III.   Mechanical Properties and Crystallinity

Mechanical properties of several foods are affected by solids that exist in the crystalline state within the food matrix. Such food materials include food fats and oils, which often contain a certain portion of the lipid fraction in the crystalline state and frozen foods, which contain ice within the freeze-concentrated solute matrix. The mechanical properties of amorphous foods are also affected by crystallinity, e.g., crystallinity of starch or a crystalline phase within an amorphous solid material.

### A. Food Lipids

Most common fats are partially crystalline materials. Crystallinity is highly dependent on composition and temperature and the mechanical properties are often related to the solid fat content. The mechanical and rheological properties of fats are important to their behavior during processing, packaging, and storage as well as to their spreadability and oral melting properties (Hoffmann, 1989).

1. Mechanical properties and firmness

Mechanical properties and firmness of fats are mainly defined by their solid fat content. According to Bailey (1950) the viscosity of a fat increases rapidly with increasing solids content. However, the mechanical properties and firmness are characterized by other properties rather than viscosity.

The mechanical properties of fats are affected by several properties of the crystalline phase. Factors that affect rheological properties include the polymorphic form of the crystals and the shape of the crystals. Large crystals tend to form firm, rigid structures. Also the presence of water in an emulsion structure affects fat consistency, especially at temperatures close to the melting temperature. Most crystals in rapidly cooled fats are small, but the size and shape of the crystals vary significantly between various fats and they are highly dependent on the thermal history of the material (Walstra, 1987). In fat mixtures the crystals generally form a network structure which is responsible for the solid properties.

According to Walstra (1987) a crystalline phase of 10% is sufficient to allow formation of solid properties in food fats. Partially crystalline fats have an elastic modulus, which applies to small deformations of less than 1%. Higher stresses break bonds between fat crystals and the material begins to flow. Breaking of bonds in the network structure occurs also during working of fats, which results in work softening and flow. After the work ceases the fat begins to recrystallize and recovers the solid-like properties. Hoffmann (1989) pointed out that work softening is an important quality characteristic of fats. Work softening is low for butter, low for puff-pastry margarines, and high for table margarines to allow easy spreading. However, depending on the crystalline form and the thermal and working history among other factors the firmness of a fat may vary by almost an order of magnitude for the same proportion of crystalline material (Walstra, 1987).

2. Plasticity

Plastic fats contain a crystalline phase and a liquid phase, which define their response to an external stress. Plastic fats are characterized by an ability to possess elasticity when they are exposed to a small external stress. However, a stress that exceeds the yield value of the fat results in viscous flow (Bailey, 1950).

Hoffmann (1989) pointed out that the plastic state of fats is characterized by (1) the co-existence of the two phases; (2) proper portions of the liquid and solid materials at a given temperature; and (3) an interconnection between

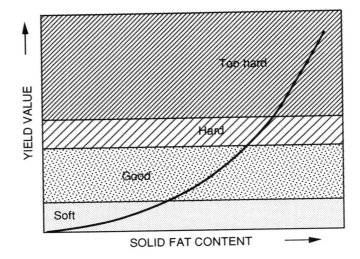

**Figure 8.6** A schematic representation of yield value (hardness) of fats as a function of solid fat content.

the fat crystals. At similar conditions hardness or the yield point of fats has a high correlation with the solid fat content. A schematic representation of the relationships between the solid fat content, yield value, and mechanical properties is shown in Figure 8.6. It is obvious that the solid fat content and thereby crystallinity is the primary determinant of the mechanical properties of fat products.

The solid fat content and crystallinity are used as indicators of mechanical properties of fat products. According to Hoffmann (1989) the solid fat content is low in soft margarines such as those that contain high amounts of polyunsaturated fatty acids, which may contain only 10% solids at 5 to 10°C. In regular soft margarines the solid fat content is 30 to 40% at 5°C and it decreases to below 20% at room temperature. Tropical margarines contain more than 30% solid fat at room temperature and they become soft at temperatures above 30°C. The mechanical properties of margarines may change during storage due to post-crystallization and recrystallization phenomena depending on the manufacturing process and storage conditions (Hoffmann, 1989).

## B. Crystallinity in Amorphous Foods

Crystallinity in amorphous foods may refer to the presence of crystalline compounds such as ice or sugars within an amorphous food matrix or partial crystallinity of polymer molecules. Mechanical properties of frozen foods are affected by the extent of ice formation at storage temperature, while polymer crystallinity may significantly affect mechanical properties of starch-containing foods. The presence of a crystalline phase within an amorphous solid structure may also affect the mechanical behavior of confectionery and various food powders.

## 1. Partially amorphous components

Starch is probably the most important partially crystalline food component. Its mechanical properties define the structure of several foods and the effect of crystallinity on the mechanical properties of starch gels has been well established. Changes in the mechanical properties of starch gels, which may occur during storage, reflect changes in crystallinity that are often caused by the slow crystallization of amylopectin.

Crystallinity in synthetic polymers affects their mechanical properties. Depending on the extent of crystallinity the effect of glass transition may be observed from modulus curves (Ferry, 1980). Miles *et al.* (1985) studied changes in shear storage modulus, $G'$, during storage of starch and amylose gels. The modulus in amylose gels developed rapidly during cooling after gelatinization and remained constant thereafter. The modulus of starch gels developed also during cooling, but the modulus continued to increase during storage. However, reheating of the gel after storage decreased the modulus to the initial value, which suggested that crystallization during storage was responsible for the increase of modulus during storage. Jankowski and Rha (1986) determined apparent moduli for cooked wheat grain at different storage times using uniaxial compression between parallel plates. The modulus increased with increasing storage time at 4 and 20°C. The increase in modulus was more pronounced during storage at 4 than at 20°C. Jankowski and Rha (1986) concluded that the transformation of amorphous starch components into the crystalline state was the main cause of the increase in modulus. It is obvious that crystallinity in starch gels and probably in starch-containing foods affects elastic modulus and mechanical properties. The elastic modulus or firmness increases with increasing crystallinity. However, it should be noticed that the increase in crystallinity due to retrogradation of starch may occur differently in various starches. Kalichevsky *et al.* (1990) studied the retrogradation behavior of amylopectins from barley, canna, maize, pea, potato, and wheat starches. The shear modulus of the gels increased with increasing storage time. The increase in the modulus with storage time was most rapid for pea starch and slowest for wheat starch. The association of the chains in cereal amylopectins occurred over shorter segment lengths. Cereal starches exhibited lower rates of crystallization, probably due to their shorter average chain lengths. This was confirmed by Keetels and van Vliet (1993), who reported that the increase in Young's modulus of concentrated potato starch gels during storage was more pronounced than in wheat starch gels.

Crystallinity is known to affect the $T_g$ of amorphous regions of partially crystalline polymers. According to Slade and Levine (1989) an increasing crystallinity increases the $T_g$, which is primarily due to the stiffening effect of microcrystalline cross-links. Such cross-links decrease the mobility of chain segments in the amorphous regions. Therefore, it may be assumed that the increase in modulus due to crystallization in polymeric food components results from the reduced molecular mobility in the crystalline regions and increased $T_g$ of the amorphous regions. The changes in the mechanical properties of starch gels and bread which occur during storage are obviously, at least to some extent, caused by the increasing crystallinity within the starch fraction.

## 2. Freeze-concentrated materials

Frozen foods contain crystalline ice within freeze-concentrated solids. The amount of ice is dependent on temperature and the ability of the solutes to depress the melting temperature of ice. In frozen foods the amount of ice affects also mechanical properties by changing the effective glass transition temperature of partially freeze-concentrated solute matrices and by probable interconnections within the ice network.

Experimental data on the mechanical properties for frozen foods are almost nonexistent. Studies on the rheological properties of ice cream have suggested that the material in the frozen state has viscoelastic characteristics (Berger, 1990). Most studies reporting mechanical properties for freeze-concentrated materials have used sugar solutions as model systems. Le Meste and Huang (1992) studied thermomechanical properties of frozen sucrose solutions. The results showed that the freeze-concentrated solute matrix had a glass transition around -46°C, but significant softening and flow in the solutions was observed to occur only above -32°C due to ice melting. Simatos and Blond (1993) used mechanical spectroscopy in the analysis of mechanical properties of freeze-concentrated model solutions. The mechanical tests included a compression test for samples with a large elastic component, an annular shearing test for viscous liquids, and a pumping shear test for viscous liquids. The results for 50% sucrose solutions showed a decrease in storage modulus and decrease in loss modulus above -45°C and increase in mobility above -32°C. A slow discontinuous increase of the phase angle associated with ice melting above -32°C was considered to indicate some reorganization of ice crystals. Simatos and Blond (1993) suggested that the mechanical properties above the onset of ice melting were dominated by ice crystals. MacInnes (1993) used DMTA in a thorough study of mechanical properties of sucrose solutions during freezing and melting. The loss modulus was found to have a peak at -41°C. Both the loss modulus and storage modulus decreased non-monotonically above -41°C and they were dependent on frequency.

Obviously, mechanical properties of frozen food materials are affected by the $T_g'$ in the maximally freeze-concentrated state and by ice melting above $T_m'$. The main cause for changes in mechanical properties of frozen foods is probably related to ice melting. The most dramatic changes in mechanical properties of frozen foods are likely to occur in the vicinity of the melting temperature as most of the ice is transformed into the liquid state.

# References

Aguilera, J.M., Levi, G. and Karel, M. 1993. Effect of water content on the glass transition and caking of fish protein hydrolyzates. *Biotechnol. Prog. 9*: 651-654.

Andersson, Y., Drake, B., Granquist, A., Halldin, L., Johansson, B., Pangborn, R.M. and Åkesson, C. 1973. Fracture force, hardness and brittleness in crisp bread, with a generalized regression analysis approach to instrumental-sensory comparisons. *J. Texture Stud. 4*: 119-144.

Arvanitoyannis, I., Kalichevsky, M.T. and Blanshard, J.M.V. 1994. Study of diffusion and permeation of gases in undrawn and uniaxially drawn films made from potato and rice starch conditioned at different relative humidities. *Carbohydr. Polym. 24*: 1-15.

Attenburrow, G. and Davies, A.P. 1993. The mechanical properties of cereal based foods in and around the glassy state. Chpt. 16 in *The Glassy State in Foods*, ed. J.M.V. Blanshard and P.J. Lillford. Nottingham University Press, Loughborough, pp. 317-331.

Attenburrow, G., Barnes, D.J., Davies, A.P. and Ingman, S.J. 1990. Rheological properties of wheat gluten. *J. Cereal Sci. 12*: 1-14.

Bailey, A.E. 1950. *Melting and Solidification of Fats*, Interscience Publishers, New York.

Barrett, A.H., Cardello, A.V., Lesher, L.L. and Taub, I.A. 1994. Cellularity, mechanical failure, and textural perception of corn meal extrudates. *J. Texture Stud. 25*: 77-95.

Berger, K.G. 1990. Ice cream. Chpt. 9 in *Food Emulsions*, ed. K. Larsson and S.E. Friberg. Marcel Dekker, New York, pp. 367-444.

Bruns, A.J. and Bourne, M.C. 1975. Effects of sample dimensions on the snapping force of crisp foods. *J. Texture Stud. 6*: 445-458.

Chattopadhyay, P.K., Hammerle, J.R. and Hamann, D.D. 1979. Time, temperature, and moisture effects on the failure strength of rice. *Cereal Foods World 24*: 514-516.

Chirife, J. and Karel, M. 1974. Effect of structure disrupting treatments on volatile release from freeze-dried maltose. *J. Food Technol. 9*: 13-20.

Chirife, J., Karel, M. and Flink, J. 1973. Studies on mechanisms of retention of volatile in freeze-dried food models. The system PVP-n-propanol. *J. Food Sci. 38*: 671-674.

Ferry, J.D. 1980. *Viscoelastic Properties of Polymers*. 3rd ed. John Wiley & Sons, New York.

Fish, B.P. 1958. Diffusion and thermodynamics of water in potato starch gel. In *Fundamental Aspects of Dehydration of Foodstuffs*, Society of Chemical Industry. Metchim & Son Ltd., London, pp. 143-157.

Flink, J.M. and Karel, M. 1972. Mechanisms of retention of organic volatiles in freeze-dried systems. *J. Food Technol. 7*: 199-211.

Hammerle, J.R. and Mohsenin, N.N. 1970. Tensile relaxation modulus of corn horny endosperm as a function of time, temperature and moisture content. *Trans. ASAE 13*: 372-375.

Hoffmann, G. 1989. *The Chemistry and Technology of Edible Oils and Fats and Their High Fat Products*. Academic Press, San Diego, CA.

Hopfenberg, H.B. and Stannett, V. 1973. The diffusion and sorption of gases and vapours in glassy polymers. Chpt. 9 in *The Physics of Glassy Polymers*, ed. R.N. Haward. John Wiley & Sons, New York, pp. 504-547.

Hsieh, F., Hu, L., Huff, H.E. and Peng, I.C. 1990. Effects of water activity on textural characteristics of puffed rice cake. *Lebensm.-Wiss. u. -Technol. 23*: 471-473.

Jankowski, T. and Rha, C.K. 1986. Retrogradation of starch in cooked wheat. *Starch 38*: 6-9.

Kalichevsky, M.T., Orford, P.D. and Ring, S.G. 1990. The retrogradation and gelation of amylopectins from various botanical sources. *Carbohydr. Res. 198*: 49-55.

Kalichevsky, M.T., Jaroszkiewicz, E.M., Ablett, S., Blanshard, J.M.V. and Lillford, P.J. 1992. The glass transition of amylopectin measured by DSC, DMTA, and NMR. *Carbohydr. Polym. 18*: 77-88.

Kalichevsky, M.T., Jaroszkiewicz, E.M. and Blanshard, J.M.V. 1993. A study of the glass transition of amylopectin-sugar mixtures. *Polymer 34*: 346-358.

Karathanos, V.T., Villalobos, G. and Saravacos, G.D. 1990. Comparison of two methods of estimation of the effective moisture diffusivity from drying data. *J. Food Sci. 55*: 218-223, 231.

Karel, M. and Saguy, I. 1991. Effects of water on diffusion in food systems. In *Water Relationships in Foods*, ed. H. Levine and L. Slade. Plenum Press, New York, pp. 157-173.

Kärki, M.-K., Roos, Y.H. and Tuorila, H. 1994. Water plasticization of crispy snack foods. Paper No. 76-10. Presented at the Annual Meeting of the Institute of Food Technologists, Atlanta, GA, June 25-29.

Katz, E.E. and Labuza, T.P. 1981. Effect of water activity on the sensory crispness and mechanical deformation of snack food products. *J. Food Sci. 46*: 403-409.

Keetels, C.J.A.M. and van Vliet, T. 1993. Mechanical properties of concentrated starch systems during heating, cooling, and storage. In *Food Colloids and Polymers: Stability and Mechanical Properties*, ed. E. Dickinson and P. Walstra. The Royal Society of Chemistry, Cambridge, pp. 266-271.

Labrousse, S., Roos, Y. and Karel, M. 1992. Collapse and crystallization in amorphous matrices with encapsulated compounds. *Sci. Aliments 12*: 757-769.

Le Meste, M. and Huang, V. 1992. Thermomechanical properties of frozen sucrose solutions. *J. Food Sci. 57*: 1230-1233.

Levine, H. and Slade, L. 1993. The glassy state in applications for the food industry, with an emphasis on cookie and cracker production. Chpt. 17 in *The Glassy State in Foods*, ed. J.M.V. Blanshard and P.J. Lillford. Nottingham University Press, Loughborough, pp. 333-373.

Lillie, M.A. and Gosline, J.M. 1990. The effects of hydration on the dynamic mechanical properties of elastin. *Biopolymers 29*: 1147-1160.

MacInnes, W.M. 1993. Dynamic mechanical thermal analysis of sucrose solutions. Chpt. 11 in *The Glassy State in Foods*, ed. J.M.V. Blanshard and P.J. Lillford. Nottingham University Press, Loughborough, pp. 223-248.

Marousis, S.N., Karathanos, V.T. and Saravacos, G.D. 1991. Effect of physical structure of starch materials on water diffusivity. *J. Food Process. Preserv. 15*: 183-195.

Miles, M.J., Morris, V.J., Orford, P.D. and Ring, S.G. 1985. The roles of amylose and amylopectin in the gelation and retrogradation of starch. *Carbohydr. Res. 135*: 271-281.

Multon, J.L., Bizot, H., Doublier, J.L., Lefebvre, J. and Abbott, D.C. 1981. Effect of water activity and sorption hysteresis on rheological behavior of wheat kernels. In *Water Activity: Influences on Food Quality*, ed. L.B. Rockland and G.F. Stewart. Academic Press, New York, pp. 179-198.

Nakamura, S. and Tobolsky, A.V. 1967. Viscoelastic properties of plasticized amylose films. *J. Appl. Polym. Sci. 11*: 1371-1386.

Peleg, M. 1993. Mapping the stiffness-temperature-moisture relationship of solid biomaterials at and around their glass transition. *Rheol. Acta 32*: 575-580.

Peleg, M. 1994a. A model of mechanical changes in biomaterials at and around their glass transition. *Biotechnol. Prog. 10*: 385-388.

Peleg, M. 1994b. Mathematical characterization and graphical presentation of the stiffness-temperature-moisture relationship of gliadin. *Biotechnol. Prog. 10*: 652-654.

Peleg, M. 1994c. A mathematical model of crunchiness/crispness loss in breakfast cereals. *J. Texture Stud. 25*: 403-410.

Peppas, N.A. and Brannon-Peppas, L. 1994. Water diffusion and sorption in amorphous macromolecular systems and foods. *J. Food Eng. 22*: 189-210.

Quast, D.G. and Karel, M. 1972. Effects of environmental factors on the oxidation of potato chips. *J. Food Sci. 37*: 584-588.

Roos, Y. and Karel, M. 1991. Water and molecular weight effects on glass transitions in amorphous carbohydrates and carbohydrate solutions. *J. Food Sci. 56*: 1676-1681.

Sauvageot, F. and Blond, G. 1991. Effect of water activity on crispness of breakfast cereals. *J. Texture Stud. 22*: 423-442.

Shimada, Y., Roos, Y. and Karel, M. 1991. Oxidation of methyl linoleate encapsulated in amorphous lactose-based food model. *J. Agric. Food Chem. 39*: 637-641.

Simatos, D. and Blond, G. 1993. Some aspects of the glass transition in frozen foods systems. Chpt. 19 in *The Glassy State in Foods*, ed. J.M.V. Blanshard and P.J. Lillford. Nottingham University Press, Loughborough, pp. 395-415.

Slade, L. and Levine, H. 1989. A food polymer science approach to selected aspects of starch gelatinization and retrogradation. In *Frontiers in Carbohydrate Research - 1: Food Applications*, ed. R.P. Millane, J.N. BeMiller and R. Chandrasekaran. Elsevier, Amsterdam, pp. 215-270.

Slade, L. and Levine, H. 1993. The glassy state phenomenon in food molecules. Chpt. 3 in *The Glassy State in Foods*, ed. J.M.V. Blanshard and P.J. Lillford. Nottingham University Press, Loughborough, pp. 35-101.

Slade, L., Levine, H., Ievolella, J. and Wang, M. 1993. The glassy state phenomenon in applications for the food industry: Application of the food polymer science approach to structure-function relationships of sucrose in cookie and cracker systems. *J. Sci. Food Agric. 63*: 133-176.

To, E.C. and Flink, J.M. 1978. 'Collapse,' a structural transition in freeze dried carbohydrates. III. Prerequisite of recrystallization. *J. Food Technol. 13*: 583-594.

Vickers, Z.M. 1981. Relationships of chewing sounds to judgments of crispness, crunchiness and hardness. *J. Food Sci. 47*: 121-124.

Vickers, Z.M. 1988. Instrumental measures of crispness and their correlation with sensory assessment. *J. Texture Stud. 19*: 1-14.

Vickers, Z.M. and Bourne, M.C. 1976. A psychoacoustical theory of crispness. *J. Food Sci. 41*: 1158-1164.

Vickers, Z.M. and Christensen, C.M. 1980. Relationships between sensory crispness and other sensory and instrumental parameters. *J. Texture Stud. 11*: 291-307.

Vrentas, J.S. and Duda, J.L. 1978. A free-volume interpretation of the influence of the glass transition on diffusion in amorphous polymers. *J. Appl. Polym. Sci. 22*: 2325-2339.

Vrentas, J.S., Duda, J.L. and Lau, M.K. 1982. Solvent diffusion in molten polyethylene. *J. Appl. Polym. Sci. 27*: 3987-3997.

Walstra, P. 1987. Fat crystallization. Chpt. 5 in *Food Structure and Behaviour*, ed. J.M.V. Blanshard and P.J. Lillford. Academic Press, Orlando, FL, pp. 67-85.

Wollny, M. and Peleg, M. 1994. A model of moisture-induced plasticization of crunchy snacks based on Fermi's distribution function. *J. Sci. Food Agric. 64*: 467-473.

Yannas, I.V. 1972. Collagen and gelatin in the solid state. *J. Macromol. Sci., Rev. Macromol. Chem. C7(1)*: 49-104.

Zabik, M.E., Fierke, S.G. and Bristol, D.K. 1979. Humidity effects on textural characteristics of sugar-snap cookies. *Cereal Chem. 56*: 29-33.

# CHAPTER 9

# *Reaction Kinetics*

## I. Introduction

Food deterioration is often a result of structural transformations, chemical changes, or microbial growth. Chemical reactions, enzymatic changes, and microbial growth may occur readily in foods with high water contents when their occurrence is not restricted by environmental factors such as pH or temperature.

Rates of deteriorative changes and microbial growth are related to water content and water activity. Water activity may be considered to be a measure of the availability of water as a solvent or its availability to support microbial growth. Food deterioration due to microbial growth is not likely at water activities below 0.6 $a_w$. However, chemical reactions and enzymatic activity may occur at considerably lower water activities. Typical deteriorative changes of low-moisture foods include enzyme-catalyzed changes, nonenzymatic browning, and oxidation. Enzymatic changes and nonenzymatic browning have been found to occur above a critical water content and to show a rate maximum at an intermediate-moisture level, which is followed by a decrease at higher water activities (Labuza *et al.*, 1970). Oxidation may have a high rate at low water contents, the rate may go through a minimum with an increase in water activity, and then it may decrease at higher water activities (Labuza *et al.*, 1970).

Rates of deteriorative changes at reduced water activities have been suggested to depend on molecular mobility (Duckworth, 1981). Molecular mobility is governed by the physical state of food solids (Karel, 1985; Levine and Slade, 1988; Slade and Levine, 1991) and it is obvious that rates of several deteriorative changes can be affected by diffusion and become diffusion-limited at low water activities (Simatos and Karel, 1988; Karel and Saguy, 1991). Therefore, a chemical reaction may proceed with an extremely slow rate in a glassy matrix. A significant increase in the rate may occur as the material is transformed into the liquid-like viscous state above $T_g$ due to increasing molecular mobility.

The various effects of the physical state of foods on reaction rates and deterioration have been given increasing attention. Phase transitions in foods may significantly contribute to rates of physicochemical changes which occur and contribute to food quality during food processing and storage. The effects of various changes in the physical state on reaction kinetics and rates of deteriorative changes in foods are discussed in this chapter.

## II.  Principles of Reaction Kinetics

The main importance of reaction kinetics is to evaluate the overall rate at which chemical reactions and changes in food quality may occur at various conditions. Food deterioration is often a result of complicated chemical changes. Knowledge and understanding of reaction mechanisms and the rate of the reactions at various conditions are of significant importance in product development and formulation. Kinetics in food systems have been reviewed, e.g., by Villota and Hawkes (1992).

### A.  Reaction Order

A chemical reaction in its general form is characterized by the fact that the number of moles of atoms in reactants and products is constant. The rate of a chemical reaction defines the change of concentration at a given time. The

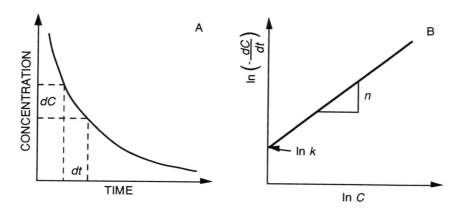

**Figure 9.1** A schematic representation of the determination of the reaction order, $n$, for a chemical change in foods. A. A plot of concentration, $C$, as a function of time, $t$, can be used to determine $dC/dt$. B. A plot showing ln $(-dC/dt)$ against ln $C$ gives a straight line. The reaction order is given by the slope of the line.

order of a chemical reaction is defined by equation (9.1), which states that the change in concentration, $C$, of a chemical compound during a chemical reaction at time, $t$, is defined by the initial concentration, the reaction rate constant, $k$, and the order of the reaction, $n$.

$$-\frac{dC}{dt} = kC^n \tag{9.1}$$

Equation (9.1) can also be written into the form of equation (9.2), which has the form of a straight line.

$$\ln\left(-\frac{dC}{dt}\right) = \ln k + n \ln C \tag{9.2}$$

According to equation (9.2) a plot showing ln $(-dC/dt)$ against $C$ gives a straight line, which can be used to obtain values for $k$ and $n$ from the intercept and slope, respectively, as shown in Figure 9.1. Such plots are often used to analyze rates of chemical changes in foods, but not the rate constant (Villota and Hawkes, 1992). A more accurate value for $k$ is obtained by applying the equation derived for the proper reaction order.

## 1. Zero-order reactions

Equation (9.1) defines that a zero-order reaction proceeds with a rate which is independent of concentration, i.e., the change in concentration in unit time is constant. For a zero-order reaction equation (9.1) is reduced to the form of equation (9.3), where $k_0$ is the zero-order reaction rate constant.

$$-\frac{dC}{dt} = k_0 \tag{9.3}$$

The integrated form of equation (9.3) is given by equation (9.4), which states that the decrease in concentration (difference between initial concentration, $C_0$, and concentration, $C$, at time, $t$) is linear with time as shown in Figure 9.2.

$$C = C_0 - k_0 t \tag{9.4}$$

Equation (9.4) defines that a reaction follows the zero-order kinetics if a plot of concentration against reaction time is linear. The reaction rate constant, which can be obtained from the slope of such a plot, gives a measure of the reactivity and defines the decrease in concentration of the reactant in unit time. According to Villota and Hawkes (1992) zero-order kinetics are not common in food materials, although some changes are often considered to follow the zero-order kinetics. However, zero-order kinetics may apply if the initial concentration, $C_0$, is high or the products of the reaction are removed in proportion to the reaction rate.

## 2. First-order reactions

A number of changes in foods follow either zero- or first-order kinetics (Labuza and Riboh, 1982). A first-order reaction follows equation (9.1) with $n = 1$. Thus, equation (9.1) for a first-order reaction is reduced to the form of equation (9.5), where $k_1$ is the first-order rate constant.

$$-\frac{dC}{dt} = k_1 C \tag{9.5}$$

Integration of equation (9.5) gives equation (9.6), which, as shown by equation (9.7), defines that a linear relationship applies between $\ln C$ and time.

$$-\ln\left(\frac{C}{C_0}\right) = k_1 t \qquad (9.6)$$

$$\ln C = \ln C_0 - k_1 t \qquad (9.7)$$

Experimental data for a first-order reaction show that the change in concentration occurs exponentially with time. Thus, a plot of $\ln C$ against $t$ shows that the rate constant, $k_1$, for the reaction is defined by the slope of the line according to Figure 9.2. Equation (9.6) allows another convenient means to analyze first-order reactions. The initial concentration of the reaction can be defined to be equal to unity, which allows definition of the half-life of the reaction with equation (9.8). Equation (9.8) defines that the half-life, $t_{1/2}$, is equal to $0.693/k_1$.

$$-\ln\left(\frac{0.5}{1}\right) = k_1 t \qquad (9.8)$$

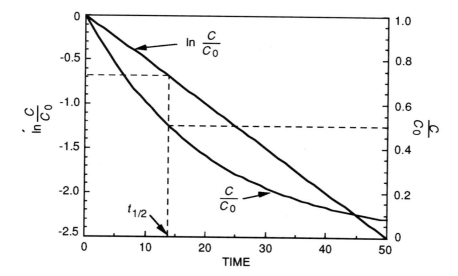

**Figure 9.2** Concentration, $C$, of a reactant during a hypothetical first-order reaction. The relative amount of the reactant, $C/C_0$, where $C_0$ is the initial concentration at time, $t = 0$, decreases with time, $t$. A plot of $\ln C/C_0$ is a linear function of $t$. The reaction rate constant, $k_1$, is obtained from the slope, which is equal to $-k_1$. The half-life, $t_{1/2}$, of the reaction is the time required to reduce the concentration by 50%.

The half-life of a change in food materials gives a measure for the time required to cause a given change during processing or storage. It is useful in evaluating the time which is required to reduce, e.g., the amount of a nutrient by 50% at a processing step or during distribution and storage. Similarly, another quantity which is critical in fulfilling product specifications or for product quality can be used instead of 50%.

### 3. Second-order reactions

The rate of a second-order reaction is defined by the concentration of the reactant according to equation (9.9), where $k_2$ is the reaction rate constant. The integrated form of the equation is given by equation (9.10).

$$-\frac{dC}{dt} = k_2 C^2 \tag{9.9}$$

$$\frac{1}{C} = \frac{1}{C_0} + k_2 t \tag{9.10}$$

The rate constant for a second-order reaction can be obtained by plotting $1/C$ against time, which gives a straight line. The slope of the line gives $k_2$. The half-life of a second-order reaction, which follows equation (9.9), is defined by equation (9.11).

$$t_{1/2} = \frac{1}{k_2 C_0} \tag{9.11}$$

Equation (9.9) is the most simple relationship between the concentration of a reactant and reaction time for a second-order reaction. It should be noticed that the rate of the reaction may be dependent on the nature of the reaction and on the reactant concentrations (Villota and Hawkes, 1992).

### B. Temperature Dependence

Deteriorative changes and chemical reactions in food materials are highly temperature-dependent. Small changes in processing or storage temperatures may significantly affect the rate of observed changes. These changes are either desired or detrimental. Changes that occur during food storage often decrease food quality. The temperature dependence of changes in foods usually follows Arrhenius-type temperature dependence, particularly in foods which have high water contents. The kinetics may show deviations from the Arrhenius kinetics at reduced water contents, which may reflect changes in the physical state and diffusional limitations.

## 1. $Q_{10}$ approach

Knowledge of the temperature dependence of reaction rates and rates of other quality changes in foods has been of great importance to food processors. Such knowledge is applied in the determination and prediction of limiting values for temperature and time in food processing and storage.

An empirical approach in studies of temperature-dependent kinetics of reaction rates and quality changes has been based on the determination of the rates at two or more temperatures and establishing relationships between the observed change, temperature, and time. Determination of the rate, $k_T$, at temperature, $T$, and the rate, $k_{T+10}$, at $T + 10$ allows definition of the ratio of the rates according to equation (9.12).

$$Q_{10} = \frac{k_{T+10}}{k_T} \tag{9.12}$$

The ratio given by equation (9.12) is commonly known as the $Q_{10}$ value of the reaction, or change in quality. The $Q_{10}$ value defines that an increase in temperature by 10 degrees increases the rate by the $Q_{10}$ factor.

## 2. Arrhenius equation

The most common relationship between rates of chemical reactions and temperature is given by the Arrhenius equation (9.13), where $k$ is the rate constant, $k_0$ is the frequency factor, $E_a$ is activation energy, $R$ is the gas constant, and $T$ is absolute temperature.

$$k = k_0 e^{-\frac{E_a}{RT}} \tag{9.13}$$

$$\ln k = \ln k_0 - \frac{E_a}{RT} \tag{9.14}$$

Equation (9.13) may also be written into the form of a straight line, which is given by equation (9.14). Thus, a plot of $k$ against $1/T$ gives a straight line with the slope $E_a/R$.

The Arrhenius equation is probably the most important relationship used to model temperature dependence of various physicochemical and chemical properties of foods. However, deviations from the Arrhenius kinetics of chemical reactions have been reported (Labuza and Riboh, 1982) and obviously a change in the physical state may change activation energy.

Labuza and Riboh (1982) listed several reasons for possible deviations from the Arrhenius kinetics, which have been observed in studies of temperature dependence of reaction rates in foods. Labuza and Riboh (1982) considered the following reasons to be the primary causes of nonlinearities in Arrhenius plots of reaction rates:

1. An increase in temperature may cause the occurrence of first-order phase transitions, e.g., melting of solid fat, which may increase mobility of potential reactants in the resultant liquid phase.

2. Crystallization of amorphous sugars may release water and affect the proportion of reactants in the solute-water phase.

3. Freeze-concentration of solutes increases concentration of reactants in the unfrozen solute matrix.

4. Reactions with different activation energies may predominate at different temperatures.

5. Increasing water activity with increasing temperature may cause an additional increase in reaction rate.

6. Partition of reactants between oil and water phases may vary with temperature due to phase transitions and solubility.

7. Solubility of gases, especially of oxygen, in water decreases with increasing temperature.

8. Reaction rates are often dependent on pH, which also depends on temperature.

9. Loss of water at high temperatures may alter reaction rates.

10. Proteins at high temperatures may become more or less susceptible to chemical reactions due to denaturation.

It is obvious that phase transitions are important causes for the observed deviations from Arrhenius kinetics in foods. Karmas *et al.* (1992) reported changes in activation energies for nonenzymatic browning reactions in low-moisture foods and food models. The reported changes suggested that the reaction rate was dependent on the physical state and that the rate increased above the glass transition temperature of the materials.

### 3. WLF equation

Molecular mobility is an important temperature-dependent factor which may affect rates of deteriorative changes in foods. It has been well established that water as a plasticizer has a significant effect on molecular mobility above a critical, temperature-dependent water activity or water content. Chemical reactions in low-moisture foods may be diffusion-limited (e.g., Simatos and Karel, 1988). Slade and Levine (1991) suggested that diffusion in amorphous foods is governed by the glass transition. Their hypothesis assumed that at temperatures below glass transition temperature rates of chemical reactions are extremely low due to the restricted molecular mobility and slow diffusion. A rapid increase in the rate is assumed to occur due to thermal or water plasticization at $T_g$. Slade and Levine (1991) pointed out that the temperature dependence of diffusion-controlled changes follows the WLF equation over the temperature range from $T_g$ to about $T_g + 100°C$.

It is likely that diffusion of reactants of bimolecular reactions in amorphous foods is related to viscosity. If the rate of a reaction is controlled by diffusion it may be assumed that below $T_g$ the rate of the reaction is extremely slow. At temperatures above $T_g$ diffusivity increases as viscosity decreases and probably follows the WLF-type temperature dependence in accordance with equation (9.15), where $D$ is the diffusion coefficient at temperature, $T$, $D_s$ is the diffusion coefficient at a reference temperature, $T_s$, and $C_1$ and $C_2$ are the WLF constants. If glass transition temperature is used as the reference temperature the diffusion coefficient, $D_g$, at $T_g$ can be used instead of $D_s$ and $T_s$.

$$\log\frac{D_s}{D} = \frac{-C_1(T - T_s)}{C_2 + (T - T_s)} \tag{9.15}$$

The use of equation (9.15) assumes that diffusion is directly related to the relaxation times of mechanical properties of amorphous solids. As reported by Peleg (1992) and Nelson and Labuza (1994), the WLF constants $C_1$ and $C_2$ can be calculated from the intercept and slope, respectively, of a straight line, which is obtained by plotting $1/(\log D_s/D)$ against $1/(T - T_s)$ according to equation (9.16).

$$\left(\log\frac{D_s}{D}\right)^{-1} = \frac{-C_2}{C_1}\left(\frac{1}{T - T_s}\right) - \frac{1}{C_1} \tag{9.16}$$

The rate constant of a diffusion-limited reaction is defined by equation (9.17), where $k'$ is the observed rate constant, $k$ is the true rate constant of a well-stirred system, and $\alpha$ is a temperature-independent constant (Karel and Saguy, 1991; Karel, 1993). If the temperature dependence of diffusion coefficients of reactants in amorphous food matrices follows equation (9.15) the rate of reactions with such reactants can be affected by glass transition.

Equation (9.16) shows that a low value of $D$ below $T_g$ decreases the observed rate constant. At temperatures above $T_g$ the reaction rate increases with increasing plasticization due to increasing diffusion and approaches the true constant, $k$, which applies when the system is well stirred. Figure 9.3 shows the ratio, $k'/k$, for a hypothetical reaction as a function of $D$ based on the assumptions that equations (9.15) and (9.17) apply and that the ratio $k/\alpha D$ is significantly greater than zero. As was pointed out by Karel (1993) these assumptions are not likely to be valid in all cases.

$$k' = \frac{k}{1 + \dfrac{k}{\alpha D}} \qquad (9.17)$$

Equation (9.17), as was also shown in Figure 9.3, defines that a dramatic change in the observed reaction rate constant may occur over a fairly narrow range of diffusion coefficients, which, depending on the probability of the

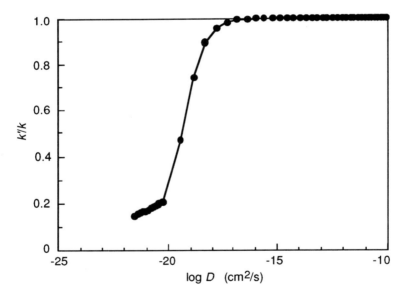

**Figure 9.3** Effect of diffusion on the ratio of observed, $k'$, and true reaction rate constants, $k$, assuming the validity of equations (9.15) and (9.17) for a hypothetical reaction. The diffusion coefficient increases with increasing temperature according to the Arrhenius-type temperature dependence below glass transition temperature, $T_g$. Exceeding $T_g$ causes a dramatic increase in the observed rate constant as the increase of the diffusion coefficient with increasing temperature becomes more pronounced and follows the WLF relationship.

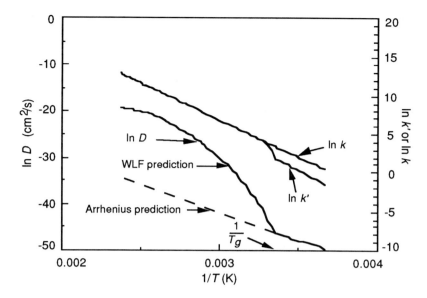

**Figure 9.4** Arrhenius plot for diffusion coefficient, $D$, true reaction rate constant, $k$, and the observed reaction rate constant, $k'$, for a hypothetical diffusion-limited reaction with the effect of glass transition temperature, $T_g$, on the temperature dependence of $D$. The diffusion coefficient was assumed to follow Arrhenius-type temperature dependence at $T < T_g$, and at $T > T_g + 100°C$ and WLF-type temperature dependence over the temperature range $T_g < T < T_g + 100°C$. The reaction rate constant for a well-stirred system is not affected by diffusion and shows no discontinuity. A diffusion-limited reaction shows a change in slope or a step change for $k'$ in the vicinity of $T_g$. The change in the reaction rate at $T_g$ may be negligible in comparison with the effect of $T_g$ on diffusion, which is significantly underestimated by the Arrhenius prediction.

reaction at $T_g$, may significantly affect the reaction rate over a narrow temperature or water content range. It is likely that in amorphous foods such a temperature or water content range is located in the vicinity of the $T_g$. Apparently, the true rate constants of deteriorative reactions at temperatures typical of food storage are relatively low and only minor changes of activation energies can be observed as food materials are transformed from the solid, glassy state into the supercooled liquid state, and only a small change in slope may be observed when $\ln k'$ is plotted against $1/T$. A representation of relationships between temperature, diffusion coefficient, and reaction rate constant in an amorphous material is shown in Figure 9.4. However, at high temperatures, which are often used in food processing, reaction rates may be substantially higher than in food storage and the effect of a glass transition on the observed reaction rate may be considerable.

The effect of a glass transition on reaction rates can be observed from experimental data for reaction rates in amorphous foods. A number of studies have reported diffusion coefficients for amorphous polymer-solvent systems. Some of the studies have found that Arrhenius plots showing the logarithm of an effective diffusion coefficient against $1/T$ exhibit a break or a step change in the vicinity of the glass transition temperature (Vrentas and Duda, 1978). The difficulties of analyzing diffusion in polymers were listed by Vrentas and Duda (1978). These included (1) inaccuracy in the determination of diffusion coefficients as a function of temperature; (2) need for extrapolations for diffusion data near $T_g$; and (3) anomalous diffusion effects near $T_g$. Perhaps these statements apply to biological materials and food solids as well. It should also be noticed that any changes in the physical state, e.g., collapse or crystallization, which are typical phenomena of amorphous food solids may significantly affect diffusion.

According to the free volume theory of diffusion (Vrentas and Duda, 1978) the temperature dependence of diffusion coefficients differs from Arrhenius behavior. Therefore, the logarithm of diffusion coefficients when plotted against $1/T$ often shows nonlinearity. Such nonlinearity becomes easier to observe with increasing molecular size of the diffusing compound or diluent (Vrentas and Duda, 1978). However, linearity over a relatively narrow temperature range may be observed which allows the determination of apparent activation energies near $T_g$. Such activation energies are useful in finding possible changes in diffusional properties which may occur within the glass transition temperature range. A schematic representation of the temperature dependence of diffusion in amorphous matrices is shown in Figure 9.4. The temperature dependence of diffusion is linear in an Arrhenius plot of $\ln D$ against $1/T$ when diffusion is not affected by other temperature-dependent factors, e.g., glass transition. Figure 9.4 shows also that a glass transition causes a fairly large change in the diffusion coefficient over the temperature range from $T_g$ to $T_g + 100°C$ when the assumption that the change in diffusion follows the WLF-type temperature dependence over that temperature range is valid.

The change in $D$ above $T_g$ may be dramatic and substantially larger than it would be according to the Arrhenius-type temperature dependence, but the difference in the observed and the true reaction rate constants is relatively small and occurs within a relatively narrow temperature range above $T_g$. In practice rates of diffusion-limited reactions have been determined for foods above $T_g$ and the rates have followed the Arrhenius-type temperature dependence. It is obvious from Figure 9.4 that a relatively small change in $D$ causes only a minor change in the observed reaction rate constant or only a change in the slope of the $\ln k'$ line as the value becomes equal to $\ln k$. It should also be noticed that the change in diffusion at a constant temperature due to water plasticization may be even more dramatic when observed as a function of water activity or water content. That may also explain the sudden increase in rates of such reactions as nonenzymatic browning above a critical water activity.

It may be concluded that diffusional properties in low-moisture foods are governed by glass transition, but the effects of diffusivity on the rates of chemical changes in such foods are more complicated. Nelson (1993) studied effects of $T_g$ on kinetic phenomena including crystallization of amorphous sugars and rates of chemical changes in food materials. She found that deteriorative reactions occurred at temperatures below $T_g$, but a large increase in the rates of several reactions in the vicinity of $T_g$ was evident. These findings were in agreement with rates of nonenzymatic browning which were reported by Karmas *et al.* (1992) for various amorphous foods and food models. Karmas *et al.* (1992) observed breaks in Arrhenius plots for the rate constant in the vicinity of $T_g$. Such breaks would be expected according to the data shown in Figure 9.4. Nelson (1993) concluded that the proper application of the WLF model in predicting kinetic data involves determination of the WLF coefficients. The coefficients are dependent on the system and also on its water content. Application of the WLF model for predicting temperature dependence of reaction rate constants should include the use of both equations (9.15) and (9.17). This was also suggested by the finding of Nelson (1993) that the rates of chemical changes in rubbery matrices instead of following the WLF model often followed the Arrhenius-type temperature dependence.

## III. Kinetics in Amorphous Foods

Molecular mobility of reactants is often considered to be one of the main factors that affect rates of deteriorative changes and shelf life of food materials. Molecular mobility may be related to the physical properties of foods and therefore also to phase transitions. It is well known that chemical reactions in foods usually occur in the solute matrix and in a liquid phase. The rates of the reactions decrease due to decreasing temperature or solvent and reactant crystallization. The effects of second-order transitions on molecular mobility have been well established (e.g., Slade and Levine, 1991), but their effects on reaction rates remain poorly understood. However, the effect of glass transition on molecular mobility is significant and a few studies on the effects of glass transition, water availability, or plasticization on rates of deteriorative reactions have been reported.

## A. Low-Moisture Foods

Molecular mobility in low-moisture foods is obviously important in defining rates at which reactants may diffuse within a food matrix. Such diffusion is necessary for any deteriorative change or enzymatic activity. Mobility may also be required for microbial growth, since transportation of nutrients from the food matrix is required to maintain activity. Food matrices at low water contents exist most probably in the amorphous state, but solutes, e.g., sugars and salts, may be present in the crystalline, amorphous, or dissolved state. They may also have dissolved and then bound to other compounds, e.g., proteins. The physical state of a number of compounds, particularly that of salts and sugars, is significantly related to the extent of water plasticization of the food and the thermal and plasticization history of the food due to the possibility of transitions between the amorphous, crystalline, and dissolved states. Such transitions may occur especially at a fluctuating storage humidity or temperature, but also during processing steps that may cause high local temperatures due to friction, e.g., in grinding.

### 1. Mobility and reaction rates

Rates of several deteriorative changes in foods are characterized by sensitivity to water. Rates often increase rapidly above a critical water content and decrease at high water contents. Therefore, a plot of reaction rate as a function of water activity may show a maximum at intermediate water activities.

Duckworth (1981) used wide-line nuclear magnetic resonance spectroscopy to determine "mobilization points" for solutes at a constant temperature in low-moisture food matrices. One of his main findings was that the mobilization point was peculiar to the system. The level of hydration needed to achieve mobility was solute-dependent. Duckworth (1981) observed that no solute mobilization occurred below the BET monolayer water content. Mobilization results for a system which contained reactants of the nonenzymatic browning reaction suggested that browning initiated at the mobilization point. An increased mobilization was apparent with increasing water activity and the rate maximum was at the water activity which corresponded to the hydration level that allowed complete mobilization. Extensive water contents caused dilution and reduced reaction rates. Mobilization of solutes according to Duckworth (1981) required sufficient amounts of hydrated reactants and the theory was referred to as the "solution scheme" by Karel *et al.* (1993). The solution scheme assumed that food materials at low water contents below the BET monolayer value were composed of an hydrated matrix with undissolved reactants and reactions did not occur. Above the monolayer value some of the reactants were dissolved, which allowed mobility in a saturated solution and an increasing rate with increasing water content. It should be noticed that mobilization through water adsorption may also depend on temperature.

Karel (1985) pointed out that water is the most important solvent and plasticizer for hydrophilic food components and therefore the mobility of food components is greatly affected by the presence of water. He also noticed that the mobility of food components affects the physicochemical and physical properties of foods, and that in most hydrophilic polymer systems diffusion of water and other small solutes as well as mechanical and thermal properties including $T_g$ are dependent on water content. Karel *et al.* (1993) considered reaction rates at low water contents to be restricted due to diffusional limitations. Restricted diffusion may decrease reaction rates as the transport of the reactants as well as transport of the products becomes reduced.

It is likely that most of the nonfat solids in low-moisture foods are amorphous and therefore mobility may be achieved by plasticization (Le Meste *et al.*, 1990). Such plasticization can be due to an increase in temperature (thermal plasticization) or addition of plasticizers such as glycerol or water. Roozen *et al.* (1991) used ESR to study molecular motions of dissolved probes in malto-oligosaccharides and maltodextrins with various water contents as a function of temperature. The rotational motions were detected from rotational correlation time, $\tau_c$, which is related to the rotational diffusion coefficient. Roozen *et al.* (1991) found that $\tau_c$ decreased linearly with increasing temperature at temperatures below $T_g$, suggesting that temperature-dependent molecular motions occurred in the glassy state. However, a dramatic decrease of the rotational correlation time occurred over the $T_g$ temperature range, which indicated the dramatic effect of $T_g$ on molecular mobility. Roozen *et al.* (1991) also noticed the decrease of the temperature at which the change in molecular mobility occurred with increasing water content due to water plasticization. The dramatic increase of molecular mobility above but in the vicinity of $T_g$ obviously has an effect on reaction rates in low-moisture foods.

## 2. Diffusion-limited reactions

Diffusion of reactants in a food matrix is probably the main requirement for the occurrence and increasing rates of chemical reactions above some critical temperature or water content. In some cases flow through pores may increase reaction rates. Such exceptions include lipid oxidation in dehydrated porous foods, since oxygen may diffuse in the material and enhance oxidation of lipids on the pore membranes.

Karel and Saguy (1991) listed a number of factors which may affect the overall effective diffusion coefficient, $D_e$. $D_e$ depends on (1) diffusion in the solid proper; (2) diffusion in gas-filled pores; (3) surface diffusion along the walls of pores; (4) capillary flow due to gradients in surface pressure; and (5) convective flow in capillaries due to total pressure differences. Therefore, it may be concluded that phase transitions and structural transformations in foods may affect diffusion and consequently observed rates of deteriorative changes. The effect of plasticization-induced mobility may significantly contribute to chemical changes in foods. Diffusion-limited reactions in foods include browning reactions, enzymatic reactions, and loss of color in dehydrated foods among others. These reactions may significantly limit the shelf life of food materials and their kinetics are important in shelf life predictions and in accelerated shelf life tests.

Diffusion in amorphous polymers has been found to be complicated and controlled by various factors. It is evident that diffusion of various compounds in biological materials is even more complicated. Effects of water content and temperature on diffusion in amorphous foods are well known, but the relationships between $T_g$ and diffusion in foods have not been well established. Unfortunately such data are not available and they may be extremely difficult and time-consuming to obtain. Some studies have reported effective diffusion coefficients for water in starch at various water contents, temperatures, and pressures (Fish, 1958; Karathanos *et al.*, 1991; Marousis *et al.*, 1991). Fish (1958) studied diffusion of water in starch gels, which, after dehydration, were clear and transparent, and obviously amorphous. The experimental data showed that the effective diffusion coefficient leveled off at about 20% water at 25°C. The same water content was also sufficient to decrease the activation energy to the value for the activation energy of self-diffusion in water. Incidentally, an estimate for the amount of water required to depress the $T_g$ of amorphous starch to 25°C is 20% (w/w) (Roos and Karel, 1991b). This coincidence suggests that diffusion in amorphous starch is probably hindered at lower water contents due to the glassy state of the matrix. Fish (1958) suggested that the rotation of water molecules was restricted at low water contents, which agrees with the fact that molecular mobility is significantly decreased in the glassy state.

Menting and Hoogstad (1967) and Menting *et al.* (1970) reported a rigorous study of diffusion of volatile compounds in amorphous maltodextrin matrices and solutions. The results indicated that diffusion was a strong function of water content. Diffusion of acetone and water in maltodextrin at low water contents was very low. Diffusion was also dependent on the size of the diffusing molecules, which suggested selective diffusion of water at low water contents. The selective diffusion was considered to be the basis for the retention of flavor compounds in dehydration (Menting *et al.*, 1970; Thijssen, 1971). Menting *et al.* (1970) found also that the diffusion coefficient decreased significantly with increasing molecular size of the diffusant, which may be important in defining reaction rates in the vicinity of $T_g$. Collapse of a dehydrated food decreases porosity. Products with high porosities, e.g., freeze-dried foods, exhibit high diffusion coefficients. Air-dried foods have a collapsed structure with low porosity and they have low diffusivity due to diffusion in the amorphous matrix, since diffusion by flow through pores is low (Karel and Saguy, 1991; Marousis *et al.*, 1991). According to Karel and Saguy (1991) apparent diffusion in porous materials may decrease with increasing water content due to collapse. However, diffusion in an amorphous food matrix increases with increasing water content. Therefore, transformation of food solids from a glassy state to the rubbery state may affect effective diffusion by causing structural changes or by increasing diffusion due to the increased molecular mobility as was described by Peppas and Brannon-Peppas (1994).

3. Water plasticization

It is obvious that mobility and diffusion in low-moisture foods are affected by water. The relative rate of deteriorative changes is traditionally related to the water content and water activity with the assumption that stability can be maintained at water contents below the BET monolayer value (e.g., Salwin, 1963; Labuza *et al.*, 1970; Labuza, 1980).

Roos (1993) analyzed water plasticization and water sorption properties of food components and food materials. The assumption that depression of the $T_g$ to below ambient temperature results in loss of stability, as was emphasized by Slade and Levine (1991), was used to establish critical water contents and $a_w$ values for stability. The critical water contents were higher than the BET and GAB monolayer values, which suggested that even water contents higher than the monolayer values could probably maintain stability. The fact that reaction rates are low below a critical water content or $a_w$ can be related to water plasticization and depression of the $T_g$ to below ambient temperature. Figure 9.5 shows viscosity of a maltodextrin (Maltrin M365) as a function of water activity. The viscosity was assumed to decrease according to the WLF equation above $T_g$. Therefore, depression of viscosity at a constant temperature is observed above the critical water content or critical $a_w$. Figure 9.5 shows also the effect of water activity on the diffusion coefficient of water and acetone in an amorphous maltodextrin matrix

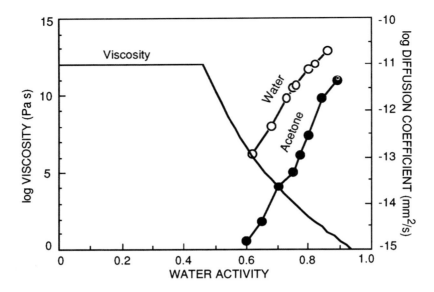

**Figure 9.5** WLF prediction of viscosity for maltodextrin (Maltrin M365) with diffusion coefficients for water and acetone in maltodextrin. Maltrin M365 (Roos, 1993) data for glass transition and water sorption were used in predicting the diffusion coefficient determined by Menting *et al.* (1970) as a function of water activity.

according to Menting and Hoogstad (1967). The composition of the maltodextrin as reported by Menting and Hoogstad (1967) was comparable with the composition of Maltrin M365, for which the $T_g$ data shown in Figure 9.5 were available (Roos and Karel, 1991a).

Enzyme activity has been found to be related to hydration. A change in heat capacity and an increase in motional freedom of enzyme molecules have coincided with the onset of enzyme activity (Careri *et al.*, 1980), which suggests that a relationship may exist between enzyme activity and glass transition. Silver and Karel (1981) studied the effect of water activity on sucrose inversion by invertase. The rate of the reaction increased with increasing water activity and the rate followed first-order kinetics. The samples were freeze-dried and the sucrose was likely to exist in the amorphous state. This was also noticed from the fact that the onset of hydrolysis occurred at water activities below the suggested mobilization point of 0.81 $a_w$ for crystalline sucrose (Duckworth, 1972). Silver and Karel (1981) observed a continuous decrease in the activation energy with increasing $a_w$, which was concluded to suggest that the reaction was diffusion-controlled. Drapron (1985) stated that not only water activity, but also the ability of water to give a certain mobility to enzymes and substrates, is important to enzyme activity. He assumed that the amount of water needed increases with increasing molecular size due to impaired diffusion. However, lipase activity was not related to the mobility provided by water. Interestingly, Drapron (1985) pointed out that in $\beta$-amylolysis the water activity at which the reaction started was lower at 30°C than at 20°C. He assumed that the mobility of the components increased with temperature.

Water plasticization may also have an impact on rates of vitamin destruction during food storage. Dennison *et al.* (1977) studied the effect of water activity on thiamine and riboflavin retention in a dehydrated food system. The product was a starch-based food model and it is obvious that the $T_g$ of the model was fairly high. They found that the retention of the vitamins was high after a storage period of 8 months at 20 and 30°C and $a_w < 0.65$. A substantial loss of thiamine and riboflavin was noticed at 45 and 37°C and $a_w > 0.24$, respectively. It may be assumed that the molecular mobility attained

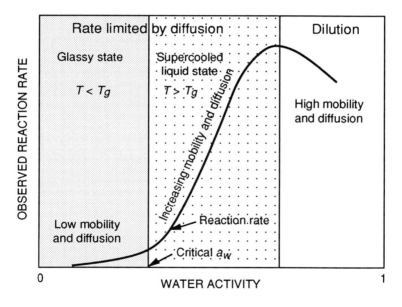

**Figure 9.6** A schematic representation of the effect of water activity, $a_w$, on mobility and diffusion, and on observed reaction rates in amorphous food materials at a constant temperature. Rates of diffusion-limited reactions may become significantly reduced in the glassy state, which is typical of food materials at low water activities. An increase in $a_w$ may enhance diffusion. The glass transition temperature, $T_g$, is depressed to below ambient temperature at a critical $a_w$, which increases diffusion and the observed reaction rate. An increasing reaction rate with increasing $a_w$ is observed until dilution of reactants decreases the rate of the reaction at high water activities.

was due to water and thermal plasticization, which occurred in the freeze-dried model and caused the observed degradation of the nutrients. The overall effects of water plasticization on rates of diffusion-limited reactions in low-moisture foods are summarized in Figure 9.6. Figure 9.6 assumes that the critical $a_w$ is defined by the $T_g$ depression to ambient temperature, e.g., to food storage temperature. The change in molecular mobility and diffusion occurs in the vicinity of the critical $a_w$. It should be remembered that $T_g$ is dependent on the experimental time scale. However, the most dramatic changes in molecular mobility and diffusion, and probably also in reaction rates, occur over a fairly narrow water activity range.

4. Observed kinetics

A number of studies have reported kinetic data for observed reaction rates in low-moisture foods. These data have often been reported as a function of water activity or water content at a constant temperature. The change in the physical state due to glass transition or other phase transitions has been considered in only a few studies. These studies have revealed that reaction rates are increased above glass transition, but some reactions have occurred also at temperatures below $T_g$ (Nelson, 1993; Karmas, 1994).

Food quality is always subject to time-dependent changes. The quality of most foods decreases with time, although some foods, e.g., some fermented products, may improve during storage. However, the extent of quality changes may be related to the time of observation and the properties of the food and its surroundings (Karel *et al.*, 1993). The change in quality, $dQ$, in time, $dt$, is defined to be a function of environmental, $E_i$, and compositional, $F_j$, effects by equation (9.18).

$$\frac{dQ}{dt} = f\left(E_i, F_j\right) \tag{9.18}$$

Rates of quality changes in foods are often considered to follow either zero- or first-order kinetics and the temperature dependence, to follow Arrhenius equation (Labuza, 1980). Thus, equation (9.19), where [$Q$] is the amount of the quality factor, may be used to predict kinetics of quality changes at various temperatures.

$$\frac{dQ}{dt} = k_0 e^{-\frac{E_a}{RT}} [Q]^n \tag{9.19}$$

It has been noticed that water influences $k_0$, $E_a$, $[Q]$, and $n$ (Labuza, 1980). The glass transition theory assumes that the molecular mobility and diffusion in the glassy state are slow and the rate of the quality change approaches zero. Above $T_g$ the observed rate of the quality change increases due to enhanced diffusion and the temperature dependence of the quality change may deviate from equation (9.19). It may be assumed that the temperature dependence of the diffusion coefficient follows equation (9.15) and that the observed rate constant, $k'$, of the quality change becomes a function of equation (9.17). If the assumption is valid a break or a step change in the Arrhenius plot for the quality change is observed in the vicinity of the $T_g$. Equations (9.15) and (9.17) do not consider the effect of water content on the quality change. The observed rate increases with increasing water activity or water content above the critical value. It is important to notice that the true rate constant, $k$, is a temperature- and water plasticization-dependent parameter. Therefore, $k'$ values that are obtained at different temperatures above $T_g$ should not be used to predict $k'$ at other temperatures by converting water plasticization to the corresponding thermal plasticization in terms of $T - T_g$.

*a. Nonenzymatic Browning.* Nonenzymatic browning is a series of condensations that can be considered to be bimolecular. The initial reactants of nonenzymatic browning in foods are often a reducing sugar and an amino acid. It is an important reaction that produces typical flavors of several foods, but it also decreases food quality during storage. The kinetics of nonenzymatic browning are related to a number of factors (Labuza and Baisier, 1992), but in low-moisture foods it may be considered to become diffusion-limited at low water activities.

Eichner and Karel (1972) published probably the first study on nonenzymatic browning, which considered the effect of mobility, viscosity, and also the glassy state on the rate of the reaction in glucose-glycine-glycerol-water model systems. They found a decrease in the browning rate especially at low water contents when the amount of glycerol in the system was low. The decrease was assumed to result from decreased mobility of the reactants and reaction products at the very high viscosity of the sugar solution, which was reported to be in the glassy state. Eichner and Karel (1972) observed that the addition of glycerol improved the mobility of the reactants due to plasticization, which increased the rate of the reaction. Although Eichner and Karel (1972) did not report relationships between $T_g$ and the browning rate their results suggested that the reaction rate decreased at low levels of plasticization as the material approached the glassy state.

Flink *et al.* (1974) studied the browning rate of nonfat milk powder, which was humidified at 0, 11, and 32% RH at 37°C and then stored at various temperatures. They observed that the rate of browning was low below a critical temperature, above which the rate of the reaction increased substantially. They also observed that the browning rate was dependent on water content and the critical temperature for the reaction decreased with increasing initial water activity. The physical state of lactose in freeze-dried nonfat milk that was used by Flink *et al.* (1974) is known to be amorphous at low water activities. The results of the study of Flink *et al.* (1974) suggested that the rate of browning at temperatures below $T_g$ was low and the increase in browning rate above the critical temperature occurred due to plasticization and increasing mobility above glass transition. In addition, the rate of browning was probably also increased due to crystallization of amorphous lactose at temperatures above $T_g$.

Nonenzymatic browning rates have been reported for several foods that are likely to exist in the glassy state at low water activities or water contents and to exhibit increasing browning rates above some critical $a_w$ values or water contents. Karmas *et al.* (1992) derived $T_g$ values for cabbage, carrots, onions, and potatoes and analyzed their browning rates as a function of $T - T_g$. The results showed that nonenzymatic browning was not likely to occur below $T_g$. Browning occurred above a critical $T - T_g$, which was dependent on water content. The critical $T - T_g$ was higher for materials with higher water contents, which probably was due to the fact that the browning

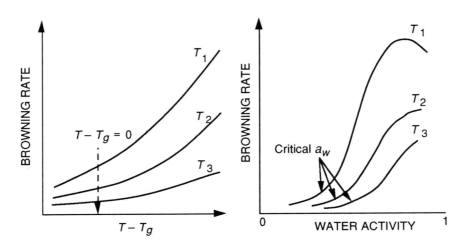

**Figure 9.7** A schematic representation of the effect of temperature, $T$, glass transition temperature, $T_g$, water activity, $a_w$, and water content on the rate of nonenzymatic browning. The $T - T_g$ at a constant temperature, $T_1 > T_2 > T_3$, increases with increasing water content and the observed browning rate increases with increasing $T - T_g$. The $T - T_g$ increases with increasing $a_w$. The observed browning rate increases with increasing $a_w$ and also with increasing temperature, $T_1 > T_2 > T_3$. Browning may occur also below $T_g$.

reaction had a different rate constant not only due to the lower $T_g$ and therefore higher $T - T_g$ but also due to the higher water content and therefore different concentration of the reactants. It would be expected that the true rate constant of the browning reaction is affected by both water content and temperature. A material with a high water content has a low $T_g$ and browning may occur at a relatively low temperature. However, the true rate constant decreases with decreasing temperature, which also decreases the observed rate constant. A schematic representation of observed relationships between browning rate, temperature, and water content is shown in Figure 9.7. It may be concluded that browning data should be obtained as a function of temperature and also as a function of water content at constant temperatures, which allows evaluation of the various effects which temperature and water content may have on the physical state and observed browning rate.

Determination of optical density is a convenient method to follow the extent of nonenzymatic browning in model systems. The method involves extraction of reaction products with an ethanol-water solution and determination of optical density at 280 or 420 nm. Optical density at 280 nm is a

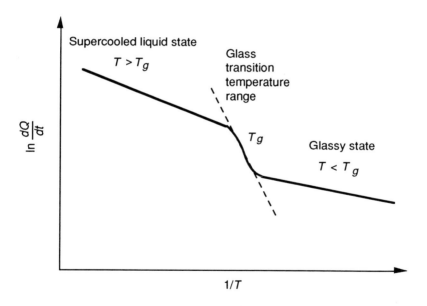

**Figure 9.8** A schematic representation of the effect of glass transition on observed rates of nonenzymatic browning. The reaction has a low activation energy, $E_a$, below glass transition temperature, $T_g$. Increased $E_a$ may be observed within the glass transition temperature range due to probable increase in diffusion. Typical $E_a$ values of the reaction but higher than below $T_g$ are observed above the glass transition temperature range.

measure of reaction products such as furfurals, which are formed at the early stages of the reaction, while optical density at 420 nm quantifies yellow and brown pigments, which are formed at the final stages of the reaction. The optical density can be used in equation (9.19) as the quality factor, [Q]. Since the reaction often follows zero-order kinetics a plot of [Q] against time, $t$, gives a straight line with a slope that is equal to the rate constant. Karmas *et al.* (1992) reported browning data as a function of temperature for several model systems that had various initial water activities at room temperature. Arrhenius plots for the materials were nonlinear with two changes. These changes were observed to occur in the vicinity of $T_g$ and at about 10°C above $T_g$ as shown schematically in Figure 9.8. Karmas *et al.* (1992) found that the activation energies for the reaction below $T_g$ (30 to 90 kJ/mol) were lower than above $T_g$. The activation energies above $T_g$ (65 to 190 kJ/mol) were typical of the nonenzymatic browning reaction. However, those within the glass transition temperature range (250 to 400 kJ/mol) were substantially higher than values commonly obtained for the reaction. Karmas *et al.* (1992) pointed out that the step change in the Arrhenius plots was similar to those found for diffusion in polymers.

The assumptions in the application of equations (9.15) and (9.17) are that diffusivity is proportional to the reciprocal of viscosity, $\eta$, which follows the WLF-type temperature dependence over the temperature range from $T_g$ to about $T_g + 100$°C. If a change in a quality factor is diffusion-limited the observed rate constant is a function of the true rate constant and diffusivity as was defined by equation (9.17). In some studies additional assumptions have led to the conclusion that the observed rate constant of diffusion-limited reactions in amorphous food matrices is also proportional to $1/\eta$, which has led to the application of the WLF equation in the form of equation (9.20), where $k'$ is the observed rate constant and $k'_s$ is the observed rate constant at a reference temperature, $T_s$ (Buera and Karel, 1993; Nelson, 1993; Roos and Himberg, 1994).

$$\log\frac{k'_s}{k'} = \frac{-C_1(T-T_s)}{C_2+(T-T_s)} \tag{9.20}$$

Application of equation (9.20) to model rates of nonenzymatic browning at several water contents has failed in most cases (Karmas, 1994; Roos and Himberg, 1994). The use of equation (9.20) may be justified when the material has a constant water content and the constants $C_1$ and $C_2$ have been derived from experimental data. Buera and Karel (1993) reported that $C_1$ for several food systems varied with water content. They found that the $C_1$ increased linearly with increasing water content at water contents above 3 g/100 g of solids. The constant, $C_2$, was found to be independent of water content. However, Buera and Karel (1993) found that reaction rate data in a plot of log $(k'/k'_s)$ against $(T - T_s)/(T - T_\infty)$, where $T_\infty$ was a fixed temperature, where the rate constant, $k'_\infty$, became vanishingly small, fell on the same straight line. The procedure used was that of Ferry (1980), where the constants were first solved for equation (9.21) and subsequently shifted to the $T_g$ as another reference temperature according to Peleg (1992).

$$\log\frac{k'_\infty}{k'_s} = \frac{-C_1(T_\infty - T_s)}{C_2 + (T_\infty - T_s)} \tag{9.21}$$

Considering that $k'_\infty$ approaches zero, $C_2 + (T_\infty - T_s)$ must also approach zero. Therefore, constant $C_1$ may be obtained from equation (9.22), which combines equation (9.20) with $C_2$ that is defined by $C_2 = T_s - T_\infty$.

$$\log\frac{k'_s}{k'} = \frac{-C_1(T - T_s)}{(T - T_\infty)} \tag{9.22}$$

Equation (9.22) suggests that a plot of log $k'_s/k'$ against $(T - T_s)/(T - T_\infty)$ is a straight line that goes through the origin. Ferry (1980) pointed out that the $T_\infty$ is often located at about 50°C below $T_g$, but several attempts with different $T_\infty$ values may be needed before $C_1$ is obtained from the slope of the line. Although equation (9.20) may have predictive value (Buera and Karel, 1993) it should be noticed that temperature and water content may have significant effects on the true reaction rate constants.

The above studies on relationships between $T_g$ and the rate of the nonenzymatic browning reaction have suggested that the reaction is diffusion-limited and that the rate is affected by $T_g$. However, the rate constant is dependent on a number of other factors that include temperature, water content, and structural transformations (Karmas, 1994). Karmas (1994) pointed out that the size of the reactants may also be an important factor that affects rates of diffusion-limited reactions. It may be assumed that the rate of diffusion decreases with increasing size of the diffusant. The temperature and water content have the most important effects due to the fact that an increasing temperature increases the true rate constant and it is also dependent on water content. Therefore, observed browning rates at the same apparent $T - T_g$ for the same material at various levels of water plasticization may be different.

*b. Other Changes.* Rates of diffusion-limited reactions other than nonenzymatic browning are likely to be affected by temperature and water content in a similar manner in addition to the change caused by glass transition. Nelson (1993) observed rates of ascorbic acid degradation within freeze-dried noncrystallizing maltodextrin matrices at various temperatures and water contents. The reaction may occur through a number of pathways, but Nelson (1993) assumed that the reaction involved diffusion of small molecules such as oxygen in the system studied. The rate of the reaction was found to follow first-order kinetics. Therefore, the reaction rate constants could be derived from the slopes of straight lines from plots that showed the logarithm of the ascorbic acid concentration against time.

Nelson (1993) found that the rate of ascorbic acid degradation generally increased with increasing temperature. The reaction occurred also at temperatures below $T_g$, probably due to the small size of the diffusing oxygen molecules. Application of the Arrhenius model to describe the temperature dependence showed a change in the activation energy of the reaction for a system that had a relatively low water content. The material with higher water contents had Arrhenius plots with a continuous line or no data were obtained below $T_g$. Nelson (1993) found that the kinetics of the reaction were affected by structural changes. The WLF equation failed to describe the temperature dependence of the reaction. Bell and Hageman (1994) found that aspartame degradation in a poly(vinylpyrrolidone) (PVP) matrix also occurred below $T_g$ and that the rate at room temperature was more dependent on water activity than on the state of the system.

5. Effects of structural transformations

Structural transformations in low-moisture foods occur at temperatures above the glass transition temperature. Such transformations include collapse of the physical structure, which decreases diffusion through pores, and crystallization of amorphous compounds, especially sugars, that may release adsorbed water and affect water activity. Crystallization of amorphous compounds may also release encapsulated compounds, which become susceptible to chemical changes. The main reactions that are affected by changes in the physical state and structural transformations include diffusion-limited reactions and oxidation.

*a. Collapse.* Collapse in low-moisture foods may significantly affect effective diffusion coefficients, mainly due to the loss of porosity and formation of a dense structure. Karmas *et al.* (1992) related some of the differences in observed rates of the nonenzymatic browning to collapse of the matrix. Collapse was observed also by Nelson (1993) to affect the rate of ascorbic acid degradation in maltodextrin matrices.

The formation of glassy carbohydrate matrices is of great importance in flavor encapsulation and in the protection of emulsified lipids in food powders from oxidation (Gejl-Hansen and Flink, 1977). Flink (1983) pointed out that the encapsulated compounds are stable as long as the physical structure of the encapsulating matrix remains unaltered. Encapsulated compounds may be released due to collapse, which results in loss of flavors and exposure of lipids to oxygen. However, Flink (1983) pointed out that collapsed structures may hold encapsulated compounds and protect them from release due to the high viscosity of a collapsed media. Labrousse *et al.* (1992) found that an encapsulated lipid was partially released during collapse, but during the collapse reencapsulation assured its stability. However, differences in the rates of diffusion within glassy carbohydrate matrices and supercooled, liquid-like, viscous matrices have not been reported. Therefore, it may be difficult to evaluate various effects of collapse on the release of encapsulated compounds or effects of diffusion on quality changes in collapsed matrices.

Nelson (1993) found that activation energies of ascorbic acid degradation in some cases were lower above $T_g$ than below $T_g$. She assumed that the decrease of the activation energy occurred due to structural collapse which was evident above $T_g$. Nelson (1993) pointed out that ascorbic acid could not be released from a maltodextrin matrix, which often occurs in the case of flavors and lipids. Therefore, she concluded that a collapsed matrix probably protected ascorbic acid from degrading due to decreased ability of oxygen to penetrate the surface of the collapsed matrix.

*b. Crystallization.* Crystallization of amorphous sugars is known to result in serious quality losses in food powders. The crystallization of amorphous lactose in dehydrated milk products has been observed to result in acceleration of the nonenzymatic browning reaction as well as other deteriorative changes and caking. Crystallization of amorphous sugars as defined by Levine and Slade (1986) is a collapse phenomenon that is a consequence of the glass transition. Glass transition may be considered to be the primary cause of an increased molecular mobility that allows the molecular reorganization which results in the formation of the crystalline phase.

Herrington (1934) pointed out that lactose crystallization in spray-dried milk powder must be prevented in order to avoid caking and decreased solubility of the powder. Lea and White (1948) reported that the storage life of milk powders ranged from 2 days for milk powder that was packed in air and kept at 37°C to 2 years or longer for milk powders that had a low water content and were packed in nitrogen and stored at 20°C. They also noticed that an increase in temperature caused a greater deleterious effect at a high water content of 7.6%. In addition lactose crystallization was observed to release water, which increased water activity and accelerated deterioration of the residual protein-sugar mixture. Therefore, milk powder with the high water content became rapidly unpalatable, discolored, and insoluble in water. King (1965) reported that crystallization of lactose coincides with an increase in free fat, which presumably facilitates lipid oxidation in milk powders.

Crystallization of amorphous lactose occurs also in other dairy powders such as whey powder above a critical water activity (Saltmarch and Labuza, 1980). Lactose crystallization in dairy powders results in increasing rates of nonenzymatic browning and loss of lysine (Saltmarch *et al.*, 1981). Saltmarch *et al.* (1981) found that the rate of browning at 45°C increased rapidly above 0.33 $a_w$ and showed a maximum between 0.44 and 0.53 $a_w$. The rate maximum for browning occurred at a lower $a_w$ than those found for other foods. That would be expected, since crystallization of lactose occurs within the reported water activity range and the $a_w$ at which crystallization

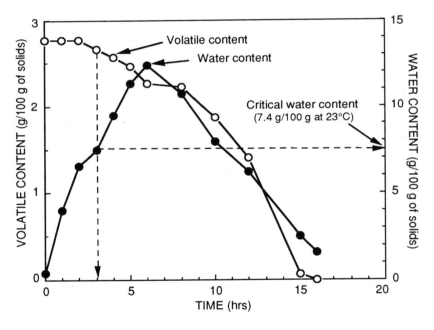

**Figure 9.9** Effect of water adsorption at 61% RH and 23°C on the amount of 2-propanol encapsulated in amorphous lactose. Amorphous lactose is hygroscopic and adsorbs water rapidly. Crystallization of lactose is observed as the glass transition temperature, $T_g$, is depressed to 23°C, which occurs at the critical water content. Further increase in water content accelerates the crystallization process. The volatile is lost due to lactose crystallization with a rate that corresponds to the rate of crystallization. Data from Flink and Karel (1972) and Jouppila and Roos (1994).

occurs decreases with increasing temperature. Saltmarch *et al.* (1981) observed that the rate maximum occurred at the same $a_w$ at which extensive lactose crystallization was observed from scanning electron micrographs. The rate of browning was significantly lower in a whey powder that contained most of the lactose in the precrystallized form. The loss of lysine was also found to be most rapid at water activities that corresponded with the occurrence of lactose crystallization. Saltmarch *et al.* (1981) found that the rate of nonenzymatic browning deviated from the Arrhenius kinetics. Such behavior is probable due to the first-order phase transition that may change the rate of the reaction as the temperature is increased to above $T_g$. Crystallization of the amorphous lactose increased water activity and accelerated the browning reaction in comparison with the rate of the reaction at the same temperature but at a constant water activity (Kim *et al.*, 1981).

Crystallization in low-moisture carbohydrate matrices which contain encapsulated volatiles or lipids results in a complete loss of flavor and release of lipids from the matrix. Flink and Karel (1972) studied the effect of crystallization on volatile retention in amorphous lactose. Crystallization was observed from the loss of adsorbed water at relative humidities, which allowed sufficient water adsorption to induce crystallization. At low relative humidities the encapsulated compound was retained at water contents below the BET monolayer value and probably also below $T_g$. The retention was high until the water content reached its maximum value. At higher relative humidities the amount of adsorbed water decreased and the rate of the volatile loss increased. The results showed that crystallization of amorphous lactose resulted in loss of both adsorbed water and encapsulated volatiles. It is obvious that the crystalline structure was not able to entrap volatile compounds.

The relationships between volatile loss or release of lipids that are encapsulated in amorphous matrices, water content, and crystallization at a constant storage relative humidity and temperature are shown in Figure 9.9. The results of Flink and Karel (1972) on the release of 2-propanol from amorphous lactose during storage at 61% RH and 23°C suggested that the release was initiated concomitantly with crystallization. It should be noticed that crystallization is possible only when water adsorption is sufficient to depress $T_g$ to below storage temperature. However, an uneven water distribution during the sorption process at a fairly high relative humidity may allow local crystallization and volatile release before the average water content is higher than the critical water content.

The effect of glass transition on the rate of oxidation of methyl linoleate that was encapsulated in an amorphous lactose matrix was studied by Shimada *et al.* (1991). Shimada *et al.* (1991) observed that oxidation did

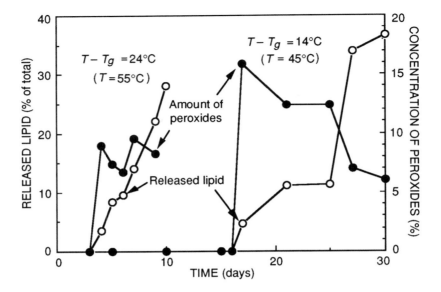

**Figure 9.10** Effect of crystallization on the release of encapsulated methyl linoleate from a lactose-based food model. The rate of crystallization is defined by temperature difference between storage temperature, $T$, and glass transition temperature, $T_g$. Crystallization releases the encapsulated lipid, which becomes susceptible to atmospheric oxygen and shows rapid oxidation. Data from Shimada *et al.* (1991).

not occur in encapsulated methyl linoleate. Lactose crystallization was observed at temperatures above $T_g$ and the rate increased with increasing $T - T_g$. Shimada *et al.* (1991) did not observe oxidation above $T_g$ until crystallization released the encapsulated compound. Methyl linoleate that was released from amorphous lactose became accessible to atmospheric oxygen and oxidized rapidly as shown in Figure 9.10. It is obvious that nonencapsulated lipids are susceptible to oxidation in low-moisture foods. Encapsulated lipids in foods may become protected from oxidation (Matsuno and Adachi, 1993), but crystallization of the encapsulating matrix releases such compounds and causes rapid deterioration.

6. Stability maps

Food stability is often related to water activity, $a_w$, which has been used as a common measure of stability of low- and intermediate-moisture foods at various storage conditions. The effect of $a_w$ on relative rates of various deteriorative changes has often been described with stability maps. Stability maps may be used to relate reactivity to $a_w$ at a constant temperature. However, such maps do not always consider the occurrence of phase transitions and time-dependent nonequilibrium phenomena (Karel *et al.*, 1993).

Stability maps are based on reaction rate data which have been obtained for various food materials. They commonly show the relative rate of enzymatic changes, nonenzymatic browning, oxidation, microbial growth, and overall stability as a function of water activity (e.g., Labuza *et al.*, 1970; Rockland and Nishi, 1980). The rate of the various reactions may also be related to the physical state and mobility in amorphous foods. Figure 9.11 shows a modified stability map that includes information on the critical water activity which determines the glass transition and molecular mobility at a constant temperature. It should be noticed that the critical $a_w$ has a material-specific value. Therefore, water activity values for observed rates may vary significantly depending on food composition and temperature.

Temperature has a significant effect on the reaction kinetics and the physical state of food materials. The assumption that stability is related to glass

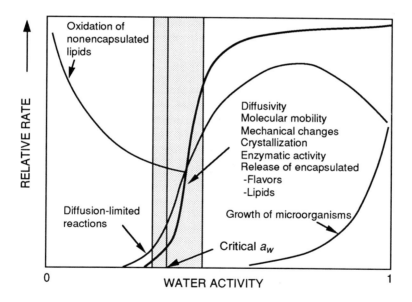

**Figure 9.11** Stability map that shows relative rates of physicochemical changes in foods as a function of water activity, $a_w$. Molecular mobility and rates of diffusion-controlled changes increase in the vicinity of a critical $a_w$, which is defined by water plasticization which is sufficient to depress $T_g$ to below ambient temperature.

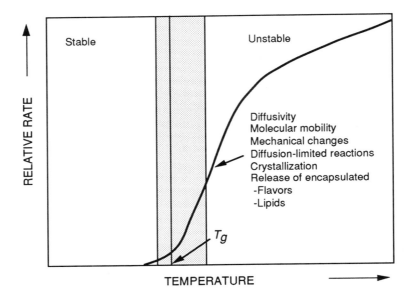

**Figure 9.12** Stability map for amorphous foods that describes the effect of temperature on relative rates of changes which are controlled by the glass transition temperature, $T_g$.

transition allows establishment of another stability map, which describes the effect of temperature on changes that depend on the relaxation times of mechanical changes in foods. Figure 9.12 relates deteriorative changes, which are governed by $T_g$, to temperature. In addition state diagrams and sorption isotherms are useful as stability maps. They may also be used to obtain material-specific data for the glass transition temperature at various water contents and relationships between water content and water activity.

### B. Frozen Foods

The physical state of frozen foods is probably more sensitive to temperature than low-moisture foods due to phase transitions of both water and food solids. At low temperatures food solids exist in the freeze-concentrated amorphous matrix that is plasticized by the unfrozen water. At temperatures typical of frozen food storage the physical state is defined by the characteristics of the solids and their effect on ice melting.

## 1. Quality changes in frozen foods

Relationships between composition and temperature in frozen foods can be characterized by state diagrams which describe the physical state as a function of temperature. It is obvious that ice formation occurs in foods as the temperature is depressed to below the equilibrium melting temperature of ice. Therefore, freeze-concentration of solutes to the unfrozen solute matrix also changes the concentration of potential reactants.

Fennema (1973a) pointed out that during freeze-concentration several chemical properties of the unfrozen phase are changed. Such properties include (1) pH; (2) titratable acidity; (3) ionic strength; and (4) oxygen-reduction potential. It is obvious that these changes in addition to the changes in concentration, mechanical properties, and temperature may affect reaction rates. Deteriorative changes in frozen foods include (1) autoxidation of ascorbic acid; (2) degradation of vitamins; (3) enzymatic browning; (4) lipid oxidation; (5) pigment degradation; (6) reactions that produce off-odors and off-flavors; and (7) textural changes (Powrie, 1973a,b; Fennema, 1973b). Some of these reactions may become diffusion-limited at low temperatures as the freeze-concentration of the unfrozen solute matrix increases viscosity and decreases the effective $T - T_g$ for relaxation times of mechanical properties.

Different food products have different storage lives and the effect of changes in storage temperature on the rate of quality changes is specific to each material. It is well known that foods with high amounts of lipids have a short shelf life due to oxidation. Fruits and vegetables have fairly long shelf lives at -18°C, but they are more sensitive to increases in storage temperature. The sensitivity of sugar-containing food products to temperature is probably due to the higher unfrozen water content and lower melting temperature. A decrease in storage temperature below -18°C increases shelf life, which is probably caused by the decreasing temperature and the high viscosity of the freeze-concentrated solute matrix. According to Levine and Slade (1986) stability is maintained during storage at temperatures below the glass transition of the maximally freeze-concentrated solute matrix.

## 2. Arrhenius and WLF kinetics

Quality changes in frozen foods are often determined with sensory techniques or chemical analysis of a specific compound such as ascorbic acid. Such studies have suggested that an increasing storage temperature increases rates of deteriorative changes. Therefore, the product shelf life decreases

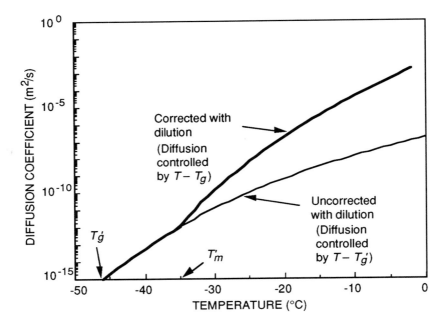

**Figure 9.13** Effect of glass transition temperature of the maximally freeze-concentrated solute matrix, $T_g'$, and ice melting at temperatures above $T_m'$ on a hypothetical diffusion coefficient in a frozen sucrose solution. The temperature dependence of the diffusion coefficient was assumed to follow the WLF equation above $T_g'$. Ice melting at temperatures above $T_m'$ decreases the $T_g$ that controls diffusivity. Therefore, diffusivity increases more than would be predicted by the temperature difference $T - T_g'$.

exponentially with increasing storage temperature. It is obvious that the increase in unfrozen water content with increasing storage temperature contributes to rates of deteriorative changes.

Levine and Slade (1986, 1988) have stressed the importance of the glass transition of the freeze-concentrated, particularly the maximally freeze-concentrated, solute matrix on rates of deteriorative changes in frozen foods. Levine and Slade (1986, 1988) have postulated that frozen foods are stable at temperatures below $T_g'$. Stability decreases exponentially above $T_g'$ with increasing $T - T_g$. According to Levine and Slade (1986) the dramatic change in rates of physicochemical changes above $T_g'$ can be described by the WLF-type temperature dependence of mechanical properties. Therefore, rates of a number of diffusion- and viscosity-related quality changes such as recrystallization processes in frozen foods may follow the WLF equation. The effect of glass transition and ice melting above $T_m'$ on a hypotheti-

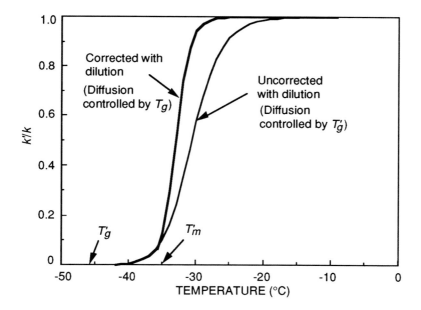

**Figure 9.14** Effect of temperature on the rate constant of a hypothetical reaction occurring in a frozen sucrose solution. The observed reaction rate constant, $k'$, approaches zero below the glass transition temperature of the maximally freeze-concentrated solute matrix, $T'_g$, due to diffusional limitations. The rate of the reaction increases above $T'_g$ due to increasing diffusion. Ice melting at temperatures above $T'_m$ increases diffusion more than would be predicted by the temperature difference $T - T'_g$, which increases the reaction rate as the observed rate constant becomes equal to the true rate constant, $k$, within a narrow temperature range.

cal diffusion coefficient in a frozen sucrose solution is shown in Figure 9.13. It is assumed that the diffusion coefficient is related to viscosity. Therefore, it follows equation (9.15) and decreases according to the WLF-type temperature dependence in the maximally freeze-concentrated solute matrix at temperatures above $T'_g$. However, ice melting occurs above $T'_m$, which causes dilution of the freeze-concentrated solute matrix. The $T_g$ that controls diffusion decreases with increasing temperature due to the plasticizing effect of water that is released from the ice fraction. It is obvious that the diffusion coefficient is controlled by $T - T'_g$ at temperatures below $T'_m$, but at temperatures above $T'_m$ the increase in the diffusion coefficient is orders of magnitude higher than would be predicted by the WLF equation with $T'_g$ as the reference temperature. It should be noticed that the $T'_m$ corresponds to the $T'_g$ reported by Levine and Slade (1986, 1988; Slade and Levine, 1991).

**Table 9.1** Glass transition temperatures (or onset of ice melting) of maximally freeze-concentrated solute matrices, $T'_g$ (or $T'_m$) for selected food materials according to Levine and Slade (1989).

| Material | $T'_g$    (°C) |
|---|---|
| Dairy foods | |
| Cottage cheese | -21 |
| Cream | -23 |
| Ice cream | -41 to -27.5 |
| Skim milk | -27.5, -32 [a] |
| Skim milk with hydrolyzed lactose | -40 [a] |
| Whipped toppings | -34.5 to -31.5 |
| Whole milk | -22 |
| Fruits | |
| Apple | -42 to -41 |
| Banana | -35 |
| Blueberry | -41 |
| Peach | -36.5 |
| Strawberry | -41 to -33.5 |
| Fruit juices | |
| Apple | -40.5 |
| Lemon | -43±1.5 |
| Orange | -37.5±1.0 |
| Pear | -40 |
| Pineapple | -37.5 |
| Prune | -41 |
| Strawberry | -41 |
| White grape | -42.5 |
| Vegetables | |
| Broccoli | -26.5 |
| Cauliflower | -25 |
| Potato | -16 to -11 |
| Spinach | -17 |
| Sweet corn | -14.5 to -8 |
| Tomato | -41.5 |

[a] Jouppila and Roos (1994)
*Source*: Levine and Slade (1989)

Karel and Saguy (1991) made shelf life predictions for stability of frozen foods, which showed that the shelf life was insensitive to temperature fluctuations below $T'_g$, but temperature fluctuations above $T'_g$ resulted in a dramatic decrease of shelf life. They assumed that deteriorative changes in frozen foods may show diffusional limitations and pointed out that reaction rates may also depend on the size of the diffusants among other factors. Diffusion in frozen foods seems to be affected by the physical state of the freeze-concentrated unfrozen solute matrix and the extent of water plasticization above $T'_m$. Assuming that equation (9.17) defines relationships between the observed rate constant, $k'$, and the true rate constant, $k$, the observed rate constant may be estimated with predicted diffusion data above $T'_g$ and $T'_m$. The effect of the physical state and ice melting on the rate constant of a hypothetical reaction in a frozen sucrose solution is shown in Figure 9.14. Again the assumption that the rate constant is defined by $T - T'_g$ would underestimate the temperature dependence of the reaction. The observed rate constant approaches the true rate constant over a relatively narrow temperature range above $T'_m$. It is obvious that quality changes in frozen foods may become diffusion-limited as the extent of freeze-concentration approaches its maximum value. At temperatures above $T'_m$ the temperature dependence probably follows the Arrhenius equation due to ice melting and the sensitivity of the diffusion coefficient to water plasticization. Moreover, rates of quality changes in a number of frozen foods have been found to follow Arrhenius kinetics (Simatos *et al.*, 1989; Simatos and Blond, 1991). However, it should be remembered that rates of quality changes are also affected by the changes in reactant concentration and other temperature-dependent factors which may affect the rate constant of deteriorative changes in frozen foods.

The effects of glass transition on reaction kinetics in frozen foods were studied by Lim and Reid (1991). Reactions studied included kinetics of enzymatic hydrolysis of disodium-p-nitrophenylphosphate, loss of actomyosin solubility, and oxidation of ascorbic acid. The $T'_g$ values were those determined and reported by Levine and Slade (1986), which may be considered to correspond to the onset of ice melting, $T'_m$, in the materials. The enzymatic hydrolysis was followed in maltodextrin matrices that had various dextrose equivalents. Lim and Reid (1991) reported that the rate of the hydrolysis was almost zero at temperatures below $T'_g$ and increased rapidly with increasing temperature above $T'_g$. Lim and Reid (1991) suggested the use of the WLF equation in predicting the temperature dependence of the reaction rate above $T'_g$, but no proof of the applicability of the equation was reported. The solubility of actomyosin in a maltodextrin matrix was also related to storage temperature. The solubility was maintained by storage below the $T'_g$ of the maltodextrin, but storage at temperatures above $T'_g$ resulted in loss of solubility with a faster rate. However, the solubility of actomyosin was maintained in a sucrose solution that was stored above $T'_g$. Ascorbic acid was found to be stable in a maltodextrin matrix at temperatures below $T'_g$. Increasing the temperature to above $T'_g$ resulted in oxidation of ascorbic acid. Ascorbic acid was oxidized in sucrose solutions, which accounted for the low $T'_g$ of sucrose. The increase in the unfrozen water content above $T'_g$ was suggested to be another cause for the observed rapid rate of oxidation of ascorbic acid with sucrose.

It is obvious that stability of frozen foods is related to the physical state of the unfrozen solute matrix. $T'_g$ values which correspond to $T'_m$ for selected foods are given in Table 9.1. Food storage below the $T'_g$ may significantly increase shelf life and decrease the severity of temperature fluctuations. The rapid decrease of viscosity and increase of diffusivity which occur above $T'_g$ result in dilution of food solutes and cause a dramatic increase in reaction rates. It is likely that rates of deteriorative changes above $T'_g$ are affected by diffusional limitations in the vicinity of $T'_g$. There are only a limited number of data on diffusivity and reaction rates in frozen foods, but the temperature dependence of deteriorative changes in common frozen foods is more likely to be of Arrhenius type than of WLF type (Lai and Heldman, 1982; Kerr *et al.*, 1993). However, diffusional limitations below $T'_m$ and the changes in the physical state which occur at $T'_g$ and above $T'_m$ should produce a step change in Arrhenius plots of rates of deteriorative changes of frozen foods.

# References

Buera, M.P. and Karel, M. 1993. Application of the WLF equation to describe the combined effects of moisture and temperature on nonenzymatic browning rates in food systems. *J. Food Process. Preserv. 17*: 31-45.

Careri, G., Gratton, E., Yang, P.H. and Rupley, J.A. 1980. Correlation of IR spectroscopic, heat capacity, diamagnetic susceptibility and enzymatic measurements of lysozyme powder. *Nature 284*: 572-573.

Dennison, D., Kirk, J., Bach, J., Kokoczka, P. and Heldman, D. 1977. Storage stability of thiamin and riboflavin in a dehydrated food system. *J. Food Process. Preserv. 1*: 43-54.

Drapron, R. 1985. Enzyme activity as a function of water activity. In *Properties of Water in Foods*, ed. D. Simatos and J.L. Multon. Martinus Nijhoff Publishers, Dordrecht, the Netherlands, pp. 171-190.

Duckworth, R.B. 1972. The properties of water around the surfaces of food colloids. *Proc. Inst. Food Sci. Technol. 5*: 60.

Duckworth, R.B. 1981. Solute mobility in relation to water content and water activity. In *Water Activity: Influences on Food Quality*, ed. L.B. Rockland and G.F. Stewart. Academic Press, New York, pp. 295-317.

Eichner, K. and Karel, M. 1972. The influence of water content and water activity on the sugar-amino browning reaction in model systems under various conditions. *J. Agric. Food Chem. 20*: 218-223.

Fennema, O.R. 1973a. Nature of the freezing process. Chpt. 4 in *Low-Temperature Preservation of Foods and Living Matter*, ed. O.R. Fennema, W.D. Powrie and E.H. Marth. Marcel Dekker, New York, pp. 150-239.

Fennema, O.R. 1973b. Freeze-preservation of foods - Technological aspects. Chpt. 11 in *Low-Temperature Preservation of Foods and Living Matter*, ed. O.R. Fennema, W.D. Powrie and E.H. Marth. Marcel Dekker, New York, pp. 504-550.

Ferry, J.D. 1980. *Viscoelastic Properties of Polymers*, 3rd ed. John Wiley & Sons, New York.

Fish, B.P. 1958. Diffusion and thermodynamics of water in potato starch gel. In *Fundamental Aspects of Dehydration of Foodstuffs*, Society of Chemical Industry. Metchim & Son Ltd., London, pp. 143-157.

Flink, J.M. 1983. Structure and structure transitions in dried carbohydrate materials. Chpt. 17 in *Physical Properties of Foods*, ed. M. Peleg and E.B. Bagley. AVI Publishing Co., Westport, CT, pp. 473-521.

Flink, J.M. and Karel, M. 1972. Mechanism of retention of organic volatiles in freeze-dried systems. *J. Food Technol. 7*: 199-211.

Flink, J.M., Hawkes, J., Chen, H. and Wong, E. 1974. Properties of the freeze drying "scorch" temperature. *J. Food Sci. 39*: 1244-1246.

Gejl-Hansen, F. and Flink, J.M. 1977. Freeze-dried carbohydrate containing oil-in-water emulsions: Microstructure and fat distribution. *J. Food Sci. 42*: 1049-1055.

Herrington, B.L. 1934. Some physico-chemical properties of lactose. I. The spontaneous crystallization of supersaturated solutions of lactose. *J. Dairy Sci. 17*: 501-518.

Jouppila, K. and Roos, Y. 1994. Glass transitions and crystallization in milk powders. *J. Dairy Sci. 77*: 2907-2915.

Karathanos, V.T., Vagenas, G.K. and Saravacos, G.D. 1991. Water diffusivity in starches at high temperatures and pressures. *Biotechnol. Prog. 7*: 178-184.

Karel, M. 1985. Effects of water activity and water content on mobility in food components, and their effect on phase transitions in food systems. In *Properties of Water in Foods*, ed. D. Simatos and J.L. Multon. Martinus Nijhoff Publishers, Dordrecht, the Netherlands, pp. 153-169.

Karel, M. 1993. Temperature-dependence of food deterioration processes. *J. Food Sci. 58*: ii.

Karel, M. and Saguy, I. 1991. Effects of water on diffusion in food systems. In *Water Relationships in Foods*, ed. H. Levine and L. Slade. Plenum Press, New York, pp. 157-173.

Karel, M., Buera, M.P. and Roos, Y. 1993. Effects of glass transitions on processing and storage. Chpt. 2 in *The Glassy State in Foods*, ed. J.M.V. Blanshard and P.J. Lillford. Nottingham University Press, Loughborough, pp. 13-34.

Karmas, R. 1994. The effect of glass transition on non-enzymatic browning in dehydrated food systems. Ph.D. thesis, Rutgers - The State University of New Jersey, New Brunswick, NJ.

Karmas, R., Buera, M.P. and Karel, M. 1992. Effect of glass transition on rates of nonenzymatic browning in food systems. *J. Agric. Food Chem. 40*: 873-879.

Kerr, W.L., Lim, M.H., Reid, D.S. and Chen, H. 1993. Chemical reaction kinetics in relation to glass transition temperatures in frozen food polymer solutions. *J. Sci. Food Agric. 61*: 51-56.

Kim, M.N., Saltmarch, M. and Labuza, T.P. 1981. Non-enzymatic browning of hygroscopic whey powders in open versus sealed pouches. *J. Food Process. Preserv. 5*: 49-57.

King, N. 1965. The physical structure of dried milk. *Dairy Sci. Abstr. 27*: 91-104.

Labrousse, S., Roos, Y. and Karel, M. 1992. Collapse and crystallization in amorphous matrices with encapsulated compounds. *Sci. Aliments 12*: 757-769.

Labuza, T.P. 1980. The effect of water activity on reaction kinetics of food deterioration. *Food Technol. 34(4)*: 36-41, 59.

Labuza, T.P. and Baisier, W.M. 1992. The kinetics of nonenzymatic browning. Chpt. 14 in *Physical Chemistry of Foods*, ed. H.G. Schwartzberg and R.W. Hartel. Marcel Dekker, New York, pp. 595-649.

Labuza, T.P. and Riboh, D. 1982. Theory and application of Arrhenius kinetics to the prediction of nutrient losses in foods. *Food Technol. 36(10)*: 66, 68, 70, 72, 74.

Labuza, T.P., Tannenbaum, S.R. and Karel, M. 1970. Water content and stability of low-moisture and intermediate-moisture foods. *Food Technol. 24*: 543-544, 546-548, 550.

Lai, D.-J. and Heldman, D.R. 1982. Analysis of kinetics of quality change in frozen foods. *J. Food Process Eng. 6*: 179-261.

Lea, C.H. and White, J.C.D. 1948. Effect of storage on skim-milk powder. Part III. Physical, chemical and palatability changes in the stored powders. *J. Dairy Res. 15*: 298-340.

Le Meste, M., Viguier, L., Lorient, D. and Simatos, D. 1990. Rotational diffusivity of solutes in concentrated caseinates: Influence of glycosylation. *J. Food Sci. 55*: 724-727.

Levine, H. and Slade, L. 1986. A polymer physico-chemical approach to the study of commercial starch hydrolysis products (SHP's). *Carbohydr. Polym. 6*: 213-244.

Levine, H. and Slade, L. 1988. Principles of "cryostabilization" technology from structure/property relationships of carbohydrate/water systems - A review. *Cryo-Lett. 9*: 21-63.

Levine, H. and Slade, L. 1989. A food polymer science approach to the practice of cryostabilization technology. *Comments Agric. Food Chem. 1*: 315-396.

Lim, M.H. and Reid, D.S. 1991. Studies of reaction kinetics in relation to the $T'_g$ of polymers in frozen model systems. In *Water Relationships in Foods*, ed. H. Levine and L. Slade. Plenum Press, New York, pp. 103-122.

Marousis, S.N., Karathanos, V.T. and Saravacos, G.D. 1991. Effect of physical structure of starch materials on water diffusivity. *J. Food Process. Preserv. 15*:183-195.

Matsuno, R. and Adachi, S. 1993. Lipid encapsulation technology techniques and applications to food. *Trends Food Sci. Technol. 4*: 256-261.

Menting, L.C. and Hoogstad, B. 1967. Volatiles retention during the drying of aqueous carbohydrate solutions. *J. Food Sci. 32*: 87-90.

Menting, L.C., Hoogstad, B. and Thijssen, H.A.C. 1970. Diffusion coefficients of water and organic volatiles in carbohydrate-water systems. *J. Food Technol. 5*: 111-126.

Nelson, K.A. 1993. Reaction kinetics of food stability: Comparison of glass transition and classical models for temperature and moisture dependence. Ph.D. thesis, University of Minnesota, St. Paul, MN.

Nelson, K.A. and Labuza, T.P. 1994. Water activity and food polymer science: Implications of state on Arrhenius and WLF models in predicting shelf life. *J. Food Eng. 22*: 271-289.

Peleg, M. 1992. On the use of the WLF model in polymers and foods. *Crit. Rev. Food Sci. Nutr. 32*: 59-66.

Peppas, N.A. and Brannon-Peppas, L. 1994. Water diffusion and sorption in amorphous macromolecular systems and foods. *J. Food Eng. 22*: 189-210.

Powrie, W.D. 1973a. Characteristics of food myosystems and their behavior during freeze-preservation. Chpt. 6 in *Low-Temperature Preservation of Foods and Living Matter*, ed. O.R. Fennema, W.D. Powrie and E.H. Marth. Marcel Dekker, New York, pp. 282-351.

Powrie, W.D. 1973b. Characteristics of food phytosystems and their behavior during freeze-preservation. Chpt. 7 in *Low-Temperature Preservation of Foods and Living Matter*, ed. O.R. Fennema, W.D. Powrie and E.H. Marth. Marcel Dekker, New York, pp. 352-385.

Rockland, L.B. and Nishi, S.K. 1980. Influence of water activity on food product quality and stability. *Food Technol. 34(4)*: 42, 44-46, 48-51, 59.

Roos, Y.H. 1993. Water activity and physical state effects on amorphous food stability. *J. Food Process. Preserv. 16*: 433-447.

Roos, Y.H. and Himberg, M.-J. 1994. Nonenzymatic browning behavior, as related to glass transition, of a food model at chilling temperatures. *J. Agric. Food Chem. 42*: 893-898.

Roos, Y. and Karel, M. 1991a. Phase transitions of mixtures of amorphous polysaccharides and sugars. *Biotechnol. Prog. 7*: 49-53.

Roos, Y. and Karel, M. 1991b. Water and molecular weight effects on glass transitions in amorphous carbohydrates and carbohydrate solutions. *J. Food Sci. 56*: 1676-1681.

Roozen, M.J.G.W., Hemminga, M.A. and Walstra, P. 1991. Molecular motion in glassy water--malto-oligosaccharide (maltodextrin) mixtures as studied by conventional and saturation-transfer spin-probe e.s.r. spectroscopy. *Carbohydr. Res. 215*: 229-237.

Saltmarch, M. and Labuza, T.P. 1980. Influence of relative humidity on the physicochemical state of lactose in spray-dried sweet whey powders. *J. Food Sci. 45*: 1231-1236, 1242.

Saltmarch, M., Vagnini-Ferrari, M. and Labuza, T.P. 1981. Theoretical basis and application of kinetics to browning in spray-dried whey food systems. *Prog. Food Nutr. Sci. 5*: 331-344.

Salwin, H. 1963. Moisture levels required for stability in dehydrated foods. *Food Technol. 17*: 1114-1116, 1118-1120.

Shimada, Y., Roos, Y. and Karel, M. 1991. Oxidation of methyl linoleate encapsulated in amorphous lactose-based food model. *J. Agric. Food Chem. 39*: 637-641.

Silver, M. and Karel, M. 1981. The behavior of invertase in model systems at low moisture contents. *J. Food Biochem. 5*: 283-311.

Simatos, D. and Blond, G. 1991. DSC studies and stability of frozen foods. In *Water Relationships in Foods*, ed. H. Levine and L. Slade. Plenum Press, New York, pp. 139-155.

Simatos, D. and Karel, M. 1988. Characterization of the condition of water in foods - Physico-chemical aspects. In *Food Preservation by Water Activity Control*, ed. C.C. Seow. Elsevier, Amsterdam, pp. 1-41.

Simatos, D., Blond, G. and Le Meste, M. 1989. Relation between glass transition and stability of a frozen product. *Cryo-Lett. 10*: 77-84.

Slade, L. and Levine, H. 1991. Beyond water activity: Recent advances based on an alternative approach to the assessment of food quality and safety. *Crit. Rev. Food Sci. Nutr. 30*: 115-360.

Thijssen, H.A.C. 1971. Flavour retention in drying preconcentrated food liquids. *J. Appl. Chem. Biotechnol. 21*: 372-377.

Villota, R. and Hawkes, J.G. 1992. Reaction kinetics in food systems. Chpt. 2 in *Handbook of Food Engineering*, ed. D.R. Heldman and D.B. Lund. Marcel Dekker, New York, pp. 39-144.

Vrentas, J.S. and Duda, J.L. 1978. A free-volume interpretation of the influence of the glass transition on diffusion in amorphous polymers. *J. Appl. Polym. Sci.* 22: 2325-2339.

# CHAPTER 10

# Food Processing and Storage

## I. Introduction

The main importance of phase transitions in foods is their tremendous impact on food behavior in processing and storage. As would be expected the main applications of the information on material properties include proper control of processing and storage conditions to achieve and maintain optimum food quality and desired properties, and to provide stability. However, processing and storage conditions are often chosen by trial and error without knowledge of the physicochemical changes occurring within food solids.

Much of the work on the effects of phase transitions on food processing and storage has been done on dehydration and stability of low-moisture foods. Dehydration in most cases results in structural changes, which often decrease food quality. In addition to the removal of water the process may cause loss of flavor. Changes during storage of dehydrated foods include deteriorative changes that are strongly dependent on water content as well as loss of structure and volatile components. Such food processes as agglomeration and extrusion may be considered to be based on the proper control of phase transitions and thermal as well as water plasticization of food solids. The success of these processes is therefore related to rigorous control of temperature, time, and water content. Other examples of phase transition-related structure formation in foods include various baking processes. The formation of the structure of bread during baking may be considered to be due to starch gelatinization and thermosetting of gluten to the form of a soft, elastic bread crumb. The formation of the crispy texture of low-moisture breakfast cereals, crisp bread products, and a number of snacks, as well as cookies and crackers is also a result of a series of controlled phase and state transitions defined by temperature, time, and water relationships during various processing steps.

The food polymer science approach to understanding of phase transitions as well as time, temperature, and water relationships in foods has provided valuable information, which can be used in predicting food behavior in processing and storage (Slade and Levine, 1991). This chapter describes the importance of time, temperature, and water content-related phenomena, which may be advantageous or disadvantageous to physicochemical changes in food processing and storage. The main emphasis here is to analyze dynamic phenomena in food processing and storage and to provide information of various means which can be used to achieve desired product properties and stability.

## II.  Food Processing

A number of food processes involve phase transitions. The phase transitions of water between the solid, liquid, and gaseous states are present in almost all food processing operations. Water is often an important component which causes evaporative cooling and controls food temperature during heating and dehydration. However, processes that are affected by phase transitions of food solids are characterized by either low temperatures, which remove water by ice formation, or high temperatures, which remove water by evaporation or cause thermal plasticization of food solids.

### A.  Dehydration and Agglomeration

The most serious problems in dehydration of food materials are caused by the difficulty of water removal from sugar-containing foods and product stickiness at late stages of dehydration processes. Brennan *et al.* (1971) considered dehydration to be one of the most difficult preservation techniques of fruit juices, because of their heat sensitivity and hygroscopicity. They also noticed that such materials were thermoplastic, which resulted in difficult wall deposition problems and difficulties in product handling. The problems have often been reduced by adding compounds which improve drying behavior or by using equipment of special design. Problems in material handling are often accompanied by loss of food quality due to collapse and loss of flavor.

1. Quality changes in dehydration

Dehydration of food materials is commonly accomplished by using air-drying techniques, which allow a relatively efficient and economic means of water removal. Phase transitions of food solids and their time-dependent characteristics during various stages of the dehydration process may contribute to the extent of collapse of the physical structure, the extent of stickiness, and other quality changes which occur due to changes in temperature and water content. Common final products include low-moisture viscoelastic materials such as raisins, vegetables with a glassy structure, and free-flowing powders, e.g., dairy powders, which are produced by spray-drying. It is well known that air-dried berries, fruits, vegetables, and other food materials are reduced in size in comparison with the fresh produce. The dry materials are stable due to the low water content and they can be rehydrated when used as food ingredients. However, a collapsed structure and impaired flavor may be disadvantageous properties due to impaired rehydration properties and reduced overall quality, which decrease the utility of dehydrated food products as such or as components of dry food mixtures.

Water contents of fresh, solid foods are high. The reduction of the water content in common air-drying techniques is achieved by applying heat with an air flow, which provides the latent heat for evaporation of water and removes the produced vapor. The reduction of water content results in concentration of solutes, which may affect rates of deteriorative changes that possibly occur during various stages of the dehydration process. It would be expected that rates of such reactions as nonenzymatic browning, depending on temperature, may cause quality changes as the water content is reduced to an intermediate level at which reaction rates are particularly sensitive to temperature and water activity. Thijssen (1971) pointed out that at temperatures between 40 and 70°C enzyme-catalyzed reactions may alter functional, nutritional, and flavor properties during food processing. Other changes included nonenzymatic browning, protein denaturation, and vitamin destruction. Thijssen (1971) also pointed out that typical activation energies of nonenzymatic browning are between 60 and 125 kJ/mol and that a 10°C increase in processing temperature results in an increase of the reaction rate by a factor of 2.5 to 5. In dehydration the activation energy for such reactions was considered to be higher than that for water removal and therefore a decrease in drying temperature was predicted to result in an improved retention of quality, which was suggested also by Leniger and Bruin (1977).

Eichner *et al.* (1985) studied the effect of drying conditions on the formation of Amadori compounds (intermediate compounds of the browning reaction) in carrots. A strong influence of temperature and water content on the formation of the intermediates was found. Obviously, browning could be decreased by lowering the dehydration temperature during final stages of drying as the activation energy increased with decreasing water content. The results of Eichner *et al.* (1985) reflect the strong influence of the reactant concentration and of the physical state on the rate of deteriorative changes during food processing, particularly at intermediate water activities, and the

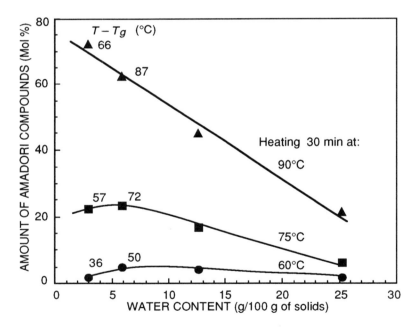

**Figure 10.1** Formation of Amadori compounds (browning intermediates) in freeze-dried carrots during heating at various water contents. The glass transition temperatures, $T_g$, indicated as $T - T_g$ data were derived from the data of Karmas *et al.* (1992) with the Gordon and Taylor equation. The rate of the reaction at the lower $T - T_g$ values decreased probably due to decreasing diffusion. Data from Eichner *et al.* (1985).

importance of reduced water plasticization and formation of a glassy structure to food stability.

Direct relationships between the physical state of food materials and rates of nonenzymatic browning have not been established. Eichner *et al.* (1985) found that the rate of nonenzymatic browning at relatively low temperatures showed a maximum at intermediate water contents. The data shown in Figure 10.1 suggested that the maximum shifted to lower water contents with increasing temperature. The rate of browning at 90°C increased linearly with no maximum. Eichner *et al.* (1985) concluded that at the higher temperature the diffusion resistance for the reactants within the product matrix was decreased and the reaction became independent of water content. The reduced reaction rate at 60°C was also evident from a higher $Q_{10}$ value and from the high activation energy of the reaction at low water contents. Published $T_g$ data for carrots suggest that the anhydrous $T_g$ is about 52°C (Karmas *et al.*, 1992). The material is significantly plasticized by water and therefore small increases in water content may dramatically decrease $T_g$. The $T - T_g$ values shown in Figure 10.1 were derived from the $T_g$ data of Karmas *et al.* (1992). The predicted $T - T_g$ values suggest that the rate of the nonenzymatic browning reaction decreased at 60 and 75°C as the $T - T_g$ became less than 50°C. It may be concluded that the nonenzymatic browning reaction during dehydration of food materials becomes extremely sensitive to temperature at low water contents. The rate is probably decreased at low $T - T_g$ values due to decreasing diffusion. It is also important to notice that the formation of browning intermediates during dehydration may affect the shelf life of dehydrated foods. Eichner *et al.* (1985) pointed out that nonenzymatic browning may occur faster during storage if the reaction has initiated during drying.

Stickiness is probably the most important property in establishing criteria for the suitability of heat-sensitive food materials to spray-drying. Several sugar-containing materials such as fruit juices have proved to be extremely difficult to dehydrate due to their tendency to stick on drier surfaces and cake inside the processing equipment. Lazar *et al.* (1956) determined "sticky points" for spray-dried tomato powder. The *sticky point* was defined as the temperature at which the force needed to turn a propeller in a test tube containing the powder increased sharply. Lazar *et al.* (1956) found that the sticky point decreased with increasing temperature. It was shown that tomato powder with 2% water could not be exposed to temperatures higher than 60°C. The data obtained suggested that at low water contents the drying air often increased product temperature, which resulted in sticking and caking. However, cooling of drying particles was suggested to allow formation of a nonsticky particle surface and free-flowing powder. Lazar *et al.* (1956) pointed out that such powders may rapidly adsorb water, which causes stickiness during product handling and storage.

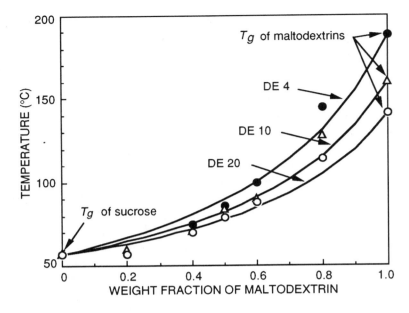

**Figure 10.2** Effect of added maltodextrin on the glass transition temperature, $T_g$, of sucrose. The $T_g$ of the mixture increases with increasing molecular weight of the additive and therefore with decreasing dextrose equivalent, DE. The effect of an added component on the $T_g$ of a food product which decreases hygroscopicity and stickiness can be predicted with the Gordon and Taylor equation. Data from Roos and Karel (1991c).

Brennan *et al.* (1971) pointed out that two approaches may be used to decrease the thermoplasticity and hygroscopicity and therefore to solve wall deposition problems in spray-drying of fruit juices. These methods were the use of additives as drying aids and the use of specially-designed equipment. The common additives, which included alginates, corn syrups, dextrins, glyceryl monostearate, natural gums, soya proteins, and sucrose, were reported to produce physical changes in the materials. Brennan *et al.* (1971) found that liquid glucose with DE of 39 to 43 was an effective additive. It reduced wall deposition markedly and the product was of acceptable flavor with free-flowing properties. It should be noticed that the additive was probably equivalent to such glucose polymers as maltodextrins, which are commonly used as additives in spray-drying of fruit products due to their efficiency in decreasing stickiness during drying, product handling, and storage. It is obvious that such additives are miscible with low molecular weight sugars and increase the sticky point by increasing the glass transition temperature as shown in Figure 10.2. The sticky point of amorphous food solids is located at an isoviscosity state above $T_g$ (Downton *et al.*, 1982; Roos and Karel,

1991a) and the measurement of the sticky point with the method of Lazar *et al.* (1956) can be considered to be a method which, in fact, locates the glass transition of the food solids (Chuy and Labuza, 1994). Thus, the sticky point of food solids with various compositions can be predicted with the same principles that are used in evaluating effects of composition on glass transition temperatures of food solids and polymers. However, the amounts of such additives as maltodextrins that are needed to give a sufficient increase of the $T_g$ are fairly high. The more expensive method to avoid wall deposition in spray-drying is achieved by using relatively low air temperatures and cooling of drier walls to avoid exposures of the particles to temperatures above $T_g$. Unfortunately, avoiding stickiness with a special equipment design does not decrease stickiness and hygroscopicity of the final powder.

Collapse of macroscopic structure during dehydration is one of the most detrimental changes, which (1) reduces drying rate; (2) causes slow and difficult rehydration; and (3) decreases the quality of dehydrated foods. Water removal from the cellular structure of fresh food products is accompanied by loss of the cellular volume. Lozano *et al.* (1980) found a significant change in the three-dimensional arrangement of the cellular tissue of apples at water contents below 1.5 g/g of solids. They concluded that the open structure that characterizes fresh products changed into a structure with locked-in pores. Lozano *et al.* (1983) showed that a water loss-based bulk shrinkage coefficient, $s_b$, could be related to initial water content and composition of vegetables, in particular to the amount of starch and sugars. The bulk shrinkage coefficient was defined to be the ratio of bulk volume, $V_b$, of the material with water content, $m_x$, and the volume of the material with the initial water content, $m_i$. The results of these studies suggested that collapse or loss of porosity during dehydration occurred as the water content was reduced to below a critical value and that the extent of shrinkage was related to composition. The strong correlation between the extent of shrinkage and the initial water content has led to the conclusion that collapse during dehydration can be decreased by increasing the initial content of solids, e.g., by such pretreatments as osmotic dehydration.

Karathanos *et al.* (1993) postulated that the extent of collapse, defined as the tissue bulk volume, is related to the glass transition temperature of food solids. Therefore, drying temperature was assumed to affect the extent of collapse of celery during air-drying. Karanthanos *et al.* (1993) found that the $T_g$ of celery was very low and significantly lower than the drying temperature during most of the drying process due to the high initial water content of the material. They observed that the extent of shrinkage increased proportionally with the decrease in volume that was caused by the loss of water. However, air-drying at 5°C resulted in a product with higher porosity than drying

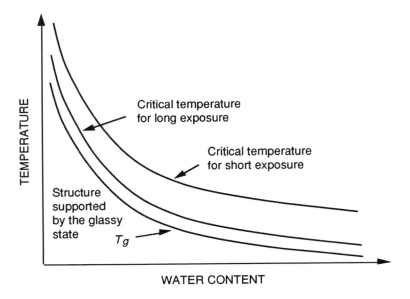

**Figure 10.3** Relationships between glass transition temperature, $T_g$, dehydration temperature, structural collapse, and water content. At temperatures below $T_g$ structural stability is provided by the solid, glassy state of the solids. Collapse occurs at temperatures above $T_g$. The relaxation time of collapse is defined by $T - T_g$ and drying in the vicinity of $T_g$ is sufficient to produce a noncollapsed material. Short exposures obtained with pulsed heating techniques to a critical temperature defined by the relaxation time may be allowed to improve drying rate without significant collapse of food structure. Adapted from Anglea *et al.* (1993).

at 60°C although drying occurred well above $T_g$ even at the lower temperature. According to Anglea *et al.* (1993) collapse during dehydration is due to structural mobility, which is a function of viscosity. Viscosity depends on the temperature difference $T - T_g$ and therefore collapse is governed by glass transition (Levine and Slade, 1986). It is likely that, as shown in Figure 10.3, minimizing the exposure time of a material to temperatures above $T_g$ and the magnitude of $T - T_g$ decreases collapse during dehydration and improves quality of the dehydrated material.

Anglea *et al.* (1993) found that drying with pulsed hot and cold air at 5°C produced a product with higher porosity and better quality than drying at higher temperatures. Unfortunately, the water contents of common food materials are high and shrinkage cannot always be avoided in air-drying until the drying temperature is decreased and the $T_g$ of the solids is increased due to the removal of most of the water. Therefore, if feasible, a significant reduction of shrinkage due to water removal may require addition of high molecular weight solids, use of low dehydration temperatures, and application of hot and cold air or microwave pulsed techniques similar to those used by Anglea *et al.* (1993). However, the theory is applicable and product collapse is minimized in freeze-drying, which removes the solid ice phase from the highly viscous, freeze-concentrated solute matrix. It is also obvious that the noncollapsed glassy materials are able to support their own weight and the porous structure when thermal and water plasticization are minimized by keeping product temperature below the glass transition temperature.

Collapse in freeze-drying occurs at a critical temperature which allows viscous flow of freeze-concentrated amorphous solutes (Bellows and King, 1973). Such viscous flow occurs due to the decreasing viscosity in the vicinity of the onset temperature of ice melting above $T_g'$ (Levine and Slade, 1986, 1988; Pikal and Shah, 1990; Franks, 1990; Roos and Karel, 1991b; Slade and Levine, 1991). For sugar solutions and other food materials the reported $T_g'$ and $T_m'$ values can be used as critical temperatures for production of noncollapsed freeze-dried materials. In vacuum freeze-drying the ice temperature of a drying material, provided that the heat and mass transfer are in equilibrium, is a function of pressure in the drying chamber. The relationships between pressure and temperature may be obtained from the phase diagram of water. Therefore, a chamber pressure of 0.1 mbar corresponds to the ice temperature of -40°C, which is too high for freeze-drying of fructose and glucose solutions. An increase of the pressure to about 0.4 mbar increases ice temperature to -30°C, which is detrimental to freeze-drying of sucrose solutions and critical to lactose or maltose solutions. A pressure of 1 mbar corresponds to about -20°C, which may be considered to be limiting pressure and temperature values for the successful freeze-drying of most foods. It should be noticed that in addition to the high viscosity of the freeze-concentrated amorphous solutes, the physical structure in tissue foods is also supported by cell walls and polymers.

Karel *et al.* (1993) compared the effects of freeze-drying on structural properties of plant tissues when drying was accomplished either below or above $T_g'$. The estimated $T_g'$ for samples of apple, celery, and potato was -46°C. Dehydration was accomplished at 0.005 mbar (initial temperature -55°C) and at 1.98 mbar (initial temperature -28°C). The results showed a drastic difference in the porosity of the dehydrated materials. The porosity of apples which were dried at the lower pressure was 0.93 and that of apples dried at the higher pressure was 0.65. The results showed that ice melting during drying at the higher pressure plasticized both the unfrozen freeze-concentrated matrix and the dehydrated solids, which resulted in partial loss of structure.

2. Flavor retention and encapsulation

Dehydration of food materials allows a fairly high retention of volatile compounds within the dehydrated food solids in comparison with the amount of evaporated water. Several studies have shown that the diffusion of volatile compounds is affected by water content and that volatile compounds often become encapsulated within amorphous carbohydrate matrices. Freeze-drying and spray-drying have proved to be particularly useful techniques in allowing a high retention of flavors and providing excellent product characteristics in comparison with most other food dehydration techniques.

Menting *et al.* (1970a) found that diffusion coefficients of volatile compounds in maltodextrins were dependent on water content and on the size of the diffusants, which was suggested to affect flavor retention in dehydration processes. Menting *et al.* (1970b) studied retention of volatile components in drying of liquid foods. Their results suggested that at high water contents the loss of volatile compounds was defined by their volatility. At low water contents the diffusion coefficients of volatile compounds became the rate-defining factors. It was concluded that in spray-drying most of the flavor compounds were lost during formation of the liquid droplets, since at later stages of the dehydration process the slow diffusion of flavors to the droplet surface facilitated retention.

The retention of flavors in food materials was considered by Thijssen (1971) to be based on selective diffusion. Thijssen (1971) suggested that the loss of flavors from food liquids during evaporation was independent of the relative volatility and controlled by the rate at which the compounds were able to migrate to the evaporating surface. The theory postulated that below a critical water content the food matrix became selectively permeable to water and completely impermeable to flavor compounds. The critical water content was considered to be dependent on the dissolved solids and it increased with increasing size of the flavor molecule as well as with decreasing temperature. It is obvious that these criteria of molecular size, temperature, and water content for selective diffusion established by Thijssen (1971) are related to the decreasing rate of diffusion, which occurs as amorphous food solids approach the glassy state during water removal. It was also pointed out by Thijssen (1971) that the selective diffusivity for water at low water concentrations is a property that all noncrystalline hydrophilic organic systems have in common.

Thijssen (1971) assumed that the diffusion of a flavor compound and water within a food matrix followed Fick's first law, which is given by equation (10.1), where $J$ is flux of the diffusant and $\partial C / \partial x$ is the concentration gradient. Thijssen (1971) used the ratio of the diffusivities of a flavor compound, $D_f$, and that of water, $D_w$, as a measure of selective diffusivity.

He defined a critical water content for the occurrence of selective diffusion to be that at which the ratio of the diffusion coefficients was 0.01.

$$J = -D\frac{\partial C}{\partial x}$$ (10.1)

According to the definition of selective diffusion by Thijssen (1971) the drying surface of a food material during dehydration became increasingly impermeable to the flavor compound as the water content at the surface decreased to below the critical water content and the ratio of $D_f/D_w < 0.01$ was achieved. The critical water content was a function of temperature and the size of the diffusant as shown in Figure 10.4. The critical water content increased with decreasing temperature, or inversely, the critical temperature for selective diffusion decreased with increasing water content. The decrease of the critical temperature with increasing water content suggests that the diffusion of flavor compounds and flavor retention in dehydration are

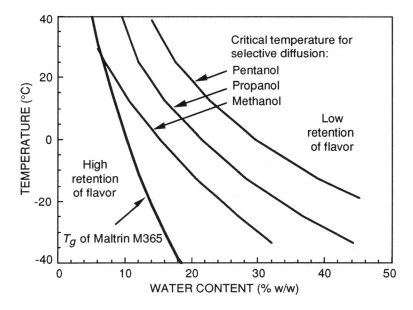

**Figure 10.4** Critical temperature for selective diffusion within an amorphous maltodextrin matrix for methanol, propanol, and pentanol. The temperature at which the ratio of the diffusion coefficient of the volatile compound, $D_f$, and that of water, $D_w$, has the value $D_f/D_w = 0.01$ decreases with increasing water content. The decrease is probably related to the glass transition temperature, $T_g$, which is shown for Maltrin M365, which has a DE of 36. Data from Thijssen (1971) and Roos and Karel (1991c,d).

affected by the glass transition temperature. Figure 10.4 shows the $T_g$ curve for Maltrin M365, which is a maltodextrin with a DE of 36, and the critical temperatures for selective diffusion of methanol, propanol, and pentanol within a corresponding amorphous maltodextrin matrix. The critical temperature for selective diffusion seems to be defined by $T_g$, particularly at low water contents. The $T - T_g$ at which selective diffusion occurs increases with increasing size of the diffusant. However, the $T - T_g$ seems to increase with increasing water content, which may be related to additional effects of decreasing temperature on diffusivity.

Relationships between diffusion, flavor retention, and the physical state of the material have also been observed to exist in sugar matrices. Chandrasekaran and King (1972) found that the activation energy for diffusion of ethyl alcohol in sucrose solutions increased with increasing solute concentration and the increase followed the increase of the activation energy for viscosity. At high solute concentrations the activation energy for diffusion of water decreased less than the activation energy for diffusion of ethyl alcohol, which indicated delayed diffusion of the organic component in comparison with diffusion of water. Flavor retention mechanisms in dehydration processes are evidently based on the effect of increasing viscosity on diffusion of flavor compounds within highly viscous amorphous matrices. Mechanisms of similar type probably apply both in air dehydration and in freeze-drying. In spray-drying the retention occurs due to the very rapid removal of water and the slow diffusion of the flavor compounds within the partially dehydrated solids. In freeze-drying the diffusion of the flavor compounds within the freeze-concentrated solute matrix must be very slow.

Flink and Karel (1970) postulated that the separation of water during freezing of carbohydrate solutions resulted in the formation of microregions within the ice crystals that contained a highly concentrated solution of the carbohydrates and volatile compounds. Loss of water from the microregions during freeze-drying decreased its permeability to volatile compounds and water. They suggested that below a critical water content the loss of volatile compounds ceased while loss of water continued. During freeze-drying organic volatiles were lost until the water content decreased to below the critical value. Flink and Karel (1970) pointed out that carbohydrates promoted the retention of organic compounds and that mono- and disaccharides were particularly efficient in retaining volatiles. It should be noticed that during air-drying processes most of the dehydration occurs at temperatures at which the solute matrix is in the supercooled liquid state above $T_g$. Therefore, retention is based on the low diffusion of flavor compounds in comparison to diffusion of water and to the dramatically decreasing diffusion of the flavor compounds at water contents below the critical water content for selective dif-

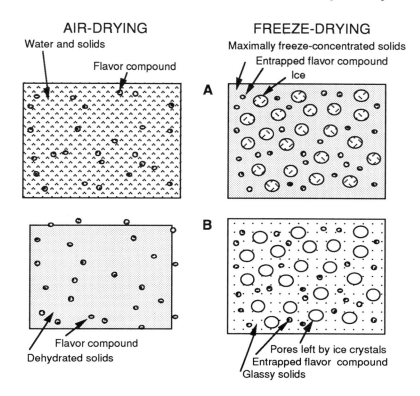

**Figure 10.5** Flavor retention mechanisms in air-drying and freeze-drying according to the results of studies by Flink and Karel (1970), Chirife and Karel (1973), Karel *et al.* (1994), and Levi and Karel (1995). A. Before dehydration the material in air-drying is composed of flavor, solids, and water. The frozen material to be freeze-dried contains ice and entrapped flavor within a glassy, maximally freeze-concentrated solute matrix. B. After dehydration the air-dried material has a reduced volume and the flavor is partially retained due to the high viscosity of the solutes and residual water. In freeze-drying the ice crystals have left empty pores and the original volume of the material is retained. The flavor compound is entrapped within the glassy solids.

fusion. In freeze-drying, however, as shown in Figure 10.5, the glassy state of the solutes may form during the prefreezing step and therefore organic compounds may become entrapped rather than adsorbed (Chirife and Karel, 1973) within the amorphous maximally freeze-concentrated solute matrix according to the microregion theory of Flink and Karel (1970). The volatiles remain entrapped within the amorphous, glassy solids until they become released. The release of such entrapped compounds may occur due to thermal or water plasticization-induced collapse and crystallization of the amorphous matrix (Flink and Karel, 1972; Karel and Flink, 1973; Chirife and Karel, 1974; Flink, 1983; Levi and Karel, 1995).

Omatete and King (1978) related data for loss of volatiles in dehydrated food matrices to collapse during rehumidification. The diffusion coefficient at low relative humidities was low and did not allow loss of volatile compounds. At high relative humidities diffusion coefficients of volatiles were high and a complete loss occurred rapidly. At intermediate relative humidities loss of volatiles was dependent on the extent of collapse and diffusion within the collapsed matrix. According to Levi (1994; Levi and Karel, 1995) flavor compounds in amorphous food matrices are retained at temperatures below $T_g$, which may also be considered to be the primary basis for the high retention of flavor compounds typical of freeze-drying. At temperatures above $T_g$ of dehydrated food solids, retained flavor compounds may become released due to crystallization of the encapsulating matrix. In noncrystallizing matrices that are typical of most foods, flavor compounds may be lost due to collapse of the amorphous dehydrated matrix. However, if such collapse is sufficiently rapid the loss of flavor becomes controlled by diffusion of the flavor compounds within the collapsed food matrix (Levi and Karel, 1995). Obviously, $T_g$ controls flavor retention within food solids during dehydration and in the dehydrated state. Loss of retained flavors in dehydrated foods can be avoided by proper packaging and storage at low temperatures. The main criterion is to avoid thermal or water plasticization and therefore exposures of the dehydrated materials to conditions at which the $T_g$ becomes lower than ambient temperature.

## 3. Agglomeration

Agglomeration is a common technique that is used to increase the particle size of fine food powders. The increase of particle size improves the wettability and solubility and therefore gives instant properties to such products as baby food powders, cocoa-sugar mixes, dairy powders, and fruit powders. Agglomeration of food powders is often referred to as *instantizing*, because of the improved rewetting characteristics of the agglomerated products. According to Kessler (1981) particles with a diameter less than 0.1 mm are difficult to rehydrate, but good rewetting properties are achieved when granules of 1 to 3 mm in diameter are produced by agglomeration. It should also be noticed that such granules are much easier to handle in processing and packaging equipment and by the consumer due to less dusty properties of the agglomerated product.

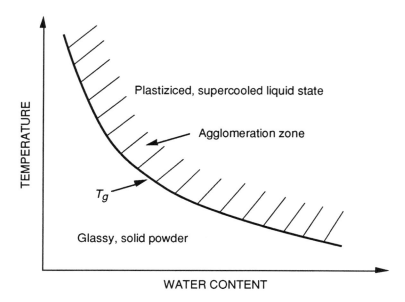

**Figure 10.6** Plasticization of food particles in agglomeration. Agglomeration occurs at temperatures above the glass transition temperature, $T_g$. Agglomeration is a result of a controlled caking process which occurs due to stickiness that is produced below a critical isoviscosity state above $T_g$. After Masters and Stoltze (1973).

Masters and Stoltze (1973) considered food solids to be composed of mixtures of organic compounds which possess no precise melting temperature. They pointed out that at a given temperature particles of food solids begin to plasticize and agglutinate. The temperature region for such agglutination was dependent on water content as shown in Figure 10.6. According to Masters and Stoltze (1973) a successful agglomeration process required a selected treatment of the powder within or near an agglomeration zone. Agglomeration occurred due to plasticization achieved by increasing temperature and surface wetting. Processing time was reported to be critical, since overwetting resulted in an unacceptable agglomerate.

The common agglomeration methods are based on rewetting of fine powders or on agglomeration during and after spray-drying using a straight-through process (Masters and Stoltze, 1973; Jensen, 1975; Schuchmann *et al.*, 1993). Agglomeration in the rewetting methods occurs in four stages (Masters and Stoltze, 1973). At the first stage the powder is wetted uniformly with water or another plasticizer which may be introduced in the liquid state or as vapor. The use of a liquid plasticizer is referred to as *droplet*

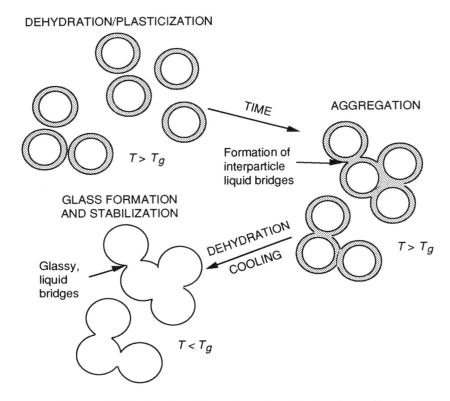

DEHYDRATION/PLASTICIZATION

**Figure 10.7** Controlled caking in agglomeration and production of instant food powders. Agglomeration requires plasticization of particle surfaces, which may be a result of rewetting and increasing of temperature or decrease of water content by spray-drying to produce sticky particles. Particles with sticky, plasticized surfaces are allowed to form aggregates by liquid bridging. Removal of the plasticizer (dehydration) and cooling are used to stabilize the agglomerated material into the glassy state. Thus, particle surfaces and interparticle liquid bridges are solidified and stickiness is lost.

*agglomeration*, in which the powder is exposed to a fine mist of the liquid. *Surface agglomeration* uses the plasticizer in the vapor form and plasticization is often accomplished with the use of steam or moist air. At the second stage the wetted and therefore plasticized powder is allowed to form clusters over a predetermined residence time in the wetting zone of the equipment. At the third stage the agglomerated powder is dehydrated to a desired water content and cooled. At the final processing step the agglomerated powder is screened and fine particles may be returned to the agglomeration process. The straight-through process is accomplished by producing plasticized particles with temperature and water content that are within the agglomeration zone. These particles are agglomerated in the drying chamber. The plasticized agglomerates enter a vibrating fluid-bed drier, which completes dehydration and allows sufficient cooling of the final product. The straight-through process is common in the production of instant dairy powders.

Agglomeration by both the rewetting and straight-through processes requires that amorphous food solids are allowed to exist a sufficiently long time in the plasticized state within the agglomeration zone. As shown in Figure 10.7 the aggregation of the fine particles may be assumed to occur through caking and the formation of interparticle liquid bridges (Peleg, 1983). The agglomerated material is stabilized by the removal of excess plasticizer and cooling to a temperature below $T_g$. It may be concluded that the agglomeration and production of instant food powders are based on controlled manipulation of the physical state. The proper agglomeration conditions are defined by the $T_g$ of the material and its dependence on water content. The time needed for the formation of liquid bridges between the fine particles is governed by viscosity and therefore by $T - T_g$ at the agglomeration zone. It should be noticed that state diagrams established for food powders may provide important information in predicting relationships between temperature, time, and water content for agglomeration processes.

## 4. Size reduction

Size reduction of dehydrated food materials is often needed in the production of powdered food materials. Such materials may include dehydrated fruits and vegetables or pure food components such as crystalline sugar.

Although size reduction of food materials is not a process that is based on a change in the physical state, phase transitions may occur during the process. Size reduction requires fairly high amounts of energy that is used to break the material into particles of a desired size distribution. However, some of the energy is transformed to heat due to friction, which increases material temperature. In amorphous food materials, e.g., dehydrated fruits and vegetables, a fairly small increase in temperature or rapid water adsorption by the small particles may depress the $T_g$ to below product temperature. Therefore, several food materials become sticky during size reduction, which results in caking problems during processing and storage. Size reduction of crystalline materials such as sugars often causes melting at the crystal surfaces. If a melted surface is cooled rapidly to the glassy state the powder is stable, although caking may occur during storage (Roth, 1977). If cooling of the particles occurs slowly the powder may cake due to formation of interparticle bridges. Liquid bridges may also exist between amorphous layers located on sugar crystals (Bressan and Mathlouthi, 1994).

Problems related to size reduction and caking of amorphous particle surfaces of the powders produced can be avoided by sufficient cooling during size reduction as well as by avoiding exposures to high relative humidities during storage.


### B. Melt Processing and Extrusion

*Melt processing* refers here to food processes in which food solids are heated to temperatures above the melting temperature with or without a plasticizer. The melt produced may then be molded and rapidly cooled to below the glass transition temperature to provide stability. Extrusion cooking of cereals is a typical example of such processes, which in addition allows expansion of the melt into a porous crispy material.


1. Plasticization and melting

Melt processing of food materials often requires the presence of water as a plasticizer. The addition of water allows the use of lower processing temperatures, which is of importance to avoid drastic changes in composition and quality that may occur at high temperatures due to nonenzymatic browning and decomposition of food components.

According to Colonna *et al.* (1989) granular starch with 9 to 20% (w/w) water in an extruder is progressively compressed and transformed into a dense, solid, and compact material. The structural transformations of starch also result in loss of the granular and crystalline properties. The physical state of starch during extrusion can be considered to change from a partially crystalline polymer to a polymer melt that is homogenized by shear. Colonna *et al.* (1989) reported that extrusion of native cereal and potato starches destroys the organized crystalline structure either partially or completely. The extent of the loss of crystallinity in cereal starches is dependent on the amylose-amylopectin ratio. Starches that contain no lipids, e.g., tuber starches and waxy corn starch, have shown reduced crystallinity at an extrusion temperature as low as 70°C. Cereal starches such as commercial corn starch have been found to exhibit V-type crystallinity after extrusion at 135°C (Colonna *et al.*, 1989). Extrusion may also decrease the molecular size of starch components, which is observed from decreased melt viscosity (Lai and Kokini, 1991), and obviously a decreased molecular size results in a decreased glass transition temperature of the extrudate.

The structure of extrusion-expanded products depends on the extent of starch gelatinization and the physical state of the melt at the die (Launay and

Lisch, 1983; Colonna *et al.*, 1989). According to Launay and Lisch (1983) rheological properties of the melt affect expansion and both longitudinal and diametrical expansion are dependent on melt viscosity and elasticity. The dramatic decrease of pressure which occurs as an extruded material exits the die causes an extremely rapid loss of water and expansion of the melt. The porous structure that is formed during expansion is retained due to loss of water and depression of temperature, which presumably allow solidification of the extruded material into the glassy state. Extrusion may also be used to produce unexpanded amorphous materials with water contents of 5 to 10%. Such materials can be expanded by increasing the temperature to above $T_g$. According to Colonna *et al.* (1989) expansion is obtained by heating to 160 to 190°C by frying or heating in an oven for 10 to 20 seconds. The rapid heating of the material evaporates water and the high pressure causes expansion of the remelted extrudate. The remelted extrudates can be cooled to below $T_g$ to provide stability.

It may be concluded that melting of starch components and formation of a melt with a sufficiently high viscosity and elasticity can be used to produce amorphous structures that have a high porosity. It is obvious that the rapid removal of water and cooling result in the formation of a glassy product that has such typical mechanical properties as crispness of low-moisture snack foods (Kaletunc and Breslauer, 1993).

2. Structural properties

Rheological properties of melts and the structure of extruded foods are affected by several factors and particularly by processing and storage conditions. Rheological properties of the resultant glassy materials are in large part governed by the molecular weight and the extent of thermal and water plasticization. It is also interesting to notice that porosity may slightly affect the glass transition temperature (Schüller *et al.*, 1994).

The molecular weight of extruded food components may change during the extrusion process (Lai and Kokini, 1991). It has been shown that the extent of macromolecular degradation is a function of extrusion parameters and in particular screw speed, temperature, and water content. Both the increase in temperature and the decrease in water content increase molecular degradation (Colonna *et al.*, 1989). The increase in molecular degradation can be assumed to decrease $T_g$, increase hygroscopicity, and increase stickiness of extrudates. One of the main desired characteristics of extruded snacks and breakfast cereals is a crispy texture. Launay and Lisch (1983) pointed out that crispness of extruded snacks is caused by their high elastic modulus and low rupture strength. These properties are typical of glassy materials and crispness may be considered to be a property that is a consequence of the glassy structure produced by extrusion.

The effects of the physical state of food materials on their behavior in extrusion and in extruded foods have not been studied in detail. Colonna *et al.* (1989) pointed out that the main limitations of using polymer mechanistic models for predicting biopolymer properties in extrusion have been the lack of knowledge of their thermophysical and flow properties. These properties depend on food composition, shear rate, temperature, thermomechanical history, and water content. Considering that the same factors define the physical state and mechanical properties of amorphous foods may significantly improve understanding of physicochemical principles of the various changes that affect material behavior in extrusion and end product characteristics.

## 3. Flavor encapsulation

Manufacturing of encapsulated flavors can be accomplished by rapid dehydration of carbohydrate solutions which contain emulsified flavors by spray-drying or by extrusion among other techniques (Jackson and Lee, 1991). Extrusion is used to produce carbohydrate melts that contain encapsulated flavor compounds. Encapsulation occurs as the melt is rapidly cooled without expansion in a cold solvent such as isopropanol. The cold solvent solidifies the carbohydrate matrix and removes flavor from surfaces (Jackson and Lee, 1991). Encapsulated flavors that are not miscible with glassy carbohydrate melts become stable and protected from release. The materials in a granulated form have free-flowing properties and their use as components of low-moisture foods is convenient.

Encapsulation of flavor compounds occurs due to selective diffusion in carbohydrate matrices. Therefore, entrapped flavor compounds that are susceptible to oxidation become protected from atmospheric oxygen due to decreased diffusion of gases through the glassy structure below $T_g$. Water plasticization of the matrices results in the release of encapsulated flavors. Such plasticization and release of the flavor may occur when the product is added in a food processing step to an aqueous material or the flavors are released from added flavor matrices during food preparation and consumption. Carbohydrates that are used in encapsulation include sugars or maltodextrins, which have a sufficiently high glass transition temperature. The material should be stable during food storage and become plasticized at a temperature which can be used in extrusion without changes in color or flavor.

## III.   Food Formulation and Storage

Phase transitions that occur in food storage include both first-order and second-order phase transitions. These transitions are often detrimental to food quality and they may accelerate food deterioration. Storage temperatures above the glass transition temperature may cause crystallization of amorphous sugars or increase rates of deteriorative changes in low-moisture foods. In frozen foods ice melting decreases viscosity and causes ice recrystallization. Ice melting together with a high storage temperature may result in a rapid deterioration of frozen foods.

### A. Prediction of Stability

The main objectives in using proper storage conditions for food materials are to decrease rates of deteriorative changes and to provide stability. Therefore, information on factors which may affect reaction rates and growth of microorganisms is needed in establishing criteria for appropriate and economical storage conditions. The information may also be used in predicting the shelf life of various foods at various storage conditions. It is well known that the main parameters of such predictions are temperature and water content.

### 1. Low-moisture foods

Low-moisture foods are considered to be stable when they are stored at cool and dry conditions. An increase in temperature or in water content may result in a significant change in rates of deteriorative changes. Knowledge of the effects of temperature and water on the physical state and diffusion in amorphous food matrices may be used to establish relationships between food composition and storage conditions.

   Traditional shelf life predictions of low-moisture foods have been based on the information on rates of deteriorative changes and loss of nutrients at various temperatures and water contents. The main assumption for stability is often that water contents close to the BET monolayer value allow maximum stability. An increase in water content at a constant storage temperature results in rapid deterioration as the reaction rates increase at intermediate water contents. It may be assumed that the rates of a number of deteriorative changes in low-moisture foods are affected by diffusional limitations. Reaction rates in the solid, glassy state at low water contents approach zero, but an increase in temperature or in water content probably increases diffusion and therefore the reaction rate. It is probable that at temperatures above $T_g$ reaction rates at a constant water content follow the Arrhenius-type temperature dependence.

Sorption isotherms of low-moisture foods provide important information that can be used in predicting shelf life. The GAB model has been shown to be particularly useful in fitting sorption data. The model gives a numerical value for the monolayer value and it may be used to predict effects of water content on water activity. The relationship between $a_w$ and water content can be directly used in evaluating effects of water content on stability. It is obvious that the relationships between mechanical properties, temperature, and water content of amorphous foods provide improved criteria for shelf life predictions. The sorption isotherm combined with water plasticization data allows the establishment of critical values for water content that can be related to stability. Water contents that are higher than the critical values decrease the $T_g$ of the material to below ambient temperature, which results in stickiness, caking, and probable crystallization of amorphous sugars. Rates of chemical changes are likely to decrease rapidly at temperatures below $T_g$ and increase at temperatures above $T_g$. The critical values may be directly used in establishing criteria for maximum water vapor permeability values for packaging materials that can maintain stability and reduce water plasticization during storage.

Accelerated shelf life tests are used to obtain information on reaction rates and activation energies of deteriorative changes in low-moisture foods. However, it should be noticed that an increase in temperature to above $T_g$ may result in faster deterioration than would be expected due to changes in the physical structure and diffusion as well as to collapse and crystallization. Therefore, combined sorption isotherms and state diagrams are extremely useful in locating critical water contents at various temperatures and in evaluating validity of extrapolated shelf life data.

## 2. Frozen foods

Stability of frozen foods is related to several factors that are dependent on temperature and composition. Food components that decrease ice melting temperature decrease also the glass transition temperature of freeze-concentrated solids and affect the unfrozen water content at food storage temperatures. Shelf life predictions of frozen foods need to consider the effect of freeze-concentration and temperature on deteriorative changes.

Levine and Slade (1986) suggested that rates of deteriorative changes in frozen foods increase by orders of magnitude with small increases in temperature above the glass transition temperature of the maximally freeze-concentrated solute matrix. The increase in the rates of the deteriorative changes was considered to follow the WLF equation and therefore an increase in temperature was reported to be more detrimental than would be predicted by the Arrhenius equation. Karel and Saguy (1991) compared the results of simulated shelf life test using the Arrhenius and WLF predictions of rates of deteriorative changes in frozen foods under steady and fluctuating storage temperatures. It was shown that according to the Arrhenius kinetics shelf life was affected only slightly by temperature fluctuations of 5°C. The WLF temperature dependence of rates of deteriorative changes suggested that stability was maintained at temperatures below $T'_g$. The assumption was that diffusion above $T_g$ increased with increasing $T - T_g$, which increased the rate constant. The WLF model predicted that the temperature fluctuation of 3°C decreased shelf life from 12 months to less than 10 months and when the fluctuation was 5°C shelf life was predicted to be less than 4 months.

The assumptions for the increase of deteriorative changes in frozen foods at temperatures above $T'_g$ are that the relaxation times of mechanical properties decrease according to the WLF temperature dependence and that the $T'_g$, which occurs at the onset temperature of ice melting, $T'_m$, can be used as a reference temperature below which stability is maintained. Therefore, diffusion of reactants occurs slowly at temperatures below $T'_g$ and the rate of diffusion above $T'_g$ increases according to the WLF relationship. However, existing data on rates of deteriorative changes in frozen foods are not sufficient to show that shelf life predictions could be based on the WLF temperature dependence. Temperatures above $T'_g$ and $T'_m$ increase water plasticization of food solids due to ice melting and the temperature difference, $T - T'_g$ is probably not sufficient to predict the effect of the temperature increase on the rates of deteriorative changes. Kerr and Reid (1994) observed that the temperature dependence of viscosity data for frozen maltodextrin solutions followed the WLF equation when the reference temperature was the $T_g$ of the partially freeze-concentrated solution above $T'_g$ and not $T'_g$. The use of $T'_g$ as a reference temperature was considered to be practical, as was pointed out by Slade and Levine (1991), although the decreases of relaxation times of mechanical properties and viscosity were most probably governed by $T - T_g$. However, the use of $T_g$ as the reference temperature is more complicated and requires that the $T_g$ of the material at various water contents and the effect of temperature on the amount of unfrozen water are known.

Although relating changes in relaxation times of partially freeze-concentrated food solids to $T'_g$ is theoretically not justified its temperature value provides an important single criterion for the evaluation of the effect of temperature on the physical state and shelf life of frozen foods. It is obvious that diffusion in frozen foods is slow at temperatures below $T'_g$ due to maximum ice formation and the glassy state of the freeze-concentrated solutes. The rapid decreases of relaxation times of mechanical properties above $T'_g$ and particularly above $T'_m$ that are caused by thermal and water plasticization result in a dramatic change in the physical state as the maximally freeze-concentrated solutes are transformed to the supercooled liquid state and become diluted. The change occurs suddenly within a relatively narrow temperature range. Therefore, it is likely that deteriorative changes in frozen foods may occur with significantly higher rates above $T'_m$ than below $T'_g$.

It may be concluded that deteriorative changes in frozen foods occur relatively slowly at temperatures below $T'_m$. Above $T'_m$ dilution may decrease reaction rates, which are likely to follow the Arrhenius-type temperature dependence. Therefore, shelf life predictions of frozen foods should consider that $T'_g$ and $T'_m$ define temperatures which govern the physical state of frozen foods. The change in the physical state may be observed in Arrhenius plots, which probably suggest a change of activation energy at the transition temperature range. When storage below $T'_m$ is not feasible the effect of temperature on the amount of unfrozen water should be taken into account as a physical state and rate-defining factor.

## B. Food Formulation

Knowledge of the physical state and physicochemical properties of food components is advantageous in food design and formulation. The information on rates of various kinetic processes can be used to manipulate and control rates of changes that occur during food processing and to develop food products which are less sensitive to detrimental changes during food storage.

### 1. Food composition

The composition of food solids determines food behavior in processing and storage. The presence, quality, and relative amounts of carbohydrates, lipids, proteins, and water affect phase transitions and transition temperatures that govern rates of chemical changes and the temperature and water content dependence of mechanical properties. Food composition is therefore the main variable that can be used to control time-dependent changes during processing and storage.

*a. Effects in Food Processing.* Several time-dependent changes in food processing are affected by the change in physical state that occurs over the glass transition temperature range. These changes may include viscoelastic properties of melts and formation of structure in extrusion, stickiness in dehydration, formation of color and flavor in baking and frying, and other changes which are governed by or related to the relaxation times of mechanical properties of supercooled liquid materials. The effect of food composition on the transition temperatures and required processing conditions can be derived from data for the effects of temperature and water content on the glass transition temperature.

It is obvious that a change in the effective molecular weight of food solids affects glass transition temperature. In extrusion or dehydration at a constant temperature such changes may result in changes in melt viscosity, expansion, or stickiness. It is probable that a decrease in the effective molecular weight increases problems which are caused by stickiness and discoloration in spray-drying, while an increase in the effective molecular weight would decrease stickiness and improve processability. Such changes in the effective

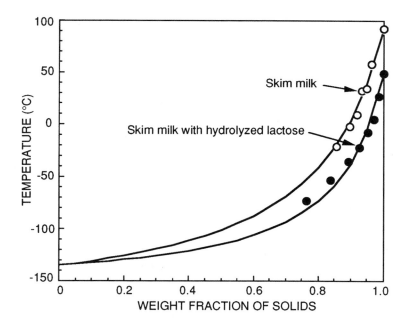

**Figure 10.8** Effect of lactose hydrolysis on the glass transition temperature, $T_g$, and water plasticization of skim milk powder. The $T_g$ of anhydrous skim milk powder is almost equivalent to that of amorphous lactose. An enzymatic hydrolysis of lactose before dehydration decreases the $T_g$ due to the compositional change of lactose to galactose and glucose, which decreases the effective molecular weight, and the $T_g$ approaches the lower $T_g$ values of the component sugars. Data from Jouppila and Roos (1994).

molecular weight may occur in food processing due to variation in relative amounts of sugars in the carbohydrate fraction of food solids. Differences in solids composition may originate from variations in the composition of raw materials, decomposition of carbohydrate polymers during processing, or changes in formulation. Valid examples of the effects of composition on the physicochemical properties are provided by dairy products. Dehydration of milk that contains lactose can in most cases be completed without problems due to stickiness and the powders produced are stable at low water contents. However, an enzymatic hydrolysis of the lactose fraction may result in serious problems in dehydration due to stickiness, caking, and discoloration. The effect of lactose hydrolysis on the glass transition temperature of skim milk powder is shown in Figure 10.8.

Jouppila and Roos (1994) found that the $T_g$ of lactose-containing anhydrous skim milk powder was 92°C. The $T_g$ decreased with increasing water content and the critical water content that decreased the $T_g$ to 24°C was 7.6 g/100 g of solids. Powder produced from skim milk that contained lactose, which was enzymatically hydrolyzed to galactose and glucose had anhydrous $T_g$ at 49°C and a water content of 2.0 g/100 g of solids decreased the $T_g$ to 24°C. According to Roos (1993a) the $T_g$ of amorphous lactose is 101°C, while those of galactose and glucose are 30 and 31°C, respectively. Although Kalichevsky *et al.* (1993) found that sugars had only a little effect on the $T_g$ of casein, the $T_g$ of milk powder with hydrolyzed lactose seems to be higher than is suggested by the $T_g$ values of the component sugars. However, the $T_g$ of milk powder is significantly decreased by lactose hydrolysis, which presumably is the main cause of stickiness during processing and storage as well as of hygroscopicity and higher susceptibility of the powder to the nonenzymatic browning reaction. It should also be noticed that although lactose is a reducing sugar the hydrolysis of 1 mol of lactose produces 2 mol of the reducing sugars, i.e., 1 mol of galactose and 1 mol of glucose.

The effect of molecular weight on the $T_g$ of food homopolymers can be estimated with the Fox and Flory equation, while the effect of composition on the $T_g$ of food solids can be estimated with the Gordon and Taylor equation or with other equations that predict effects of composition on the $T_g$ of polymer mixtures. In food formulation the $T_g$ of the solids may be manipulated by using food components with different $T_g$ values. A typical approach for improving food dehydration characteristics is to increase the effective molecular weight by the addition of maltodextrins. The required amount of the added material may also be assumed to depend on its molecular weight. Roininen (1994) used various maltodextrins with wheat flour in formulating an extruded snack food model with desired physical properties. The amount and DE of the maltodextrin that gave the desired product characteristics for the model were based on the determination of the effect of the maltodextrin on the $T_g$ at various water contents. The requirements for the material were that dry ingredients could be used to produce an extruded snack with an expanded structure and crispy texture, and that the $T_g$ of the product was depressed to below $T_g$ at intermediate water contents. Most snack foods contain starch as the major component and their $T_g$ values are fairly high. The addition of maltodextrins in such products may decrease the $T_g$ and increase hygroscopicity and eventually increase product sweetness.

Food formulation by using ingredients that affect product $T_g$ is probably most important in the development of snack foods, baked foods, and confectioneries. It may be assumed that an increase in the $T_g$ is achieved by the addition of polymeric components, e.g., carbohydrates and proteins, which act as antiplasticizers within food solids. Such materials also increase resistance to water plasticization. The addition of low molecular weight carbohydrates decreases $T_g$, which may be important in the manufacturing of sweet foods and confectionery.

*b. Food Composition and Stability.* Food composition affects the chemical and physical stability. Phase transitions may significantly affect the physical state and long-term stability of low-moisture and frozen foods. It is obvious that the main factors which control stability of low-moisture foods are water activity and composition and those of frozen foods are temperature and composition.

The physical state of low-moisture foods may be affected by phase transitions of lipids and glass transition of water-plasticized nonfat food solids. Water plasticization and loss of stability of low-moisture foods may be related to the critical values for water activity and water content. In food formulation the linearity between $T_g$ and $a_w$ provides a rapid method for the prediction of the effects of storage relative humidity on physical state. Therefore, determination of $T_g$ values for amorphous food products that have been equilibrated at three or more relative humidities allows prediction of the critical $a_w$ as was suggested by Roos (1987, 1993b). The critical $a_w$ indicates the maximum relative humidity during storage. A low critical $a_w$ suggests low storage stability without a highly protective package. A high critical $a_w$ suggests low hygroscopicity and good storage stability. Therefore, an increase or a decrease of the critical $a_w$ may be necessary in food formulation to obtain product characteristics that also maximize storage stability. The possibilities are the addition of high $T_g$ compounds to increase the $T_g$ of the solids or the addition of low molecular weight components (Ollett *et al.*, 1991), which reduces $T_g$.

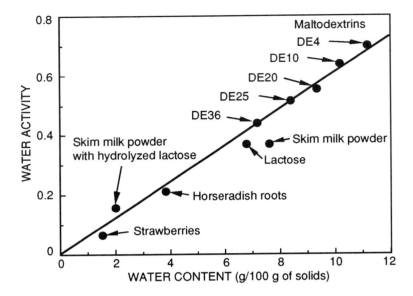

**Figure 10.9** Relationships between critical water activity, $a_w$, and critical water content. The critical $a_w$ increases linearly with increasing critical water content. The critical water content increases also with increasing molecular weight as shown for maltodextrins. Data from Roos (1993b) and Jouppila and Roos (1994).

The effects of composition on the $T_g$ of polymers and food solids have been well established (Roos and Karel, 1991a-e; Slade and Levine, 1991). The relationship between critical water content and critical $a_w$ as shown in Figure 10.9 reflects the effect of molecular weight at a constant temperature on water plasticization and food stability. The linear relationship between critical $a_w$ and water content can be used in food formulation with sorption isotherms to locate safe storage conditions. It is obvious from Figure 10.9 that a food material which is designed to maintain stability up to a storage relative humidity of 60% should have a steady state water content of about 10 g/100 g of solids or higher at 25°C. Similarly, the steady state water content of a food product may be used to determine the critical $a_w$ and the limiting storage relative humidity. Therefore, criteria for food storage conditions may be established in food formulation with the information that is obtained from the relationships between composition, phase transitions, and water content.

The physical state of frozen foods is extremely sensitive to temperatures above $T'_g$ and $T'_m$. The $T'_g$ and $T'_m$ values of food components increase with increasing molecular weight as was shown by Levine and Slade (1986) and

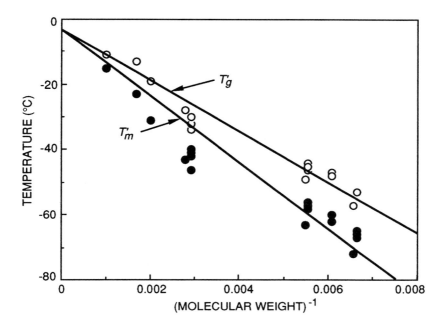

**Figure 10.10** Effect of molecular weight of maltodextrins and sugars on the glass transition temperature of the maximally freeze-concentrated solute matrix, $T'_g$, and onset temperature of ice melting within the maximally freeze-concentrated solute matrix, $T'_m$. Both $T'_g$ and $T'_m$ increase with increasing molecular weight and approach the same temperature at a limiting molecular weight. Data from Roos and Karel (1991d) and Roos (1993a).

Roos and Karel (1991d). The effect of molecular weight on the $T'_g$ and $T'_m$ of sugars and maltodextrins is shown in Figure 10.10. It may be assumed that both $T'_m$ and $T'_g$ increase with decreasing reciprocal of the effective molecular weight. It is obvious that the $T'_g$ and $T'_m$ approach the same temperature that is located slightly below the melting temperature of water at a high molecular weight. The $T'_m$ value is depressed to below -20°C as the molecular weight of the solute becomes lower than about 500. Therefore, it may be assumed that food materials which have an effective molecular weight higher than 500 exist in the stable maximally freeze-concentrated state at typical temperatures of frozen food storage. In food formulation the relationship between molecular weight and $T'_m$ value can be used to evaluate stability of frozen foods at various storage temperatures. Materials with effective molecular weights lower than 500 are likely to show decreased stability in comparison with food solids with higher effective molecular weights due to partial freeze-concentration of solutes at temperatures above $T'_m$.

Slade and Levine (1991) suggested that germination of mold spores is a mechanical relaxation process that is governed by the translational mobility of water. The effect of glass transition on the heat resistance of bacterial spores was studied by Sapru and Labuza (1993). They found that the inactivation of spores followed the WLF equation, which fitted to the data above $T_g$ better than the Arrhenius equation. Sapru and Labuza (1993) found also that the heat resistance of bacterial spores increased with increasing $T_g$ of the spores. These findings emphasize the importance of water plasticization for microbial growth and heat inactivation. However, the growth of microorganisms has been observed to occur within glassy food materials (Chirife and Buera, 1994) and other factors in food formulation, including water activity and pH, should be considered in addition to the physical state for the increase of microbial stability.

## 2. Application of state diagrams

State diagrams are effective tools in establishing relationships between the physical state of food materials, temperature, and water content. State diagrams show the glass transition temperature as a function of water content and the effect of ice formation on $T_g$ and on the equilibrium ice melting temperature, $T_m$. State diagrams may also show solubility as a function of temperature and information on various changes that may occur due to the metastable state of amorphous food solids and their approach towards the equilibrium state. In food formulation and design state diagrams allow evaluation of the effects of food composition and water content on the physical state and physicochemical properties during processing and storage.

State diagrams may be used in food formulation as "maps" that show the physical state at various water contents and temperatures (Levine and Slade, 1988; Slade and Levine, 1991). A state diagram that locates various food processes and food materials according to temperature and water content in processing or storage is shown in Figure 10.11. It is obvious that the typical temperature and water content conditions of baking and of food extrusion locate starch well above $T_g$, where the material may exist as a melt. Baked bread exists above $T_g$ at conditions that support its viscoelastic properties, although small changes in water content may significantly affect $T - T_g$ and probably also rheological properties and the rate of starch retrogradation. Most extruded materials belong to dry crispy foods which are located at temperature and water conditions in which starch exists in the glassy state well below $T_g$. Dry foods that contain high amounts of low molecular weight sugars, such as raisins, exist in the supercooled liquid state above

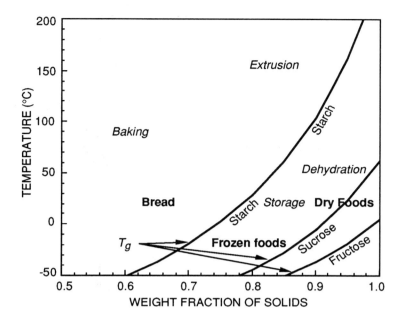

**Figure 10.11** State diagram for fructose, sucrose, and starch at typical water contents showing the physical state of food materials in baking, dehydration, extrusion, and storage. Low-moisture and frozen food solids exist either in the glassy or in the supercooled liquid state above the glass transition temperature, $T_g$, depending on composition.

$T_g$ and they exhibit rheological properties that are typical of viscoelastic materials (Karathanos *et al.*, 1994). Frozen foods that contain high amounts of polymeric compounds are likely to exist in the maximally freeze-concentrated state, while foods with high amounts of low molecular weight carbohydrates, e.g., fruit products and ice cream, probably exist in the partially freeze-concentrated state. In food formulation state diagrams may be established to show the $T_g$ of the main components. Such state diagrams explain the physical state and therefore characteristics of the material at conditions used in processing and storage. The state diagrams may also be used to evaluate feasibility of temperatures and water contents used in processing and storage.

The various effects of phase transitions on food properties are summarized in the state diagram shown in Figure 10.12. State diagrams may be used in food formulation to evaluate the effects of water content on the physical state during food processing. State diagrams provide also information on the physical state of dehydrated foods and the amount of unfrozen water at

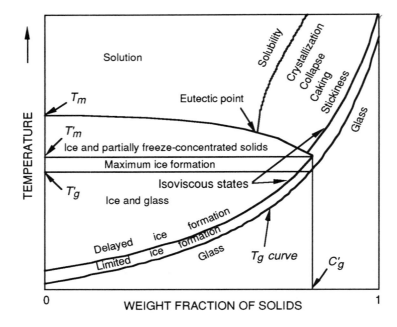

**Figure 10.12** A schematic state diagram typical of food solids. The glass transition temperature, $T_g$, decreases with increasing water content. Rates of various changes increase with increasing temperature difference to $T_g$, which is described by the isoviscous states above $T_g$. The amount of ice formed in frozen foods is defined by the equilibrium melting temperature, $T_m$. Maximum ice formation occurs at temperatures above the glass transition temperature of the maximally freeze-concentrated solute matrix, $T'_g$, but below the onset temperature of ice melting within the maximally freeze-concentrated solutes, $T'_m$. The solute concentration in the maximally freeze-concentrated solute matrix is given by $C'_g$.

various frozen storage conditions. The isoviscous states at temperatures above $T_g$ are useful in predicting rates of various changes that affect food quality. Such changes include structural changes of dehydrated foods as well as ice recrystallization and solute crystallization in frozen foods. The isoviscous states may also be related to time-dependent quality changes and therefore they are useful in establishing time limits for allowed exposures to $T - T_g$ conditions which are not likely to cause unacceptable changes in product characteristics.

# References

Anglea, S.A., Wang, J. and Karel, M. 1993. Quality changes of eggplant due to drying regime. Paper No. 192. Presented at the Annual Meeting of the Institute of Food Technologist, Chicago, July 10-14.

Bellows, R.J. and King, C.J. 1973. Product collapse during freeze drying of liquid foods. *AIChE Symp. Ser. 69(132)*: 33-41.

Brennan, J.G., Herrera, J. and Jowitt, R. 1971. A study of some of the factors affecting the spray drying of concentrated orange juice, on a laboratory scale. *J. Food Technol. 6*: 295-307.

Bressan, C. and Mathlouthi, M. 1994. Thermodynamic activity of water and sucrose and the stability of crystalline sugar. *Zuckerindustrie (Berlin) 119*: 652-658.

Chandrasekaran, S.K. and King, C.J. 1972. Multicomponent diffusion and vapor-liquid equilibria of dilute organic components in aqueous sugar solutions. *AIChE J. 18*: 513-519.

Chirife, J. and Buera, M.P. 1994. Water activity, glass transition and microbial stability in concentrated/semimoist food systems. *J. Food Sci. 59*: 921-927.

Chirife, J. and Karel, M. 1973. Contribution of adsorption to volatile retention in a freeze-dried food model containing PVP. *J. Food Sci. 38*: 768-771.

Chirife, J. and Karel, M. 1974. Effect of structure disrupting treatments on volatile release from freeze-dried maltose. *J. Food Technol. 9*: 13-20.

Chuy, L.E. and Labuza, T.P. 1994. Caking and stickiness of dairy-based food powders as related to glass transition. *J. Food Sci. 59*: 43-46.

Colonna, P., Tayeb, J. and Mercier, C. 1989. Extrusion cooking of starch and starchy products. Chpt. 9 in *Extrusion Cooking*, ed. C. Mercier, P. Linko and J.M. Harper. American Association of Cereal Chemists, St. Paul, MN, pp.247-319.

Downton, D.P., Flores-Luna, J.L. and King, C.J. 1982. Mechanism of stickiness in hygroscopic, amorphous powders. *Ind. Eng. Chem. Fundam. 21*: 447-451.

Eichner, K., Laible, R. and Wolf, W. 1985. The influence of water content and temperature on the formation of Maillard reaction intermediates during drying of plant products. In *Properties of Water in Foods*, ed. D. Simatos and J.L. Multon. Martinus Nijhoff Publishers, Dordrecht, the Netherlands, pp. 191-210.

Flink, J.M. 1983. Structure and structure transitions in dried carbohydrate materials. Chpt. 17 in *Physical Properties of Foods*, ed. M. Peleg and E.B. Bagley. AVI Publishing Co., Westport, CT, pp. 473-521.

Flink, J. and Karel, M. 1970. Effects of process variables on retention of volatiles in freeze-drying. *J. Food Sci. 35*: 444-446.

Flink, J. and Karel, M. 1972. Mechanisms of retention of organic volatiles in freeze-dried systems. *J. Food Technol. 7*: 199-211.

Franks, F. 1990. Freeze drying: From empiricism to predictability. *Cryo-Lett. 11*: 93-110.

Jackson, L.S. and Lee, K. 1991. Microencapsulation and the food industry. *Lebensm.-Wiss. u. -Technol. 24*: 289-297.

Jensen, J.D. 1975. Agglomeration, instantizing, and spray drying. *Food Technol. 29(6)*: 60, 62, 64-65, 68, 70-71.

Jouppila, K. and Roos, Y. 1994. Glass transitions and crystallization in milk powders. *J. Dairy Sci. 77*: 2907-2915.

Kaletunc, G. and Breslauer, K.J. 1993. Glass transitions of extrudates: Relationship with processing-induced fragmentation and end-product attributes. *Cereal Chem. 70*: 548-552.

Kalichevsky, M.T., Blanshard, J.M.V. and Tokarczuk, P.F. 1993. Effect of water content and sugars on the glass transition of casein and sodium caseinate. *Int. J. Food Sci. Technol. 28*: 139-151.

Karathanos, V., Anglea, S. and Karel, M. 1993. Collapse of structure during drying of celery. *Drying Technol. 11*: 1005-1023.

Karathanos, V.T., Kostaropoulos, A.E. and Saravacos, G.D. 1994. Viscoelastic properties of raisins. *J. Food Eng. 23*: 481-490.

Karel, M. and Flink, J.M. 1973. Influence of frozen state reactions on freeze dried foods. *J. Agric. Food Chem. 21*: 16-21.

Karel, M. and Saguy, I. 1991. Effects of water on diffusion in food systems. In *Water Relationships in Foods*, ed. H. Levine and L. Slade. Plenum Press, New York, pp. 157-173.

Karel, M., Buera, M.P. and Roos, Y. 1993. Effects of glass transitions on processing and storage. Chpt. 2 in *The Glassy State in Foods*, ed. J.M.V. Blanshard and P.J. Lillford, Nottingham University Press, Loughborough, pp. 13-34.

Karel, M., Anglea, S., Buera, P., Karmas, R., Levi, G. and Roos, Y. 1994. Stability-related transitions of amorphous foods. *Thermochimica Acta 246*: 249-269.

Karmas, R., Buera, M.P. and Karel, M. 1992. Effect of glass transition on rates of nonenzymatic browning in food systems. *J. Agric. Food Chem. 40*: 873-879.

Kerr, W.L. and Reid, D.S. 1994. Temperature dependence of the viscosity of sugar and maltodextrin solutions in coexistence with ice. *Lebensm.-Wiss. u. -Technol. 27*: 225-231.

Kessler, H.G. 1981. *Food Engineering and Dairy Technology*. Verlag A. Kessler, Freising, Germany.

Lai, L.S. and Kokini, J.L. 1991. Physicochemical changes and rheological properties of starch during extrusion. *Biotechnol. Prog. 7*: 251-266.

Launay, B. and Lisch, J.D. 1983. Twin-screw extrusion cooking of starches: Flow behaviour of starch pastes, expansion and mechanical properties of extrudates. *J. Food Eng. 2*: 259-280.

Lazar, M., Brown, A.H., Smith, G.S., Wong, F.F. and Lindquist, F.E. 1956. Experimental production of tomato powder by spray drying. *Food Technol. 10*: 129-134.

Leniger, H.A. and Bruin, S. 1977. The state of the art of food dehydration. In *Food Quality and Nutrition Research. Research Priorities for Thermal Processing*, ed. W.K. Downey. Elsevier, London, pp. 265-295.

Levi, G. 1994. The effects of glass transitions on release of organic volatiles from amorphous carbohydrates. Ph.D. thesis, Rutgers - The State University of New Jersey, New Brunswick, NJ.

Levi, G. and Karel, M. 1995. The effect of phase transitions on release of n-propanol entrapped in carbohydrate glasses. *J. Food Eng. 24*: 1-13.

Levine, H. and Slade, L. 1986. A polymer physico-chemical approach to the study of commercial starch hydrolysis products. *Carbohydr. Polym. 6*: 213-244.

Levine, H. and Slade, L. 1988. Principles of "cryostabilization" technology from structure/property relationships of carbohydrate/water systems - A review. *Cryo-Lett. 9*: 21-63.

Lozano, J.E., Rotstein, E. and Urbicain, M.J.1980. Total porosity and open-pore porosity in the drying of fruits. *J. Food Sci. 45*: 1403-1407.

Lozano, J.E., Rotstein, E. and Urbicain, M.J. 1983. Shrinkage, porosity and bulk density of foodstuffs at changing moisture contents. *J. Food Sci. 48*: 1497-1502, 1553.

Masters, K. and Stoltze, A. 1973. Agglomeration advances. *Food Eng. 45(2)*: 64-67.

Menting, L.C., Hoogstad, B. and Thijssen, H.A.C. 1970a. Diffusion coefficients of water and organic volatiles in carbohydrate-water systems. *J. Food Technol. 5*: 111-126.

Menting, L.C., Hoogstad, B. and Thijssen, H.A.C. 1970b. Aroma retention during the drying of liquid foods. *J. Food Technol. 5*: 127-139.

Ollett, A.-L., Parker, R. and Smith, A.C. 1991. Deformation and fracture behavior of wheat starch plasticized with glucose and water. *J. Mater. Sci. 26*: 1351-1356.

Omatete, O.O. and King, C.J. 1978. Volatiles retention during rehumidification of freeze dried food models. *J. Food Technol. 13*: 265-280.

Peleg, M. 1983. Physical characteristics of food powders. In *Physical Properties of Foods*, ed. M. Peleg and E.B. Bagley. AVI Publishing Co., Westport, CT, pp. 293-323.

Pikal, M.J. and Shah, S. 1990. The collapse temperature in freeze drying: Dependence on measurement methodology and rate of water removal from the glassy phase. *Int. J. Pharm. 62*: 165-186.

Roininen, M.-K. 1994. Water plasticization of crispy snack foods. M.Sc. thesis, University of Helsinki, Helsinki, Finland. In Finnish.

Roos, Y.H. 1987. Effect of moisture on the thermal behavior of strawberries studied using differential scanning calorimetry. *J. Food Sci. 52*: 146-149.

Roos, Y. 1993a. Melting and glass transitions of low molecular weight carbohydrates. *Carbohydr. Res. 238*: 39-48.

Roos, Y.H. 1993b. Water activity and physical state effects on amorphous food stability. *J. Food Process. Preserv. 16*: 433-447.

Roos, Y. and Karel, M. 1991a. Plasticizing effect of water on thermal behavior and crystallization of amorphous food models. *J. Food Sci. 56*: 38-43.

Roos, Y. and Karel, M. 1991b. Amorphous state and delayed ice formation in sucrose solutions. *Int. J. Food Sci. Technol. 26*: 553-566.

Roos, Y. and Karel, M. 1991c. Phase transitions of mixtures of amorphous polysaccharides and sugars. *Biotechnol. Prog. 7*: 49-53.

Roos, Y. and Karel, M. 1991d. Water and molecular weight effects on glass transitions in amorphous carbohydrates and carbohydrate solutions. *J. Food Sci. 56*: 1676-1681.

Roos, Y. and Karel, M. 1991e Applying state diagrams to food processing and development. *Food Technol. 45*: 66, 68-71, 107.

Roth, D. 1977. Das Agglomerationsverhalten von frisch gemahlenem Zucker. *Zucker 30*: 464-470.

Sapru, V. and Labuza, T.P. 1993. Glassy state in bacterial spores predicted by polymer glass-transition theory. *J. Food Sci. 58*: 445-448.

Schuchmann, H., Hogekamp, S. and Schubert, H. 1993. Jet agglomeration processes for instant foods. *Trends Food Sci. Technol. 4*: 179-183.

Schüller, J., Mel'nichenko, Y.B., Richert, R. and Fisher, E.W. 1994. Dielectric studies of the glass transition in porous media. *Phys. Rev. Lett. 73*: 2224-2227.

Slade, L. and Levine, H. 1991. Beyond water activity: Recent advances based on an alternative approach to the assessment of food quality and safety. *Crit. Rev. Food Sci. Nutr. 30*: 115-360.

Thijssen, H.A.C. 1971. Flavour retention in drying preconcentrated food liquids. *J. Appl. Chem. Biotechnol. 21*: 372-377.

# Index

Ablimation, 14
Acetone, 286
Actin, 134
Activation energy, 24, 33, 68, 136, 255-
   257, 277-278, 286, 288, 293-297,
   316-317, 324, 334-336
   of chemical reactions, 277-278, 334-
   336
   of diffusion, 255-257, 286, 324
   of enzymatic reactions, 288
   of nonenzymatic browning, 293-297,
   316-317
   of protein denaturation, 136
Activity coefficient, 82-83
Adsorption, *see* Sorption
Agglomeration, 194, 203, 313, 326-329
Air-drying, 315-324
Alginates, 318
Amadori compounds, 316

Amylopectin, 49, 69, 119-133, 140,
   161-169, 187, 235-240, 265-266,
   330
Amylose, 49, 94, 119-133, 235-240,
   265
Amylose-lipid complexes, 53, 121, 128,
   132-133
Annealing, 100, 122, 129, 133, 145,
   174, 196, 200-201, 226-229
Ante-melting, 99
Antiplasticization, 130, 339
Apple, 321
Arrhenius, 22, 68, 95, 172-176, 215-
   218, 226, 233, 255-258, 277-278,
   282-283, 291-296, 307-308, 334-
   335, 342
   equation, 172-176, 215-218, 226,
   277, 290, 307
   kinetics, 233, 255-258, 290-300,
   303-308, 334-336

[Arrhenius]
    plots 176, 256, 281, 336
Ascorbic acid, *see* Degradation
Aspartame, *see* Degradation
Avrami equation, 44-47, 217-218, 228, 235-240


Baby food powders, 326
Baked foods, 132, 219, 339
Baking, 121, 138, 195, 314, 342-343
Barley, 265
BET model, *see* Sorption models
BET monolayer, *see also* Monolayer
    value, 89-90, 284, 299, 333
Biopolymer(s), 41, 43, 52, 55, 96-97, 119-142, 159-166, 182-187, 250
Birefringence, 125-126
Boiled sweets, 114, 202
Boiling temperature, 11-17, 73-76, 83-84
    elevation, 83-84
    of water, 74-76
Bragg's law, 52
Bread, 131, 137-139, 184, 194, 235-240, 266, 314, 342
    dough, 142, 184
    elastic modulus of, 235
    staling of, 235, 238-240
Breakfast cereals, 260- 262, 314, 331
Browning, see *also* Nonenzymatic
    browning, 44, 97, 202, 205, 271, 278-303, 316-317, 338
    intermediates, 317
Bulk
    shrinkage coefficient, 319-320
    volume, 319-320
Butanediol, 176
Butter, 143-147, 263


Cabbage, 292
Caking, 44, 194, 202-206, 219, 251, 258, 297, 317, 329-330, 334, 338
Caramelization, 111
Carbohydrates, 2, 19, 44, 86, 109-133, 140, 142, 159, 165, 167, 185, 195, 225, 254, 259-260, 336, 339, 343

Carbohydrate polymers, 31, 40, 110, 338
Carrots, 81, 292, 316-317
Casein, 124, 140, 161, 166, 338
Celery, 319-321
Cellulose, 35
Cereal(s), 96, 102, 119-142, 166, 170, 183-187, 194, 235, 249-253, 262, 330-331
    foods, 253
    proteins, 96, 124, 131-142, 166, 183, 185-187, 249-251
    starches, 119-133, 235
Chemical potential, 11
Clausius-Clapeyron equation, 16
Cocoa
    butter, 143-147
    sugar mixes, 326
Coffee extract, 204
Collagen, 35, 134, 249
Collapse, 163-174, 198, 202, 206-210, 213, 222, 258-260, 282, 287, 296-297, 315-326
    during freeze-drying, 171-174, 198, 322-326
    phenomena, 166, 202, 206-210, 222, 258, 282, 287, 296-297, 315-326
    temperature 163, 206-209
Complex modulus, *see also* Modulus
    shear modulus, 28-29
    Young's modulus, 28-29
Compliance, 27
Compressibility, 15, 198-200
Conalbumin, 135-136
Confectionery, 21, 110, 112, 142, 219, 265, 339
Convective heat transfer coefficient, 225
Copolymers, 164
Corn, 121, 124, 139, 146, 237, 249, 330
    endosperm, 249
    oil, 146
    proteins, 139
    starch, 121, 330
    syrups, 118, 223, 318
Couchman and Karasz equation, 41, 140, 160-166, 179
Crispness, 179, 248, 260-262, 331-332

Critical
  nuclei, 23, 111
  point, 11
  pressure, 11
  radius, 22
  relative humidity, 212, 258
  size, 22, 24, 231
  state, 11
  temperature, 11, 136, 177, 210, 292,
    321-324
  viscosity, 203-210
  volume, 11
  water activity/water content, 159,
    167-170, 219-223, 254-255, 260-
    262, 271, 283-285, 287, 290-302,
    322-324, 336-342
Cross-linking, 30, 139, 184, 266
Cryopreservation, 103, 110, 117, 224,
  229
Cryopreservative, 110
Cryoprotectant, 86, 110, 224
Cryostabilization, 117, 142, 232
Crystal
  growth, 20, 24-25, 42-47, 173-174,
    224, 233-240
  lattice, 174
  morphology, 45-47, 51, 235-238
  size, 51, 230-233
  structure, 52-55, 230
  surface, 25, 230, 329
Crystalline
  polymers, 19, 26
  structure, 194, 237
Crystallinity, 30-31, 44-53, 119-133,
    214-219, 236-239, 262-267
  of frozen foods, 266-267
  of lipids, 55, 262-264
  of polymers, 31, 265
  of starch, 119-133, 235-240
  of sugars, 214-219
Crystallization, 14, 20-24, 42-47, 85-86,
    99, 110-112, 142-149, 193-198,
    210-240, 258-260, 282-292, 297-
    302, 326, 333
  of amorphous compounds, 42-47,
    282, 292
  of amylose-lipid complexes, 132-133

[Crystallization]
  of carbohydrate matrices, 259-260,
    297-302, 326
  exotherm, 213-214, 220
  fractional, 147
  instant, 212-214
  kinetics, 44-47, 193-198, 210-240
  of lipids, 142-149
  of polymers, 42, 217
  rate, 24, 43, 237
  of solutes, 85-86, 99
  of sugars, 110-112, 210-224, 333
  temperature, 14, 43-45, 111, 219-222
  theories, 237
  time-dependent, 210-224
Curing, 138

Dairy powders, 219-223, 298, 315, 326-
  329
Deborah number, 257-258
Degradation
  of ascorbic acid, 296-308
  of aspartame, 296
  of nutrients, 290
  of vitamins, 303
Dehydration, 2, 9, 13, 16, 73, 93, 167,
    177, 195-198, 206, 219, 254,
    258, 286-287, 313-330, 336-344
Denaturation, *see also* Protein
  denaturation
  heat of, 134-136
  kinetics, 135
  temperature, 134-136
Density, 21, 33, 145, 225, 294
Desorption, see Sorption
Deterioration, 193-194, 271-272, 276,
    286, 298-300, 304, 333-334
Devitrification, 99, 225-226, 229
Dextrins, 318
Dextrose equivalent, 124, 207, 307, 318,
    338-340
Dielectric
  constant, 67-69
  loss constant 67-69
  properties, 35, 60-69
  relaxation, 114, 201

[Dielectric]
    spectroscopy, 69
    thermal analysis, 69
Differential scanning calorimetry, 60-69,
        77, 82, 94, 102-103, 111-114,
        122, 126-149, 161-167, 197-239,
        259
Differential thermal analysis, 63-66,
        214, 235-239
Diffusion 19, 24-25, 43-47, 57-60, 95,
        125, 130, 173-174, 194, 218,
        225, 233-234, 255-258, 272, 278-
        308, 317, 322-326, 333-335
    amorphous matrices, 255-258
    anomalous, 257
    chemical reactions, 272, 278-308,
        317
    coefficient, 233, 255-258, 278-308,
        322-326
    limited reactions, 277, 279, 281,
        291, 294, 296
    selective, 286, 322-326, 332
Diffusivity, 33, 255-258, 278-308, 322-
        326
Dilatation, 148
Dilatometry, 13, 60-62, 144
DMTA, *see* Dynamic mechanical
    thermal analysis
Drum-drying, 197
Dry food mixes, 219, 315
Drying
    aids, 166
    chamber, 321, 329
    rate, 319
    temperature, 316, 319
    time, 172
DSC, *see* Differential scanning
    calorimetry
DTA, *see* Differential thermal analysis
Dühring's rule, 84
Dynamic mechanical thermal analysis,
        68-69, 114, 161, 173, 267


Effective
    diffusion, 258, 282, 286
    molecular weight, 82, 185-187, 337-
        338, 341
Egg white proteins, 135-136

Elastic modulus, *see* Modulus
Elastin, 134, 137-138, 140, 183, 250
Electron diffraction, 50, 53, 213
Electron microscopy, *see* Microscopy
Electron spin resonance spectroscopy,
        50, 54, 58-60, 173, 285
Encapsulation, 119, 224, 258-260, 296-
        300, 332
Enthalpy, 5-15, 63-65, 82, 195-201,
        236-240
    relaxation, 198-201
Entropy, 7-15, 26, 34, 198-200
    relaxation, 198
Enzymatic
    changes, 301
    hydrolysis, 307, 338
Enzyme(s)
    activity, 134, 137, 142, 193, 235,
        271, 287- 289, 301
    amylolytic, 235
    structure, 134
ESR, *see* Electron spin resonance
    spectroscopy
Ethyl alcohol, 324
Ethylene glycol, 99
Eutectic
    crystallization, 197
    freezing, 97-101
    solutions, 85-86, 97
    temperature, 85-86, 98
Evaporation, 1, 7, 9, 13, 16, 26, 74-76,
        83-85, 93, 195-197, 314-315, 322
Expansion, 33-34, 62, 93, 330-332, 337
Extrudate, 255, 330-331
Extrusion, 17, 93, 95, 121, 159, 177,
        179, 195-196, 258, 262, 313,
        330-332, 336-337, 342-343


Fats, *see* Lipids
Fermi's model,  217
Fick's law, 322-323
Firmness, of food lipids, 263-265
First-order
    kinetics, 136, 274-276, 288, 290,
        296
    reaction, 274-275
    transition, 7, 13-15, 19-20, 27, 50,
        61-65, 73, 110, 170, 179, 333

Flavor, 119, 202, 258-260, 291, 297,
299, 313, 315, 318, 322-326,
332, 337
retention, 258, 322-325
Flory-Huggins
equation, 40, 92-95
polymer-diluent interaction parameter,
40, 94
Flory plot, 94
Fluid-bed drier, 329
Food design, 336
Fox and Flory equation, 41, 124, 166,
208, 338
Fox equation, 164-165
Fragile liquids, 39
Free
electron, 58
energy, 26, 230
fat, 298
flowing, 171, 179, 202-203, 253-
254, 315-318, 332
sulfhydryl groups, 138
volume 31-41, 96, 130, 159, 177,
196, 201, 256, 282
volume theory, 31, 36
Freeze-concentration, 20, 74-79, 83, 85,
97, 171-174, 197-198, 223-237,
303-308, 334-336, 341
Freeze-drying, 74-76, 83, 103, 114, 171,
173, 197-198, 224, 259, 321-326
Freezing
injury, 174
of molecules, 30, 63, 171, 196
temperature, 74, 78- 83, 85-86, 97-
98, 109, 114, 223-225, 233-234,
237
time, 224-230
Friction, 59, 284, 329
coefficient, 59
Fringed-micelle model, 130, 237
Frozen
desserts, 111, 223, 230-233
dough, 142
fish, 231
foods, 20, 51, 56, 66, 76, 83, 86, 99,
103, 110, 171-174, 194, 210,
219, 223-225, 229-234, 247, 258,
262, 265-267, 302-308, 333-336,
339, 342-344

[Frozen]
solutions, 99-101, 117, 172-174,
304-305
storage, 74, 82, 86, 99, 102, 142,
170, 172, 202, 210, 223-224,
229-234, 341
Fruit
juices, 86, 167, 175, 203, 315, 317-
318, 326
powders, 326

GAB model, *see* Sorption models
Gas
constant, 4, 21, 40, 79, 255, 277
law, 4
Gelatin, 249
Gelatinization, *see* Starch gelatinization
Gibbs energy, 8-16
Glass
formation, 26, 63, 195-197
transition theories, 31-34
Gliadin, 138-141
Glucose polymers, 123-124, 166, 318
Gluten, 124, 131, 136-142, 165-166,
183-187, 250, 314
Glutenin, 69, 95-96, 138-141, 186
Glycerol, 99, 131, 176, 224, 285, 291
Glyceryl monostearate, 318
Gordon and Taylor equation, 41, 115,
118, 138-140, 159-160, 162-170,
178-179, 207, 256, 316, 318, 338
Growth
rate, 24, 231
unit, 25

Half-life, 275-276
Hardness, of food lipids, 149, 262-264
Heat, *see also* Latent heat
of activation, 201
capacity, 1, 6, 14-16, 33-34, 39, 41,
63-66, 76-77, 82, 103, 115, 130,
137-140, 159-165, 198, 200, 288
content, 5
of denaturation, 129, 134-136
exchange, 4, 6
flow, 65, 66
of gelatinization, 129-131

[Heat]
    inactivation, 342
    of ionization, 134
    resistance, 342
    transfer, 3, 225
Helmholtz free energy, 8
Henry's law, 18
Homopolymers, 39, 166, 338
Huang equation, 165-166
Humidity, 51, 206, 211, 215, 284

Ice
    crystals, 75, 100, 172, 198, 224,
        230-234, 267, 324-325
    formation, 21, 73-74, 78-79, 85-86,
        97-103, 112-117, 171-174, 197,
        210, 224-235, 265, 303, 314,
        342-344
    melting 74, 97-103, 116, 118, 124,
        172-174, 197-198, 223, 225, 227,
        232-233, 266-267, 302-307, 321,
        333-335, 341
    network, 266
    nucleation, 234
    recrystallization, 172, 202, 223-235,
        333, 344
    sintering, 231
Ice cream, 44, 86, 99, 111-114, 172,
        219, 223, 230-233, 250, 266, 343
Inactivation, of spores, 342
Infrared spectroscopy, 54-55, 143-145
Instant dairy powders, 329
Intermediate
    moisture foods, 167, 170, 177, 179,
        300
    water activity, 284, 316
    water content, 94, 126-128, 132-133,
        165, 317, 333, 339
Interparticle
    bridges, 206, 328-329
    forces, 203
    fusion, 206
Invertase, 288
Invert sugar, 223
Isodilation, 149
Isopropanol, 332
Isosolids diagram, 148-149

Isoviscosity state, 38, 176-177, 204,
    209, 319, 327, 344

Kelvin element, 180-181
Kelvin equation, 21
Kinetic(s)
    analysis, 136, 229
    phenomena, 101, 229, 234, 283
    temperature dependence of, 276-283
    theory, 31-33
Kwei equation, 160

Lactose
    crystallization, 86, 99, 194, 214-224,
        260, 297-300
    hydrolysis, 222, 337-339
    precrystallized, 219, 299
Lard, 146
Latent heat
    of crystallization, 24
    of fusion, 82, 111-113, 143-146, 200
    of melting, 40, 64, 79-83, 93, 102-
        103, 225
    of transition, 7, 16-17, 63-66
    of vaporization, 79, 84
Leathery region, 184
Legume
    proteins, 136
    seed, 119, 121
Linear
    equation, 164
    growth rate, 43
    polymers, 30
Lipase, 289
Lipid(s)
    crystallization of, 142-149
    encapsulation of, 258-260, 297
    fraction, 248, 262
    melting of, 145-147
    oxidation, 260, 285-286, 297-298,
        300
    polymorphic forms of, 142-145
Liquid
    bridges, 205-206, 326-329
    flow, 184, 206
    flow region, 31

[Liquid]
  relaxation, 39
Loss modulus, *see* Modulus
Loss tangent
  dielectric, 67
  mechanical, 29
Lysine, loss of, 298-299
Lysozyme, 134-137

Magnetic
  dipoles, 57
  field, 55, 58
  moment, 56
  rotation, 56
  spin energy, 55
Maltodextrins, 41, 60, 92, 124, 166-
    167, 172, 204, 207-208, 210,
    286-288, 296-297, 307-308, 318-
    319, 322-326, 332-342
Maltohexose, 165
Malto-oligosaccharides, 124, 161, 285
Maltose, 96, 113-115, 117, 124, 165,
    258-259, 321
Maltotriose, 114, 165
Margarine, 143, 146, 263
Master curve, 182-184, 249, 252
Maximum freeze-concentration, 114,
    197,223-235, 303-308, 334-336
Maxwell element, 180-181
Mechanical
  loss measurements, 39
  moduli, 26-29, 39, 68
  properties, 26, 35, 37-38, 62-63, 66-
    69, 95, 125, 137, 149, 170-184,
    201, 218, 247-254, 260-267, 279,
    285, 303-304, 331-337
  relaxation processes, 342
Mechanical spectroscopy 68-69, 139,
    141, 266
Melt
  processing, 330-332
  viscosity, 337
Melting
  of crystallites, 130, 239
  of fats and oils, 145-147, 278
  temperature, 20, 27, 42-44, 92-95,
    110-112, 121, 142-147, 225, 266,
    330, 341

Methyl linoleate, 260, 299-300
Microbial
  growth, 73, 193, 271, 284, 301, 342
  stability, 342
Microregion theory, 324-326
Microscopy
  electron microscopy, 50-51, 299
  optical microscopy, 50-51
  polarized light, 126, 220
Milk
  fat, 146-147
  powders 44, 114, 178-179, 291, 297,
    337-338
  proteins, 140
Mobilization point, 284, 288
Modulus
  complex, 28-29
  curve, 183-186, 248-253, 265
  elastic, 180-182, 235-239, 249, 263-
    267, 331
  loss, 28-29, 39, 67-69, 139, 182,
    265-267
  loss factor, 39
  shear, 28, 265
  storage, 28-29, 39, 67-69, 139, 182,
    186, 265-267
  Young's, 27-28, 140, 184, 250, 260,
    266
Molecular
  arrangements, 42, 50, 213
  degradation, 331
  mobility, 4, 19-20, 30, 34-35, 50-60,
    67, 128, 130, 148, 161, 171,
    181, 194, 201, 211, 213, 220,
    226, 238, 266, 272, 278-279,
    283-308
  organization, 11
  relaxation, 33, 182, 199, 258
  rotations, 67-68
  size, 75-76, 194, 282, 287, 289, 322,
    330
  structure, 74, 258
Molecular weight
  distribution, 40
  effective, 81-82, 185, 338, 341
  of maltodextrins, 166-167, 207
  number average, 40
  weight average, 40, 131

Monolayer value, 89-91, 284, 287, 299, 333
Muscle proteins, 134-136


Native
 proteins, 40, 133-134
 starches, 51, 121-122, 130
Natural
 gums, 318
 polymers, 35, 127
Neutron diffraction, 53
NMR, *see* Nuclear magnetic resonance spectroscopy
Nonenzymatic browning, 193, 271, 291-295, 298, 301, 315-317, 330
Nuclear magnetic resonance spectroscopy, 50, 54-58, 62, 131, 139, 143-144, 148, 161, 230, 284
Nucleation, 20-24, 42-45, 195, 224, 234, 237, 240
 heterogeneous, 23
 homogeneous, 21-22, 76
 primary, 21-22
 processes, 20-24
 secondary, 23


Oil-in-water emulsions, 259
Oils, *see* Lipids
Onions, 292
Optical
 density, 293-294
 grating, 52
Optical microscopy, *see* Microscopy
Orange juice, 207
Osmotic dehydration, 319
Ostwald-Freundlich equation, 21
Ostwald ripening, 230
Ovalbumin, 135
Oxidation, 202, 259-260, 271-272, 285, 296-308, 332


Partial
 crystallinity, 120, 122, 265
 crystallization, 235
 pressure, 18

Partially,
 crystalline, 31, 43-44, 49, 93, 96, 110, 120, 122, 130, 184-185, 237, 262-266, 330
 freeze-concentrated, 172-173, 225-227, 266, 335, 343
Pasting, 125
Phase diagram, 9-11, 75, 83, 85, 99, 101, 114
Phase transition(s)
 classification of, 13-17
 first-order, 13-14, 110
 second-order, 14-16
 third-order, 16
Physical aging, 29, 194, 200
Plank equation, 224-225
Pochan-Beatty-Hinman equation, 162, 164-165
Polyamides, 35, 95
Polymer(s),
 blends, 119, 162-164
 chains, 31, 35, 67, 130
 crystallization, 43, 235, 237-238
 glassy, 27, 184, 201
 homogeneous, 125
 mechanistic models, 332
 melting theory, 128
 miscible, 41, 159
 mixtures, 40, 338
 networks, 138
 science, 26, 122, 157, 179, 314
Polymorphic forms, 19, 142
Polymorphism, 142-144
Poly(vinylpyrrolidone), 97, 258, 296
Porosity, 202, 206, 247, 255, 287, 296, 319-321, 331
Potatoes, 265, 292, 321
Power-law
 equation, 38, 174, 176, 233
 exponent, 233
Propagation, 20, 44
Propylene glycol, 226
Protein
 aggregation, 136
 cross-linking of, 138
 denaturation, 65, 134-136, 315
 films, 137
 hydrolysates, 254

[Protein]
matrices, 140
network, 139
solubility, 307
state diagrams of, 140
structure, 137, 141
PVP, *see* Poly(vinylpyrrolidone)

$Q_{10}$ value, 277, 317
Quenching, 26

Raisins, 315, 342
Raman spectroscopy, 55, 144
Raoult's law, 17-18, 78-84
Rate
of bread staling, 238
constant, 236-237, 272-283, 291-
296, 305, 307, 335
of crystallization, 25, 43, 45, 211,
217, 220-222, 237, 298, 300
of diffusion, 25, 295, 322, 335
of ice recrystallization, 232-234
maximum, 271, 284, 298-299
of nonenzymatic browning, 97, 292
of nucleation, 22-25, 43, 45
of relaxation, 183
of retrogradation, 55, 235, 237
Reaction
kinetics, 271-308
mechanisms, 272
order, 272-276
rate, 272-308, 315-317, 333-336
Recrystallization
of amylose-lipid complexes, 133
control of, 234-235
of ice, 230-235
mechanisms, 230-231
processes, 210, 304
time-dependent, 171-174, 210-224
of triglycerides, 143
Relative
humidity, 88-89, 158, 167, 179, 211-
215, 219-220, 258-260, 299, 326,
330, 339-340
stiffness, 253
volatility, 322

Relaxation
enthalpy, 198-201
measurements, 39
of molecules, 200
phenomena, 198-202, 229
rate, 183
structural, 201-202
time(s), 31-39, 56-57, 67, 95, 114,
171, 173-174, 177, 179-182, 198-
201, 218, 228, 232, 248, 252,
257-258, 279, 302-303, 320, 335-
337
Retention
flavor, 258, 287, 322-326
vitamins, 289
volatile compounds, 258, 322-326
Retrogradation, *see* Starch retrogradation
Rheological,
behavior, 249
changes, 138
properties, 141, 262-263, 266, 331,
342-343
Rice
kernels, 249
starch, 94, 121, 128-131
Roller-drying, 219
Rotational
correlation time, 58-60, 285
diffusion coefficient, 58-60, 285
freedom, 59
mobilization, 60
motions, 29, 58-60, 285
Rubbery
flow region, 30-31
matrices, 283
plateau region, 30, 39, 139, 183-187
state, 26, 32, 250, 253-254, 258, 287

Salt
solutions, saturated, 17, 89, 167, 215
Second-order
reaction, 276
transition, 14-16, 19, 29, 31, 34, 39,
50, 61-65, 124, 283, 333
Selective diffusion, *see* Diffusion
Semicrystalline polymer(s), 31, 55, 128
Shear modulus, *see* Modulus

Shelf life, 2, 44, 97, 158, 171, 219, 254, 283, 286, 303, 307-308, 317, 333-336
  accelerated tests, 334
  predictions, 307, 333
Shift factor, 182, 249
Sintering of ice, 231
Size reduction, 329-330
Snack(s), 29, 253, 260, 262, 314, 331, 338
Snap test, 260-262
Solid
  biopolymers, 55
  fat content, 56, 61, 148-149, 262-264
  fat index, 148
Solubility
  of actomyosin, 307
  of lactose, 223
  of sugars, 112-116, 219
Sorption
  behavior, 87-91
  of gases, 88
  hysteresis, 87
  isotherms, 86-92, 158, 169-170, 179, 212, 219-223, 302, 334, 340
  prediction, 88-92
  properties, 74, 86, 179, 220-222, 287
Sorption models
  BET model, 89-90
  GAB model, 90-92, 169-170, 178-179, 334
Spectroscopic techniques, 53-60
Spores
  bacterial, 342
  mold, 342
Spray-drying 197, 211, 219, 297, 314-330, 332, 337
Stability, 19, 102, 334
  of food powders, 220
  of frozen foods 308, 334-336, 341
  maps 300-302
Starch
  anhydrous, 123-124
  cereal, 330
  classification of, 120-122
  corn, 121, 330
  extrusion, 330
  gelatinization, 51, 65, 95, 110, 119-133, 184, 265, 314, 330

[Starch]
  gelatinization temperature, 94, 125-131, 236
  gelatinized, 210, 235-236, 256
  gels, 44, 124, 235-239, 255, 265, 286
  glass transition temperature of, 96, 161, 166
  granular, 119-122, 330
  hydrolysis products, 41, 124-125, 166
  legume seed, 121
  maize, 121
  matrices, 255
  mechanical properties of, 249
  melt, 133
  melting temperature of, 94
  native, 121-122, 330
  partially crystalline, 31, 184, 330
  polymers, 120
  potato, 94, 121, 126, 128-129, 237, 265, 330
  pregelatinized, 122
  retrogradation of, 44, 65, 122, 132, 210, 235-239, 265, 342
  rice, 121, 128-129, 131
  state diagram of, 123-124, 343
  sugar interactions, 130
  tuber, 121, 235, 330
  wheat, 121, 265
State
  diagram(s), 101-103, 114-117, 122-124, 140-142, 158, 170, 177, 221, 223, 225, 302-303, 329, 334, 342-344
  equation, 4
  functions, 3
  variables, 2-5, 8, 10
Stickiness, 166, 179, 194, 202-206, 211, 219, 222, 251, 254, 258, 315-321, 331, 334, 336-338
Sticky point, 163, 203-205, 207-208, 317-319
Stiffness, 27-28, 30, 217-218, 228, 248, 251-262
Storage modulus, *see* Modulus
Strawberries, 167
Strong liquids, 39
Sublimation, 14

Sugar(s),
 anomers, 111
 blends, 118-119, 223
 candies, 112, 195
 crystallization of, 21, 24, 214, 222-223
 mixtures, 86, 117-119, 124
 phase transitions of, 110-119
 snap cookies, 139
 solutions, 84, 86, 172, 174, 225, 229, 231, 266, 291, 321
 syrups, 23, 223
Supercooling, 20-25, 42-44, 74, 97, 142, 224-225, 234, 238
Supercritical
 extraction, 11
 fluid, 10
Superpositioning
 of modulus data, 184
 principle, 182-183, 249
 time-moisture, 249
Supersaturation, 20-25
Surface
 active, 133
 agglomeration, 328
 contact, 205
 diffusion, 286
 morphology, 51
 tension, 21, 172, 203
 viscosity, 203
 wetting, 206, 327
Synthetic polymers, 26, 31, 40-42, 54-55, 92, 95-96, 109, 130, 157, 177, 179, 196, 235, 251

Tensile
 stress, 27-29
Texture, 122, 143, 158-159, 231, 235, 247-248, 254, 263, 314, 331, 339
Thermal
 analysis, 60-69, 99
 conductivity, 225
 decomposition, 95, 123, 161
 expansion coefficient, 15, 31-32, 34 36, 61-64, 198, 200
 mechanical analysis, 60-69, 114, 173
 plasticization, 248, 262, 279, 285, 291, 314, 321, 325-326, 331, 336

[Thermal]
 properties, 62, 76-77, 126, 199-200
Thermodynamic(s)
 first law of, 4
 second law of, 6
 systems, 3
 theory, 31, 34, 41, 160
 third law of, 8
 zeroth law of, 4
Thermosetting, 41, 137-139, 141, 314
 proteins, 141, 184
Thiamin, 289
Three-point
 bend test, 140
 snap test, 260
Time-temperature
 superposition, 35, 181-182
 transformation curves, 229
TMA, *see* Thermal mechanical analysis
Translational
 diffusion coefficient, 59-60
 mobility, 342
Triglycerides, 142-149
Triple point, 10, 75

Unfrozen water, 56, 60, 79, 81-83, 97, 103, 173, 225, 228, 231, 233-234, 302-304, 308, 334-336, 343
UNIFAC-UNIQUAC model, 82-83

Vacuum dehydration, 76
Vaporization, 14
Viscoamylograph, 125
Viscoelastic
 five regions of behavior, 29-31, 182
 properties, 26-31, 60, 179-187, 249, 252, 336, 342
Viscosity, 1, 19-20, 28-29, 33-39, 43-44, 59-60, 67, 95, 97, 99, 101, 126, 171-180, 194, 201-211, 217-218, 222-226, 233-234, 251-254, 263, 279, 287-297, 303-308, 320-321, 324, 329- 331, 333, 335
Viscous flow, 206, 254, 260, 263, 321
Vitamin destruction, 289, 315
Vitrification, 17, 171

Vogel-Tammann-Fulcher equation, 38, 175-177
Volatile
  compounds, 49, 258, 286, 299, 322, 324, 326
  content, 298
  loss, 299
  retention, 286, 299
Volume, 3-17, 21, 24, 31-33, 36. 40, 45, 60-63, 94, 157, 195, 198-200, 206, 214, 230, 254, 319
  expansion, 31-32, 62
VTF equation, *see* Vogel-Tammann-Fulcher equation
Vulcanization, 138

Wall deposition, 318
Water, *see also* Sorption
  activity 17, 79, 87-92, 129-131, 167-170, 177-179, 183, 187, 193, 219, 221, 248-249, 251-254, 258-262, 271-272, 278-279, 283-302, 315, 334, 339, 342
  amorphous, solid, 76-77
  availability 130, 159, 167, 193, 283
  phase behavior of, 73-86
  plasticization 73-75, 91-97, 114, 118, 137-142, 158-170, 177, 179, 214, 220, 248, 250, 256, 260-262, 282, 284-285, 287, 289, 290-291, 295, 301, 307, 314, 317, 332, 334, 337, 339, 340, 342
Wheat
  flour, 139, 338
  grains 249-251, 265
  starch 121-123, 126, 128-131, 236-237, 265-266

Whey
  powder, 219, 298-299
  proteins, 135-136
Williams-Landel-Ferry
  constants, 35-39, 204, 216, 279, 294-295
  equation, 35-39, 172-182, 204, 215-218, 233, 250, 254, 278-283, 287, 294-296, 303-308, 335, 342
  kinetics, 233, 303-308
  model, 36, 175, 217, 252, 283, 335
  relationship, 101, 218, 232, 249, 280, 335
WLF equation, see Williams-Landel-Ferry
Work
  pressure-volume, 4-9
  softening, 263

X-ray
  diffraction, 27, 50, 52-53, 120-122, 125-126, 132-133, 142-144, 213, 235-238
  radiation, 53-54
  scattering, 128
  techniques, 50, 220

Yield value, 263
Young's modulus, *see* Modulus

Zein, 139-141
Zero-order
  kinetics, 274, 294
  reaction, 274

# FOOD SCIENCE AND TECHNOLOGY

*International Series*

Maynard A. Amerine, Rose Marie Pangborn, and Edward B. Roessler, *Principles of Sensory Evaluation of Food*. 1965.

Martin Glicksman, *Gum Technology in the Food Industry*. 1970.

Maynard A. Joslyn, *Methods in Food Analysis*, second edition. 1970.

C. R. Strumbo, *Thermobacteriology in Food Processing*, second edition. 1973.

Aaron M. Altschul (ed.), *New Protein Foods*: Volume 1, *Technology, Part A*—1974. Volume 2, *Technology, Part B*—1976. Volume 3, *Animal Protein Supplies, Part A*—1978. Volume 4, *Animal Protein Supplies, Part B*—1981. Volume 5, *Seed Storage Proteins*—1985.

S. A. Goldblith, L. Rey, and W. W. Rothmayr, *Freeze Drying and Advanced Food Technology*. 1975.

R. B. Duckworth (ed.), *Water Relations of Food*. 1975.

John A. Troller and J. H. B. Christian, *Water Activity and Food*. 1978.

A. E. Bender, *Food Processing and Nutrition*. 1978.

D. R. Osborne and P. Voogt, *The Analysis of Nutrients in Foods*. 1978.

Marcel Loncin and R. L. Merson, *Food Engineering: Principles and Selected Applications*. 1979.

J. G. Vaughan (ed.), *Food Microscopy*. 1979.

J. R. A. Pollock (ed.), *Brewing Science*, Volume 1—1979. Volume 2—1980. Volume 3—1987.

J. Christopher Bauernfeind (ed.), *Carotenoids as Colorants and Vitamin A Precursors: Technological and Nutritional Applications*. 1981.

Pericles Markakis (ed.), *Anthocyanins as Food Colors*. 1982.

George F. Stewart and Maynard A. Amerine (eds.), *Introduction to Food Science and Technology*, second edition. 1982.

Malcolm C. Bourne, *Food Texture and Viscosity: Concept and Measurement*. 1982.

Hector A. Iglesias and Jorge Chirife, *Handbook of Food Isotherms: Water Sorption Parameters for Food and Food Components*. 1982.

Colin Dennis (ed.), *Post-Harvest Pathology of Fruits and Vegetables*. 1983.

P. J. Barnes (ed.), *Lipids in Cereal Technology*. 1983.

David Pimentel and Carl W. Hall (eds.), *Food and Energy Resources*. 1984.

Joe M. Regenstein and Carrie E. Regenstein, *Food Protein Chemistry: An Introduction for Food Scientists*. 1984.

Maximo C. Gacula, Jr. and Jagbir Singh, *Statistical Methods in Food and Consumer Research*. 1984.

Fergus M. Clydesdale and Kathryn L. Wiemer (eds.), *Iron Fortification of Foods*. 1985.

Robert V. Decareau, *Microwaves in the Food Processing Industry*. 1985.

S. M. Herschdoerfer (ed.), *Quality Control in the Food Industry*, second edition. Volume 1—1985. Volume 2—1985. Volume 3—1986. Volume 4—1987.

F. E. Cunningham and N. A. Cox (eds.), *Microbiology of Poultry Meat Products*. 1987.

Walter M. Urbain, *Food Irradiation*. 1986.

Peter J. Bechtel, *Muscle as Food*. 1986.

H. W. -S. Chan, *Autoxidation of Unsaturated Lipids*. 1986.

Chester O. McCorkle, Jr., *Economics of Food Processing in the United States*. 1987.

Jethro Jagtiani, Harvey T. Chan, Jr., and William S. Sakai, *Tropical Fruit Processing*. 1987.

J. Solms, D. A. Booth, R. M. Dangborn, and O. Raunhardt, *Food Acceptance and Nutrition*. 1987.

R. Macrae, *HPLC in Food Analysis*, second edition. 1988.

A. M. Pearson and R. B. Young, *Muscle and Meat Biochemistry*. 1989.

Dean O. Cliver (ed.), *Foodborne Diseases*. 1990.

Marjorie P. Penfield and Ada Marie Campbell, *Experimental Food Science*, third edition. 1990.

Leroy C. Blankenship, *Colonization Control of Human Bacterial Enteropathogens in Poultry*. 1991.

Yeshajahu Pomeranz, *Functional Properties of Food Components*, second edition. 1991.

Reginald H. Walter, *The Chemistry and Technology of Pectin*. 1991.

Herbert Stone and Joel L. Sidel, *Sensory Evaluation Practices*, second edition. 1993.

Robert L. Shewfelt and Stanley E. Prussia, *Postharvest Handling: A Systems Approach*. 1993.

R. Paul Singh and Dennis R. Heldman, *Introduction to Food Engineering*, second edition. 1993.

Tilak Nagodawithana and Gerald Reed, *Enzymes in Food Processing*, third edition. 1993.

Dallas G. Hoover and Larry R. Steenson, *Bacteriocins*. 1993.

Takayaki Shibamoto and Leonard Bjeldanes, *Introduction to Food Toxicology*. 1993.

John A. Troller, *Sanitation in Food Processing*, second edition. 1993.

Ronald S. Jackson, *Wine Science: Principles and Applications*. 1994.

Harold D. Hafs and Robert G. Zimbelman, *Low-fat Meats*. 1994.

Lance G. Phillips, Dana M. Whitehead, and John Kinsella, *Structure–Function Properties of Food Proteins*. 1994.

Robert G. Jensen, Marvin P. Thompson, and Robert Jenness, *Handbook of Milk Composition*. In Preparation.